36th Winter Workshop on Nuclear Dynamics (WWND 2020)

Journal of Physics: Conference Series Volume 1602

Puerto Vallarta, Mexico
1 - 7 March 2020

Editors:

Rene Bellwied
John Harris
Claudia Ratti
Anthony Timmins

ISBN: 978-1-7138-5749-5
ISSN: 1742-6588

Printed from e-media with permission by:

Curran Associates, Inc.
57 Morehouse Lane
Red Hook, NY 12571

Some format issues inherent in the e-media version may also appear in this print version.

Institute of Physics
Dirac House, Temple Back
Bristol BS1 6BE UK

Phone: 44 1 17 929 7481
Fax: 44 1 17 920 0979

techtracking@iop.org

Additional copies of this publication are available from:

Curran Associates, Inc.
57 Morehouse Lane
Red Hook, NY 12571 USA
Phone: 845-758-0400
Fax: 845-758-2633
Email: curran@proceedings.com
Web: www.proceedings.com

TABLE OF CONTENTS

PREFACE

These are the proceedings of the 36th edition of the Winter Workshop on Nuclear Dynamics, which was held In Puerto Vallarta, Mexico, in March 2020. Our meeting was the final in-person gathering in our field before the international lockdown period due to Covid-19. As in previous years the unique character of this conference series has allowed us to bring scientists from various fields of nuclear physics together to discuss their scientific achievements.

At the high-energy frontier very exciting results not only from the heavy-ion collisions but also smaller systems at the LHC were shown. At this time the LHC experiments are upgrading their detectors for the third large run period, which is scheduled to start in early 2022. Several talks focused on the future experimental program at the LHC. At the lower energies the second RHIC beam energy scan is underway to span the gap between the SPS and RHIC and search for critical phenomena in the nuclear matter phase diagram. This program is complemented by spin and small systems measurements at the highest RHIC energies. The future programs of sPHENIX and the planned electron-ion collider (EIC) were represented as well.

Theory progress has been made across all these different energy and system size regimes, and the new data, in particular from the LHC and RHIC, are motivating more detailed modeling and a deeper understanding of the underlying physics.

These proceedings of the 36th Winter Workshop on Nuclear Dynamics again provide a snapshot of the status of the field. The articles, many of which are written by some of the most promising young scientists in the field, are conveying the present excitement that permeates all subfields of nuclear science.

All papers published in this volume have been peer reviewed through processes administered by the Editors. Reviews were conducted by expert referees to the professional and scientific standards expected of a proceedings journal published by IOP Publishing.

Rene Bellwied (University of Houston, USA)
John Harris (Yale University, USA)
Claudia Ratti (University of Houston, USA)
Anthony Timmins (University of Houston, USA)

Journal of Physics: Conference Series **1602** (2020) 011002 doi:10.1088/1742-6596/1602/1/011002

Peer review declaration

All papers published in this volume of Journal of Physics: Conference Series have been peer reviewed through processes administered by the Editors. Reviews were conducted by expert referees to the professional and scientific standards expected of a proceedings journal published by IOP Publishing.

- **Type of peer review: Single-blind / Double-blind / Triple-blind / Open / Other (please describe)** Single-blind. Editors chose experienced peer-reviewer

- **Conference submission management system:** All submissions were collected by editors and then sent out for review. Reviews received after two week period. ~30% of submissions needed revision

- **Number of submissions received: 32**

- **Number of submissions sent for review: 32**

- **Number of submissions accepted:** 22 without revision, 10 after revision

- **Acceptance Rate (Number of Submissions Accepted / Number of Submissions Received X 100):** 22/32 = 69%

- **Average number of reviews per paper: 1**

- **Total number of reviewers involved: 14**

- **Any additional info on review process: NONE**

- **Contact person for queries:** Rene Bellwied (bellwied@uh.edu)

Droplets of quark gluon plasma: PHENIX results on small systems at RHIC

R. Belmont (for the PHENIX Collaboration)

University of North Carolina at Greensboro, Greensboro, NC 27402-6170, USA

E-mail: r.belmont@cern.ch

Abstract. In these proceedings we discuss PHENIX results on small systems at RHIC, specifically those relating to collectivity and flow. We discuss the small systems geometry scan, which comprises collisions of p+Au, d+Au, and ^{3}He+Au at $\sqrt{s_{NN}}$ = 200 GeV and is designed to exploit differences in intrinsic geometry. We also discuss the small systems beam energy scan, which comprises d+Au collisions at $\sqrt{s_{NN}}$ = 200 GeV, 62.4 GeV, 39 GeV, and 19.6 GeV and is designed to explore differences in system size and lifetime while keeping the geometry fixed. We find the data in nearly all cases are well-described by hydrodynamical calculations involving a QGP phase, suggestion formation of QGP droplets in a wide variety of collisions.

1. Small systems geometry scan

In 2014, a small systems geometry scan at the relativistic heavy ion collider (RHIC) was proposed [1] to exploit intrinsic geometry of p+Au, d+Au, and ^{3}He+Au collisions. This ambitious program made use of the versatility and flexibility of the RHIC accelerator complex.

In p+Au collisions, there is no intrinsic geometry, so the n-th harmonic eccentricities ε_n come from nucleonic and subnucleonic fluctuations only. In d+Au collisions, there is intrinsic ellipticity due to the configuration of the nucleus (viz. two nucleons in the projectile nucleus), but no intrinsic triangularity—therefore, the ellipticity ε_2 has contributions both from fluctuations and intrinsic geometry while the triangularity ε_3 is generated by fluctuations only. In ^{3}He+Au collisions, the shape of the projectile nucleus contributes both intrinsic triangularity (viz. three nucleons) and intrinsic ellipticity—there's more phase space available for configurations that are elongated than for configurations resembling an equilateral triangle, both in terms of the ^{3}He wave function and in the fact that the plane of the ^{3}He can be inclined relative to the transverse plane of the collision. It is critical to understand that fluctuations contribute to all ε_n in all systems, but intrinsic geometry contributes to only some ε_n in only some collision systems.

Different models will lead to different relative contributions from fluctuations and intrinsic geometry. However, based simply on the presence or absence of contributions from intrinsic geometry, we expect the following qualitative hierarchy:

- $\varepsilon_2(p\text{+Au}) < \varepsilon_2(d\text{+Au}) \approx \varepsilon_2(^3\text{He+Au})$;
- $\varepsilon_3(p\text{+Au}) \approx \varepsilon_3(d\text{+Au}) < \varepsilon_3(^3\text{He+Au})$.

This hierarchy is generally seen in various theoretical calculations of the ε_n in these small systems, as shown in Table 1. The hierarchy is more consistent for ε_2 than for ε_3; in some cases, the ε_3 values of d+Au and ^{3}He+Au tend stand further apart from p+Au.

	PHOBOS MC Glauber [1, 2]	Multiplicity fluctuations [2, 3]	Mult. & gluon fluctuations [2, 3]	IP-Glasma Nucleons $t = 0$ [2, 4, 5]	IP-Glasma 3 quarks $t = 0$ [2, 4, 5]
$\varepsilon_2(p\text{+Au})$	0.23	0.32	0.38	0.10	0.50
$\varepsilon_2(d\text{+Au})$	0.54	0.48	0.51	0.58	0.73
$\varepsilon_2(^3\text{He+Au})$	0.50	0.50	0.52	0.55	0.64
$\varepsilon_3(p\text{+Au})$	0.16	0.24	0.30	0.09	0.32
$\varepsilon_3(d\text{+Au})$	0.18	0.28	0.31	0.28	0.40
$\varepsilon_3(^3\text{He+Au})$	0.28	0.32	0.35	0.34	0.46

Table 1. Eccentricity values from various calculations with descriptions and references given in the first row.

The completion of that geometry scan culminates in the publication of event plane method v_2 and v_3 in p+Au, d+Au, and ^{3}He+Au collisions at $\sqrt{s_{NN}}$ = 200 GeV in Nature Physics [6]. Eschewing theoretical comparisons for the moment, Figure 1 show v_2 (v_3) in the left (right) panel in p+Au (red points), d+Au (blue points), and ^{3}He+Au (black point) collisions at $\sqrt{s_{NN}}$ = 200 GeV. Indeed, we see the same hierarchy as the ε_n discussed above:

- $v_2(p\text{+Au}) < v_2(d\text{+Au}) \approx v_2(^3\text{He+Au})$;
- $v_3(p\text{+Au}) \approx v_3(d\text{+Au}) < v_3(^3\text{He+Au})$.

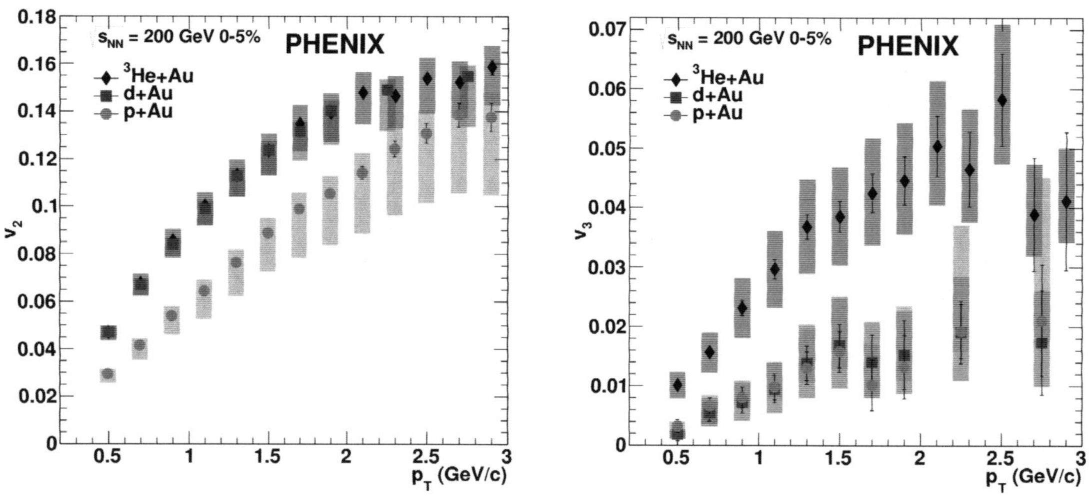

Figure 1. v_2 (left panel) and v_3 (right panel) in p+Au (red circles), d+Au (blue squares), and ^{3}He+Au (black diamonds) collisions at $\sqrt{s_{NN}}$ = 200 GeV [6]. The data exhibit a geometrical hierarchy as discussed in the text.

We now consider some theoretical comparisons. Figures 2, 3, and 4 show the v_2 (solid circles) and v_3 (solid diamonds, top and middle; sold triangle, bottom) in p+Au (left panels), d+Au (middle panels), and ^{3}He+Au (right panels) at $\sqrt{s_{NN}}$ = 200 GeV.

Two hydrodynamical calculations are shown in Figure 2: the SONIC model [1] (solid red lines) and iEBE-VISHNU [7] (dashed blue lines). Both sets of theory curves describe the data

Journal of Physics: Conference Series **1602** (2020) 012001 doi:10.1088/1742-6596/1602/1/012001

extremely well. Figure 3 shows the theoretical calculations in the color glass condensate effective field theory formalism [8]; the curves completely fail to describe the data, which seems to fully rule out the possibility for explaining the data with initial state effects alone. It is noteworthy that both the overall curves and specifically the peaks appear to decrease with increasing system size, which is suggestive of an intuitive scaling in which the strength of the correlation is inversely proportional to the number of color domains, as suggested in Ref. [6].

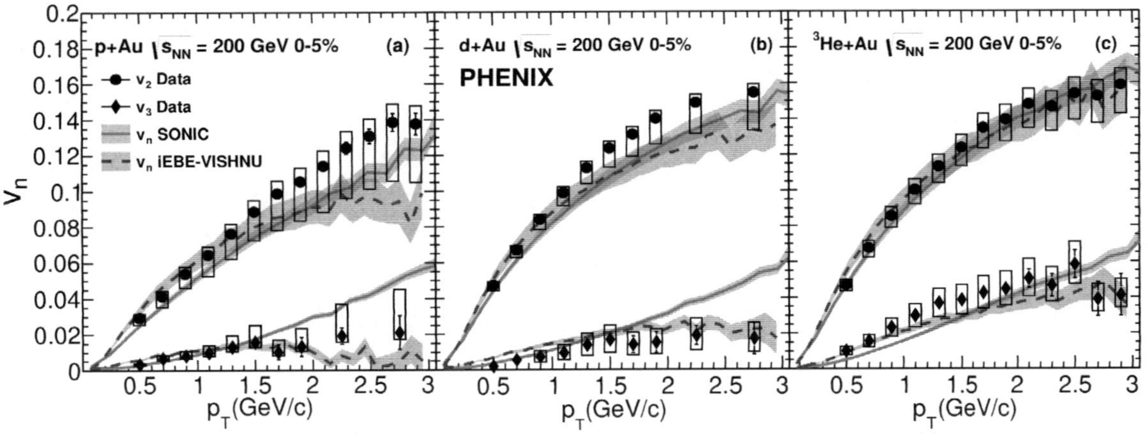

Figure 2. v_2 (black circles) and v_3 (black diamonds) as a function of p_T in p+Au (left), d+Au (middle), and ^{3}He+Au (right) collisions at $\sqrt{s_{NN}} = 200$ GeV and comparison to hydrodynamical calculations [6].

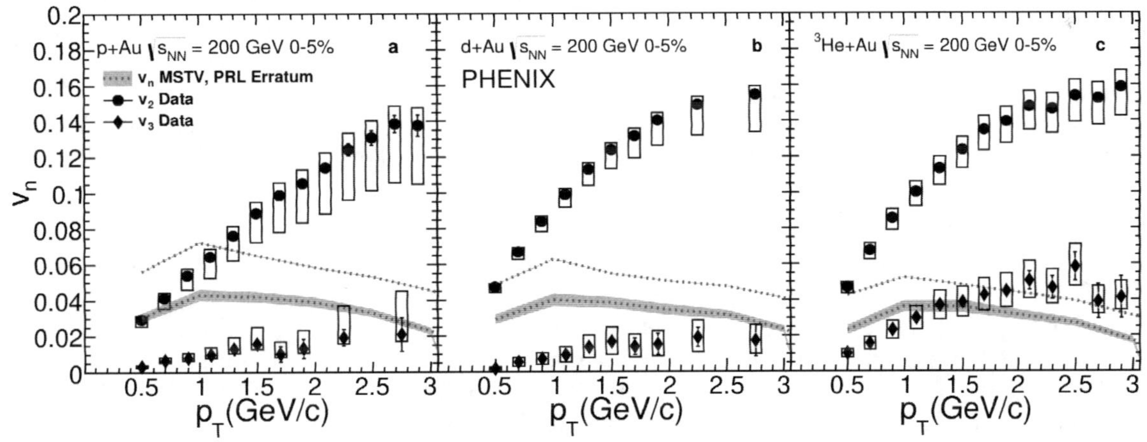

Figure 3. v_2 (black circles) and v_3 (black diamonds) as a function of p_T in p+Au (left), d+Au (middle), and ^{3}He+Au (right) collisions at $\sqrt{s_{NN}} = 200$ GeV [6] and comparison to an initial state calculation [8].

This does not, however, rule out any role for initial state effects. Recent efforts [9] have sought to include effects both from the initial state and from the hydrodynamical evolution—these are shown in Figure 4. They are generally in quite reasonable agreement with the data, especially when taking the uncertainties into account. Nevertheless, they do not describe the data quite as well as hydrodynamics alone, which could indicate a relatively small role for initial state effects, or the need for a rather different set of fluid parameters.

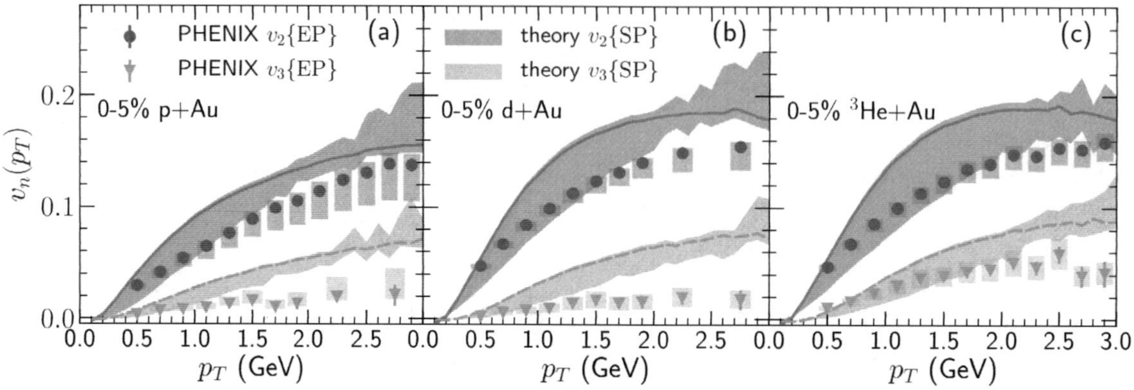

Figure 4. v_2 (red circles) and v_3 (blue triangles) as a function of p_T in p+Au (left), d+Au (middle), and ^{3}He+Au (right) collisions at $\sqrt{s_{NN}} = 200$ GeV [6] and a comparison to a combined initial state and hydrodynamical calculation [9].

Longitudinal dynamics were also explored via measurements of $dN_{ch}/d\eta$ and $v_2(\eta)$ [10]. Figure 5 shows v_2 as a function of η in green points in p+Al, p+Au, d+Au, and ^{3}He+Au collisions at $\sqrt{s_{NN}} = 200$ GeV. Also shown as blue points is $dN_{ch}/d\eta$ in each system, scaled down to meet the forward rapidity points. The v_2 and $dN_{ch}/d\eta$ follow the same trend at central and forward rapidity. At backward rapidity, we see a very large increase in the v_2 in p+Al, and a similar but smaller increase in the v_2 in p+Au; this increase is absent in d+Au and ^{3}He+Au. This increase is due to the prevalence of non-flow when the particles are near ($\Delta\eta < 1$) the event plane detector ($-3.9 < \eta < -3.1$). The relative strength of this rise is inversely proportional to the size of the collision system, which is expected based on combinatoric dilution of few-particle correlations. Also shown are calculations in a 3D hydrodynamic model [11] for p+Au, d+Au, and ^{3}He+Au. The theory curves describe the p+Au and d+Au very well; contrariwise, the theory has less forward/backward asymmetry than the experimental data in the ^{3}He+Au case.

2. d+Au beam energy scan

In 2016, RHIC performed a beam energy scan (BES) of d+Au collisions, colliding at $\sqrt{s_{NN}} =$ 200 GeV, 62.4 GeV, 39 GeV, and 19.6 GeV. The collision geometry is fixed by colliding the same species, and the system size and lifetime are varied by changing the collision energy [12].

Figure 6 shows event plane method v_2 as a function of p_T in d+Au collisions at $\sqrt{s_{NN}} =$ 200 GeV (left), 62.4 GeV (left middle), 39 GeV (right middle), and 19.6 GeV (right) [13]. Hydrodynamical calculations [12] describe the 200 GeV and predict the 62.4 GeV well. On the other hand, the 39 GeV and 19.6 GeV data are under-predicted. In the large systems BES, it has been shown [14, 15] that adjusting the equation of state and accounting for conserved charges (e.g. baryon number) is very important in getting accurate calculations. It seems likely that this is the case in the small systems BES as well.

Figure 7 shows multiparticle cumulant v_2 as a function of multiplicity in d+Au collisions at $\sqrt{s_{NN}} =$ 200 GeV (left), 62.4 GeV (left middle), 39 GeV (right middle), and 19.6 GeV (right) [16]. The $v_2\{4\}$ is real-valued in all collision systems. In the 200 GeV case, there is sufficient statistical precision to measure $v_2\{6\}$ as well, which is consistent with the $v_2\{4\}$, as seen in large systems and in p+Pb at the LHC.

Journal of Physics: Conference Series **1602** (2020) 012001 doi:10.1088/1742-6596/1602/1/012001

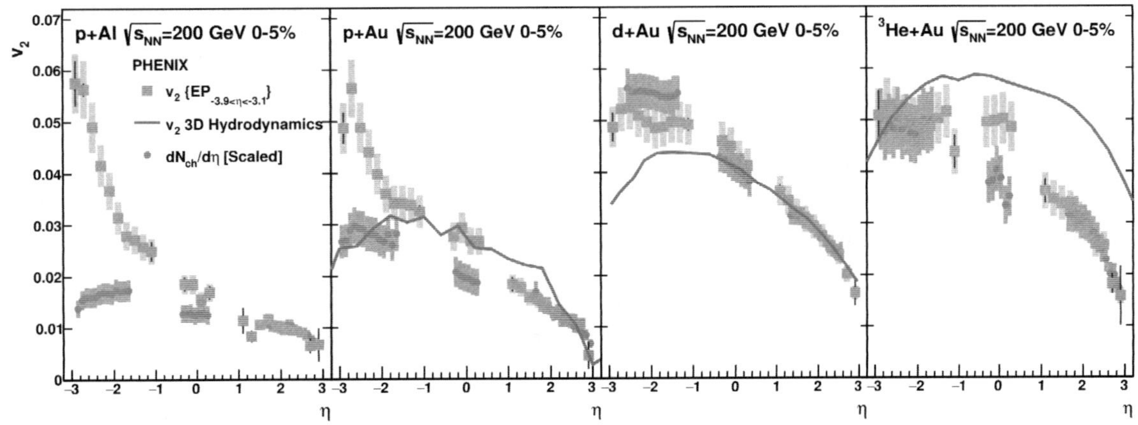

Figure 5. $v_2(\eta)$ (green squares) and $dN_{ch}/d\eta$ (blue circles, scaled downward to match $v_2(\eta)$) in p+Al (left), p+Au (left middle), d+Au (right middle), and ^{3}He+Au (right) collisions at $\sqrt{s_{NN}}$ = 200 GeV [10]. Also shown are 3D hydrodynamics calculations for p+Au, d+Au, and ^{3}He+Au [11].

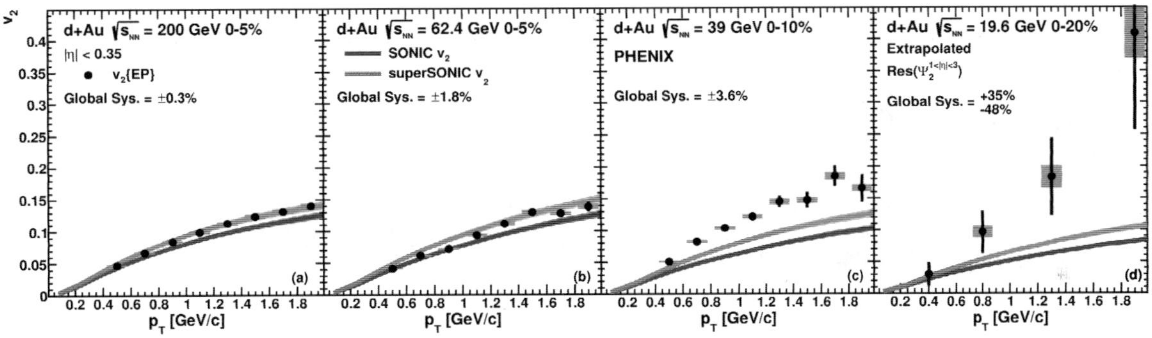

Figure 6. $v_2(p_T)$ (black circles) in d+Au collisions at $\sqrt{s_{NN}}$ = 200 GeV (left), 62.4 GeV (left middle), 39 GeV (right middle), and 19.6 GeV (right) [13]. Also shown are hydrodynamical calculations [12].

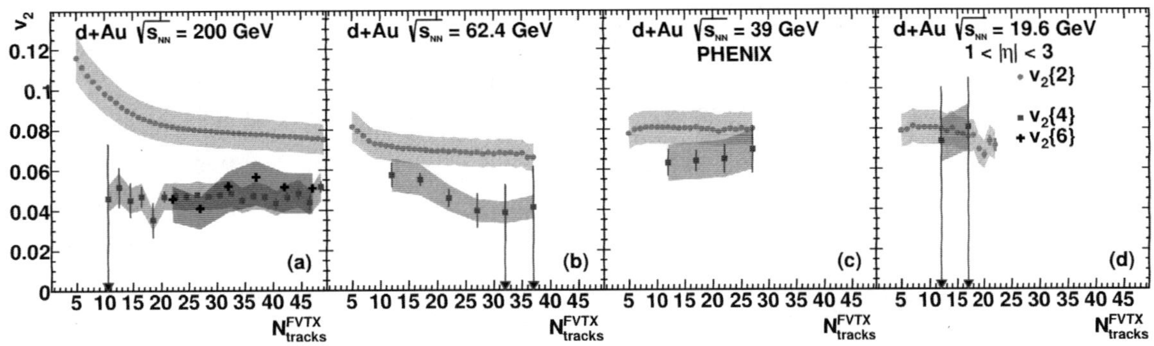

Figure 7. $v_2\{2\}$ (red circles) and $v_2\{4\}$ (blue squares) as a function of multiplicity in d+Au collisions at $\sqrt{s_{NN}}$ = 200 GeV (left), 62.4 GeV (left middle), 39 GeV (right middle), and 19.6 GeV (right), also shown for 200 GeV collisions is $v_2\{6\}$ (black crosses) [16].

3. Summary

In these proceedings we have shown the marquee results from the RHIC small systems program. The multiparticle cumulant v_2 measurements show signatures of collectivity at all four collision energies in the d+Au BES, which is a necessary (but insufficient) condition to demonstrate QGP formation. The $v_2(p_T)$ measurements in the d+Au BES show the success of hydrodynamical theory with a QGP phase at both 200 GeV and 62.4 GeV, and invite additional calculations specifically tuned for lower collision energies. The $v_2(\eta)$ and $dN_{ch}/d\eta$ in p+Al, p+Au, d+Au, and ^{3}He+Au demonstrate the importance of longitudinal dynamics in small systems and the need for three-dimensional hydrodynamical calculations. Lastly, the $v_2(p_T)$ and $v_3(p_T)$ in p+Au, d+Au, and ^{3}He+Au demonstrate a geometrical ordering, and show excellent agreement with predictions and descriptions from hydrodynamical calculations. Taken together, there is strong evidence for QGP droplet formation in small systems at RHIC.

4. References

[1] J. L. Nagle, A. Adare, S. Beckman, T. Koblesky, J. Orjuela Koop, D. McGlinchey, P. Romatschke, J. Carlson, J. E. Lynn, and M. McCumber. Exploiting intrinsic triangular geometry in relativistic ^{3}He+Au collisions to disentangle medium properties. *Phys. Rev. Lett.*, 113(11):112301, 2014.
[2] J.L. Nagle. Private communication.
[3] Kevin Welsh, Jordan Singer, and Ulrich W. Heinz. Initial state fluctuations in collisions between light and heavy ions. *Phys. Rev. C*, 94(2):024919, 2016.
[4] S. Lim. Private communication.
[5] B. Schenke. https://github.com/schenke/ipglasma.
[6] C. Aidala et al. Creation of quark–gluon plasma droplets with three distinct geometries. *Nature Phys.*, 15(3):214–220, 2019.
[7] Chun Shen, Jean-François Paquet, Gabriel S. Denicol, Sangyong Jeon, and Charles Gale. Collectivity and electromagnetic radiation in small systems. *Phys. Rev. C*, 95(1):014906, 2017.
[8] Mark Mace, Vladimir V. Skokov, Prithwish Tribedy, and Raju Venugopalan. Hierarchy of Azimuthal Anisotropy Harmonics in Collisions of Small Systems from the Color Glass Condensate. *Phys. Rev. Lett.*, 121(5):052301, 2018. [Erratum: Phys.Rev.Lett. 123, 039901 (2019)].
[9] Bjoern Schenke, Chun Shen, and Prithwish Tribedy. Hybrid Color Glass Condensate and hydrodynamic description of the Relativistic Heavy Ion Collider small system scan. *Phys. Lett. B*, 803:135322, 2020.
[10] A. Adare et al. Pseudorapidity Dependence of Particle Production and Elliptic Flow in Asymmetric Nuclear Collisions of p+Al, p+Au, d+Au, and ^{3}He+Au at $\sqrt{s_{NN}}$ = 200 GeV. *Phys. Rev. Lett.*, 121(22):222301, 2018.
[11] Piotr Bozek and Wojciech Broniowski. Collective flow in ultrarelativistic ^{3}He-Au collisions. *Phys. Lett. B*, 739:308–312, 2014.
[12] J. D. Orjuela Koop, R. Belmont, P. Yin, and J. L. Nagle. Exploring the Beam Energy Dependence of Flow-Like Signatures in Small System d+Au Collisions. *Phys. Rev. C*, 93(4):044910, 2016.
[13] C. Aidala et al. Measurements of azimuthal anisotropy and charged-particle multiplicity in d+Au collisions at $\sqrt{s_{NN}}$ = 200, 62.4, 39, and 19.6 GeV. *Phys. Rev. C*, 96(6):064905, 2017.
[14] Mauricio Martinez, Matthew D. Sievert, Douglas E. Wertepny, and Jacquelyn Noronha-Hostler. Initial state fluctuations of QCD conserved charges in heavy-ion collisions. 11 2019.
[15] Mauricio Martinez, Matthew D. Sievert, Douglas E. Wertepny, and Jacquelyn Noronha-Hostler. Toward Initial Conditions of Conserved Charges Part II: The ICCING Monte Carlo Algorithm. 11 2019.
[16] C. Aidala et al. Measurements of Multiparticle Correlations in d + Au Collisions at 200, 62.4, 39, and 19.6 GeV and p + Au Collisions at 200 GeV and Implications for Collective Behavior. *Phys. Rev. Lett.*, 120(6):062302, 2018.

Status of CME Search Before Isobar Collisions and Methods of Blind Analysis From STAR

Prithwish Tribedy for the STAR collaboration
Physics Department, Brookhaven National Laboratory, Upton, NY
E-mail: ptribedy@bnl.gov

Abstract. The STAR collaboration is currently pursuing the blind analysis of the data for isobar collisions that was performed at RHIC in the year 2018 to make a decisive test of the Chiral Magnetic Effect (CME). Why is it so difficult to detect signals of CME in the experiment? Do we really understand different sources of background? Why observing similar charge separation between $p/d + A$ and $A + A$ does not stop us from pursuing the search for CME? In this contribution, I attempt to address some of these questions and briefly outline a few recent STAR analyses based on new methods and observables to isolate the possible CME-driven signal and non-CME background contributions at the top RHIC energy. Finally, I describe the procedure for the blind analysis of the isobar data. An outstanding question remains – what happens if we go down in energy? I address this by discussing how the new event-plane detector (EPD) upgrade provides a new capability at STAR towards CME search using the data from the RHIC BES-II program.

1. Introduction

Finding a conclusive experimental signature of the Chiral Magnetic Effect (CME) has become one of the major scientific goals of the heavy-ion physics program at the Relativistic Heavy Ion Collider (RHIC). The existence of CME will be a leap towards an understanding of the QCD vacuum, establishing a picture of the formation of deconfined medium where chiral symmetry is restored and will also provide unique evidence of the strongest known electromagnetic fields created in relativistic heavy-ion collisions [1, 2]. The impact of such a discovery goes beyond the community of heavy-ion collisions and will possibly be a milestone in physics. Also, as it turns out, the remaining few years of RHIC run and analysis of already collected data probably provides the last chance for dedicated CME searches in heavy-ion collisions in the foreseeable future.

Over the past years significant efforts from the STAR as well as other collaborations have been dedicated towards developing new methods and observables to isolate the possible CME-driven signal and non-CME background contributions in the measurements of charge separation across the reaction plane. The most widely studied experimental observable in this context is the γ-correlator, defined as $\langle\cos(\phi_a^\alpha + \phi_b^\beta - 2\Psi_{RP})\rangle$, where ϕ_a and ϕ_b denote the azimuthal angles of charged particles, α and β are labels for the charge of the particles and Ψ_{RP} is the reaction plane angle [3]. The angle Ψ_{RP} is expected to be strongly correlated to the direction of the magnetic field that enables the γ-correlator to be sensitive to signals of CME, more specifically, CME leads to a difference between same sign (SS, $\alpha = \beta$) and opposite sign (OS, $\alpha \neq \beta$) charge correlations: $\Delta\gamma = \gamma_{OS} - \gamma_{SS}$. The STAR time projection chamber (TPC) has a wide acceptance at mid-rapidity ($|\eta|<1$) that is used to detect ϕ_a and ϕ_b. And, in STAR the proxy for Ψ_{RP} can be played by: 1) second-order harmonic anisotropy plane Ψ_2 of produced particles at mid-rapidity measured by TPC, 2) the first-order plane due to the spectator neutrons ($\Psi_{\rm ZDC}$) detected by the zero degree calorimeters (ZDC), 3) the forward Ψ_2 plane using the STAR beam beam counter BBCs and 4) very recently using both the first and second-order harmonic anisotropy planes using the forward Event Plane Detector (EPDs). Each of these planes are expected to have more or less measurable correlations to B-field and serves their purpose for the CME search. The first measurement of non-zero $\Delta\gamma$ by the STAR collaboration goes back to [4] where connections to several expectations from CME driven signals of charge separation was identified. Most importantly, the first measurement from STAR [4] also identified several possible contributions from non-CME effects in the experimental observation of non-zero $\Delta\gamma$. Several subsequent measurements from RHIC and LHC have confirmed this observation and provided many additional insights in that direction [4, 5, 6, 7, 8, 9, 10, 11, 12, 13, 14]. In this contribution, I will focus only on RHIC results and refer to LHC results wherever necessary.

A major challenge that the γ-correlator faces towards detecting signals of CME involves large non-CME background sources that are: 1) correlated to Ψ_{RP} and 2) independent of Ψ_{RP}. The distinction between the two sources must be carefully noted as they are crucial to the interpretation of several key measurements performed at both RHIC and LHC.

2. Major challenges in isolating background

2.1. Background sources-I: reaction plane dependent correlations

The possible background contamination due to the first source of Ψ_{RP} dependent correlation was already alluded to in the reference where γ-correlator was first proposed [3]. At that time only neutral resonance particles were identified as the major source of such background albeit thought to be sub-dominant. When a flowing neutral resonance decays it enhances the probability of a pair of opposite sign particles to move together along Ψ_{RP}. Such correlations lead to non-zero magnitudes of $\Delta\gamma$ mimicking CME. In this context it is important to mention that recently a data-driven approach to measure resonance contribution of background has been developed in STAR by studying the invariant mass dependence of $\Delta\gamma$ that we discuss later [15]. Later on, a more severe source of Ψ_{RP} background due to correlated production of a pair of opposite charged particles due to local charge conservation (LCC) was proposed [16]. Parametrically, if v_2 is the elliptic flow and N is the multiplicity the background contribution from resonance and LCC should go as $\Delta\gamma_{\rm bkg} \sim v_2/N$ [3] that is also verified by many model calculations [17]. Recently, many models that incorporate the same basic picture of particle production conserving charge locally from a flowing neutral matter, are able to very well explain measurements of $\Delta\gamma$ without invoking the physics of CME. Despite the success of background models experimental search of CME continued because of a number of reasons. Model predictions have large systematics since exact mechanism of hardronization is poorly understood, limited constraints from independent measurements are available. Above all, even the most state-of-the art background models fail to explain all qualitative features of the data (e.g. $\Delta\gamma$ in central collisions, see Fig.1). While the models continue to refine their predictive power, over many years this largely lead to a major effort in beating the background sources in the measurement of charge separation along Ψ_{RP}. It is worth to mention that pheomenological predictions based on anomalous viscous hydrodynamics are now available that include both CME signal and background contribution and can be used to test the sensitivity of different observables [18].

2.2. Background sources-II: reaction plane independent correlations

The second major sources of non-CME background to $\Delta\gamma$ arises from reaction plane independent non-flow correlations. The possibility of such background was discussed in the first publication of charge separation from STAR [4]. One possible source of such background was identified to

Journal of Physics: Conference Series **1602** (2020) 012002 doi:10.1088/1742-6596/1602/1/012002

be three-particle correlations induced by mini-jet fragmentation which is known to: 1) influence the determination of event plane, 2) introduce more opposite charge correlation than same charge correlations. The combination of these two artifacts are supposed to lead to non-zero $\Delta\gamma$ and mimic CME signals. In Ref [4], an indication of larger contribution of reaction plane independent background can already be seen in: 1) the sharp increasing strength of $\Delta\gamma$ towards peripheral events and, 2) large $\Delta\gamma$ in Cu+Cu than in Au+Au system at the same centrality. Both observations can be supported by HIJING calculation.

3. Using small systems to estimate data driven background

Small collision systems provide unique data-driven ways to measure charge separation in the background scenario. This is based on the idea that the direction of B-field is uncorrelated to the elliptic anisotropy plane of the produced particle with respect to which $\Delta\gamma$ is measured [11, 19]. In low-multiplicity or min-bias collisions of small systems such planes are dominated by non-flow correlations from di-jets or momentum conservation. However, tell-tale signatures of collectivity have been observed in high multiplicity events of small collision systems – the origin of which has been a widely discussed topic in our community. There are a few scenarios that decide whether the elliptic anisotropy plane measured in the experiment will be: 1) correlated to a geometric plane of participants if collectivity is due to hydrodynamics flow, 2) uncorrelated or less correlated to geometric plane if collectivity is due to non-hydrodynamic but other initial state momentum space correlations, e.g. from CGC or escape mechanism and, 3) dominated by non-flow from di-jets and momentum conservation if no collectivity is observed [20]. Why is this important for CME search? It is important as these scenarios determine the nature of non-CME background that will dominate the measurements of $\Delta\gamma$ in small systems. It is also important to know what kind of baseline measurement do these small systems provide because our ultimate goal is to interpret measurements in heavy-ion collisions. For example, in the first scenario hydrodynamic flow driven background combined with local charge conservation will be the dominant source, important for heavy-ion measurements in most centralities. For the second and third scenarios reaction plane independent background will be the dominant source, important for peripheral and smaller sized heavy-ion collisions. Nevertheless, the expectation is that CME signal in all such scenarios will be small as the B-field in small collision systems are weakly correlated to elliptic anisotropy plane other than some specific scenarios like what was discussed in Ref [21]. So in summary, small systems have the potential to provide baseline measurements for heavy-ion collisions where CME signals are expected to disappear but different background sources will be present. The CMS measurement was the first to show that in overlapping multiplicity $\Delta\gamma$ measurements are quantitatively similar between $p + Pb$ and $Pb + Pb$ [11]. STAR measurements performed in $p + Au$ and $d + Au$ systems show similar and in fact larger charge separation measured in terms of the scaled quantity $\Delta\gamma/v_2 \times N_{\rm ch}$ than the same in $Au + Au$ measurements [14]. Such observations are striking as they tell us that a very large value of $\Delta\gamma$ is expected even for 100% background scenario.

The following question is often asked. Does measurement in small systems completely rule out CME? Why do we still pursue the CME search? There are several reasons for not abandoning CME search in heavy-ion collisions based on the observations from small collision systems. It is already known that $\Delta\gamma$ in heavy-ion collisions suffer from major background, the possible existence of CME driven signal has become more of a quantitative question. Therefore only a quantitative baseline will serve our purpose. So a better question to ask is whether small system measurements can provide direct quantitative baseline for heavy-ions. Heavy-ion measurements for CME search are performed where the system size, multiplicity do not necessarily overlap with that of small systems. It is not straightforward to extrapolate the quantitative background baselines for $\Delta\gamma$ into such unknown territories where change of physics is eminent. For example, $\Delta\gamma$ measured for $N_{\rm ch} = 10$ in $p/d+Au$ maybe a good baseline for $A+A$ at the same multiplicity but may not serve as quantitative baselines for $\Delta\gamma$ in $Au + Au$ at $N_{\rm ch} = 100$. One may try to make a projection under some working assumptions but that will lead to a qualitative baseline and defeats the major purpose of using small systems as direct quantitative baselines. This is where isobar collisions come in – that ensures measurements in two systems with very similar size and shape are compared. It is also difficult to conclude that the case of CME is ruled out entirely based on the raw $\Delta\gamma$ measurements between $p/d + Au$ and $Au + Au$. In lieu of which several variants of $\Delta\gamma$, as well as alternative observable such as $R-$observable, signed balance function has been developed to quantify the signals of CME [22, 23]. The measurements based on $R-$observable show qualitative difference in $p/d + Au$ and $Au + Au$ [24] – that is discussed in the following section.

4. The way forward

With the aforesaid introduction on the challenges to disentangle CME from non-CME background I would like to now proceed with the possible solutions to overcome such a problem. Many cleaver ideas have been proposed and applied to existing data. The general consensus is that measurement from the isobar collisions (Ru+Ru that has $10 - 18\%$ higher B-field than Zr+Zr) provides the best solution to this problem. In following sections of this conference proceedings I would like to mention a few such recent efforts such as: 1) Differential measurements of $\Delta\gamma$ to identify and quantify backgrounds, 2) measurement of higher order harmonics of γ-correlator, 3) exploiting the relative charge separation across participant and spectator planes, 4) the use of R-observable to measure charge separation and 5) the use of signed balance function. The first three approaches are based on aforementioned three-particle correlator and the last two employ slightly different approaches to quantify charge separation. There have been many more developments in the recent times and also many LHC measurements have been performed but I will specifically focus on these five approaches because they will be explored with the isobar data. The following five sections describe these procedures in brief with comments on the outlook for isobar blind analysis.

5. Differential measurements of $\Delta\gamma$ to identify and quantify background

5.1. Invariant mass dependence of charge separation

Differential measurements of $\Delta\gamma$ with invariant mass and relative pseudorapidity provide interesting prospects to identify and quantify the sources of flow and non-flow driven backgrounds. The idea to use invariant mass is simple and was first introduced in Ref [25]. Resonances are widely identified by observing structures in the invariant mass spectra of the decay daughters. Take a pair of opposite sign pions for example, a large fraction of them come from the neutral resonances that show up in the invariant mass spectrum of $m_{inv}(\pi^+ + \pi^-)$. If we restrict the analysis to pairs of pions, differential measurements of $\Delta\gamma$ with $m_{inv}(\pi^+ + \pi^-)$ should also show similar peak like structures if background from neutral resonances dominate the charge separation. Indeed similar peak structures are observed and a careful analysis is performed by STAR collaboration to extract the possible fraction of CME signals from measurements [15]. This analysis relies on the assumption that CME signals do not show peak like structures in $m_{inv}(\pi^+ + \pi^-)$ therefore calls for more theoretical inputs in this direction.

5.2. Relative pseudorapidity dependence

The relative pseudorapidity dependence of azimuthal correlations are widely studied to identify sources of long-range components that are dominated by early time dynamics as compared to late time correlations that are prevented by causality to appear as short-range correlations. The same can be extended to charge dependent correlations that provides the impetus to explore the dependence of $\Delta\gamma$ on the pseudorapidity gap between the charge carrying particles $\Delta\eta_{ab} = |\eta_a - \eta_b|$ in $\langle \cos(\phi_a^\alpha + \phi_b^\beta - 2\Psi_{RP})\rangle$. Such measurements have been performed in STAR

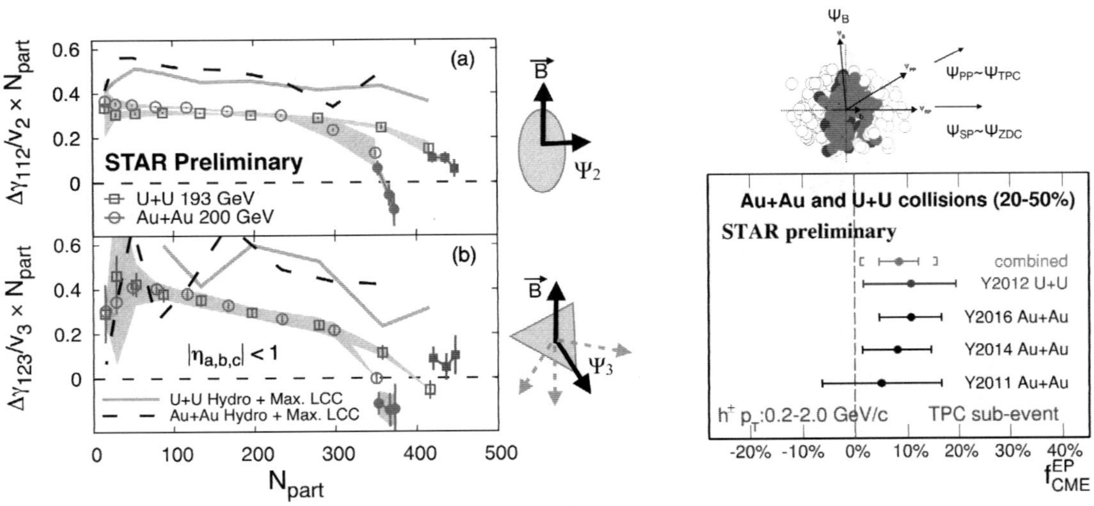

Figure 1. (Left) Measurement of charge separation along second and third order event planes in Au+Au and U+U collisions. (Right) Fraction of possible CME signal in the measurement of $\Delta\gamma$ with respect to spectator and participant planes [27].

with Au+Au and U+U data. It turns out that the possible sources of short range correlations due to photon conversion of $e^+ - e^-$, HBT and Coulomb effects can be identified and described as Gaussian peaks at small $\Delta\eta_{ab}$, the width and magnitude of which strongly depend on centrality and system size. Going to more peripheral centrality bins it becomes harder and harder to identify such components as they overlap with sources of di-jets fragmentation that dominate both same-sign and opposite sign correlations. An effort to decompose different components of $\Delta\gamma$ via study of $\Delta\eta_{ab}$ can be challenging although a clear sign of different sources of correlations are visible in change of shape of individual same-sign and opposite sign measurements of γ-correlator [26].

In any case, these differential measurements of $\Delta\gamma$ in isobar collisions provide the prospect to extract the $m_{inv}(\pi^+ + \pi^-)$ and $\Delta\eta$ dependence of CME signals that will provide much deeper insights on the origin of the effect.

6. Mixed harmonics measurements with second and third order event planes

In order to proceed in this section it is better to rewrite the conventional γ-correlator by a more general notation as $\gamma_{112} = \langle\cos(\phi_a^\alpha + \phi_b^\beta - 2\Psi_2)$. The idea is to measure charge separation across the third harmonic event plane by constructing a new correlator $\Delta\gamma_{123} = \gamma_{123}(OS) - \gamma_{123}(SS)$, where $\gamma_{123} = \langle\cos(\phi_a^\alpha + 2\phi_b^\beta - 3\Psi_3)\rangle$ that was introduced by CMS collaboration in Ref [13]. Since the Ψ_3 plane is random and not correlated to B-field direction (see Fig.1), γ_{123} is purely driven by non-CME background, the contribution of which should go as v_3/N. This is very useful to contrast signal and background scenario by comparing the measurements in two isobaric collision systems. Since Ru+Ru has larger B-field than Zr+Zr but have comparable background, the case for CME would be as follows: $(\Delta\gamma_{112}/v_2)^{\mathrm{Ru+Ru}}/(\Delta\gamma_{112}/v_2)^{\mathrm{Zr+Zr}} > 1$ and $(\Delta\gamma_{112}/v_2)^{\mathrm{Ru+Ru}}/(\Delta\gamma_{112}/v_2)^{\mathrm{Zr+Zr}} > (\Delta\gamma_{123}/v_3)^{\mathrm{Ru+Ru}}/(\Delta\gamma_{123}/v_3)^{\mathrm{Zr+Zr}}$. Fig.1 (left) shows the measurement of these observables in U+U and Au+Au collisions. Within the uncertainties of the measurements no significant difference in the trend of $\Delta\gamma_{112}/v_2$ and $\Delta\gamma_{123}/v_3$ is observed for the two collision systems except for the very central events. Predictions from hydrodynamic model calculations with maximum possible strength of local charge conservation [17] is shown

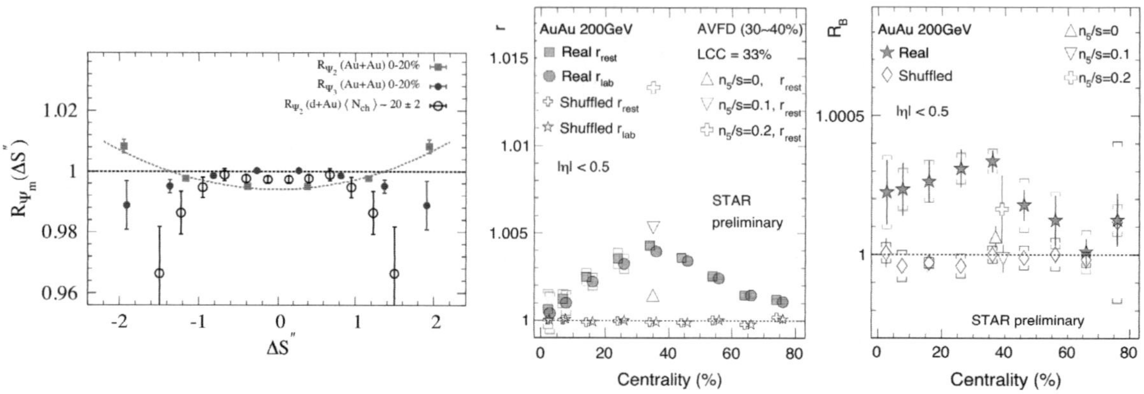

Figure 2. (Left) The R-observable shown for different collision systems, concave shape is consistent with CME expectation [24]. (Right) The two main quantities r and R_B derived from signed balance function, deviation from unity is consistent with CME expectation [30].

on the same plot. Overall observation indicates background dominate the measurements and a similar analysis of the isobar data is highly anticipated.

7. Charge separation along participant and spectator planes

This analysis makes use of the fact that B-field driven signal is more correlated to spectator plane in contrast to flow driven background which is maximum along the participant planes. The idea was first introduced in Ref [28] and later on followed up in Ref [29]. It requires measurement of $\Delta\gamma$ with respect to the plane of produced particles, a proxy for participant plane as well as with respect to the plane of spectators. In STAR the two can be done by using Ψ_2 from TPC and Ψ_1 from ZDC respectively. The approach is based on three main assumptions: 1) measured $\Delta\gamma$ has contribution from signal and background that can be expressed as $\Delta\gamma = \Delta\gamma^{\text{bkg}} + \Delta\gamma^{\text{sig}}$, 2) the background contribution to $\Delta\gamma$ should follow the scaling $\Delta\gamma^{\text{bkg}}(\text{TPC})/\Delta\gamma^{\text{bkg}}(\text{ZDC}) = v_2(\text{TPC})/v_2(\text{ZDC})$ and, 3) the signal contribution to $\Delta\gamma$ should follow the scaling $\Delta\gamma^{\text{sig}}(\text{TPC})/\Delta\gamma^{\text{sig}}(\text{ZDC}) = v_2(\text{ZDC})/v_2(\text{TPC})$. The first two have been known to be working assumptions, widely used for a long time and can be used to test the case of CME [29] if $(\Delta\gamma/v_2)(\text{ZDC})/(\Delta\gamma/v_2)(\text{TPC}) > 1$. The validity of the last one was studied and demonstrated in Ref [28]. Using all three equations one can extract [27] the fraction of possible CME signal $f_{\text{CME}} = \Delta\gamma^{\text{sig}}/\Delta\gamma$ in a fully data-driven way as shown in Fig.1(right). This analysis will be done with the isobar data and the case for CME will be $f_{\text{CME}}^{\text{Ru+Ru}} > f_{\text{CME}}^{\text{Zr+Zr}} > 0$.

8. Alternate measure: The novel R-observable

The R-observable is actually a distribution, introduced in Ref [22], and defined as the ratio of two distribution functions of the quantity ΔS parallel and perpendicular to B-field direction defined as $R_{\Psi_m}(\Delta S) = C_{\Psi_m}(\Delta S)/C^{\perp}_{\Psi_m}(\Delta S)$. Here ΔS measures the difference in the dipole moment of the positive and negative charge in an event (see Ref [22] for details). The shape of $R_{\Psi_2}(\Delta S)$ will be sensitive to CME as well as non-CME background whereas $R_{\Psi_3}(\Delta S)$ is purely driven by non-CME background and serves as a baseline. Model calculations have established several unique features of this observable: 1) presence of CME signal will lead to a concave shape of the $R_{\Psi_2}(\Delta S)$, 2) increasing strength of CME signal will increase the concavity of $R_{\Psi_2}(\Delta S)$, 3) in presence of CME, the concavity of $R_{\Psi_2}(\Delta S)$ will be larger than that of $R_{\Psi_3}(\Delta S)$. The measurement of R_{Ψ_m} is shown in Fig.2. The quantity $\Delta S''$ shown is a slight variant of

(ΔS) that incorporates correction for particle number fluctuations and event plane resolution. The observation of Fig.2 indicates more concave shape for R_{Ψ_2} compared to R_{Ψ_3} in Au+Au whereas flat or convex shapes for $p/d + Au$ indicating that the measurements are consistent with expectations of CME [24]. For isobar collisions the case of CME will be confirmed if: 1) a concave shape is observed for the ratio of the observables $R_{\Psi_2}(\Delta S)^{\mathrm{Ru+Ru}}/R_{\Psi_2}(\Delta S)^{\mathrm{Zr+Zr}}$ and 2) the concavity should be weaker for $R_{\Psi_3}(\Delta S)^{\mathrm{Ru+Ru}}/R_{\Psi_3}(\Delta S)^{\mathrm{Zr+Zr}}$.

9. Alternate measure: The signed Balance function

A very recently proposed observable to search for CME is the signed balance function (SBF) [23]. The idea is to account for the ordering of the momentum of charged pairs measured by the width of SBF that is expected to be different for out-of-plane as compared to in-plane measurement captured in the ratio r_{lab}. In addition, one can also account for the boost due to collective expansion of the system that forces all pairs to move in the same direction and measure the ratio in pair's rest frame r_{rest}. In presence of CME the individual ratios as well as the double ratio $R_B = r_{\mathrm{rest}}/r_{\mathrm{lab}}$ is expected to be greater than unity. The preliminary measurements shown in Fig.2 (right) from STAR in Au+Au 200 GeV seems to be consistent with CME expectation. This observable will be studied with the isobar data in STAR but not as a part of the blind analysis and the CME expectation will be: 1) $r(\mathrm{Ru+Ru}) > r(\mathrm{Zr+Zr})$, and 2) $R_B(\mathrm{Ru+Ru}) > R_B(\mathrm{Zr+Zr})$.

10. Steps for blind analysis of the isobar data from STAR

10.1. Modality of isobar running at RHIC

It is better to start with a short background on the activities that preceded the isobar blind analysis in STAR. The idea of colliding isobar, particularly Ru+Ru and Zr+Zr to make a decisive test of CME was proposed by Voloshin in Ref [31], the same paper which also proposed to use Uranium collisions to disentangle signal and background of CME. The possible difference in the signals relies on $10-18\%$ higher B-field in Ru+Ru compared to Zr+Zr [32] in contrast to about 4% difference in flow driven background [17]. Such estimates are sensitive to details of shapes, charge distribution and neutron skin thickness of the two isobar nuclei [32, 33, 34]. In the 2017-18 RHIC beam user request [35] STAR collaboration therefore proposed to collect data for two 3.5 week runs in the year 2018. The projection was based on the prospect of achieving five-sigma significance or better in a scenario where the measurement of $\Delta\gamma$ has 80% non-CME background. This however corresponds to the fact that the systematic uncertainty in the measurements has to be within a few percent and below the statistical significance of the measurements, something that has never been attempted before in the correlation measurements from STAR or by other heavy-ion collision experiment in recent times to the best of my knowledge. This started a large scale collaboration wide effort in synergy with the RHIC collider accelerator department to plan for the isobar running in the year 2018. Based on the studies of previous years of data from $Au + Au$ and $U + U$ collisions several major sources of systematics in the measurement of $\Delta\gamma$ were identified. The major sources include: run-to-run variation of detector response due to loss of acceptance, change in efficiency and variation in luminosity that affects the number of reconstructed tracks in the Time Projection Chamber. This eventually leads to uncorrectable systematic uncertainties in $\Delta\gamma$. In order to minimize such systematics the proposal were to: 1) switch species in RHIC between stores e.g., in orders like Ru+Ru, Zr+Zr, Ru+Ru and so on and, 2) keep long stores to level the luminosity aiming for specific rates in the coincidence measurements of beam fragments by the STAR zero-degree calorimeters. The aim was to maintain exact balance of run and detector conditions for the two species so that observations in the two systems are equally affected and can later on be largely eliminated in the ratios of observables.

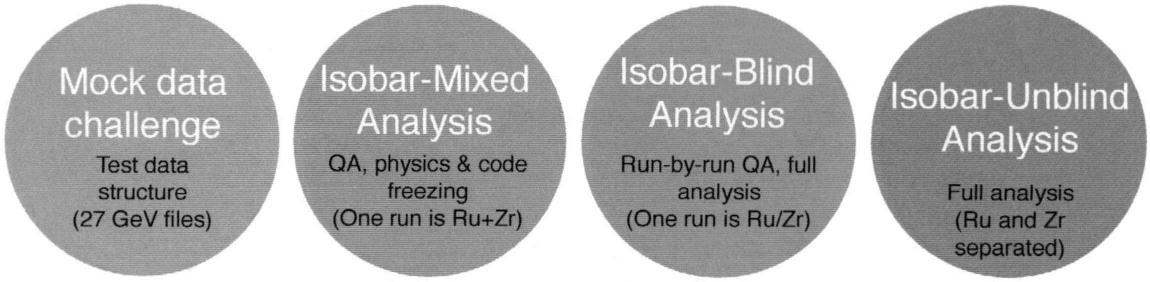

Figure 3. The steps of isobar blind analysis. This cartoon is based on the procedure for the blind analysis of isobar data that have been outlined in Ref [36].

10.2. Blinding of data sets and preparation for analysis

With the successful conclusion of the isobar run in the year 2018 STAR experiment collected more than 3 billion events for each isobar species. The next step was to develop the plans for a blind analysis, the main idea behind which is to eliminate predetermined biases. A total of five institutional groups are expected to perform the analysis of the data. The analysts from each group will focus on a specific aspect of the analysis described in the previous section although in many cases there are substantial overlap in some analyses that will help cross check the results. An important part of the blind analysis is the blinding of the data. The details of the blinding of the data structure is decided by members of a blinding-committee who are not part of the team of analysts and will work in close collaboration with STAR experts who are part of the production team. The idea is to provide the analysts the access to data in files where species-specific information are disguised or removed before the final step of unblinding. A careful consideration is taken by the blinding-committee to make sure the essential information available to do the analysis specific quality assurance of the data by the analysts. Some of the quality assurance, calibration and centrality determination that require species information are done only by STAR experts who are not a part of the team of analysts. Above all, the main goal of the committee is to make sure that under no circumstances physics analysts can access un-blinded data that can jeopardize the blind analysis. For example, all the data sets are produced with pseudo-run-number that cannot be used by the analysts to retrieve the exact species information.

10.3. Methods for the isobar blind analysis

The detailed procedure for the blind analysis of isobar data have been outlined in Ref [36]. Figure.3 is a cartoon that summarizes the four steps and the main idea.

In the zeroth step shown in (by orange circle) the extreme left of Fig.3 is the mock data challenge which is not exactly a step of the isobar data analysis but a crucial step to familiarize the analysts with the technicalities of the data structures that have been specifically designed for blind analysis.

The first step shown in Fig.3 (by green circle) as the "isobar-mixed analysis" or "mixed-blind analysis" is truly the first step of blind analysis. This is also the most challenging steps from the point of view of the analysts. In this step the analysts are provided with data sample where each run comprise of events that are "mix" samples from two species. In this step the analysts perform the full quality assurance (QA) and physics analysis of the data, document every details of steps of the procedure and freeze the codes. After the completion of this step no changes to the analysis code is permissible. Also, no changes in the analysis procedure is allowed. The only permissible change in the following step is to reject bad runs or pile-up events. However, in order to avoid predetermined bias in analysis such rejection cannot be done arbitrarily and

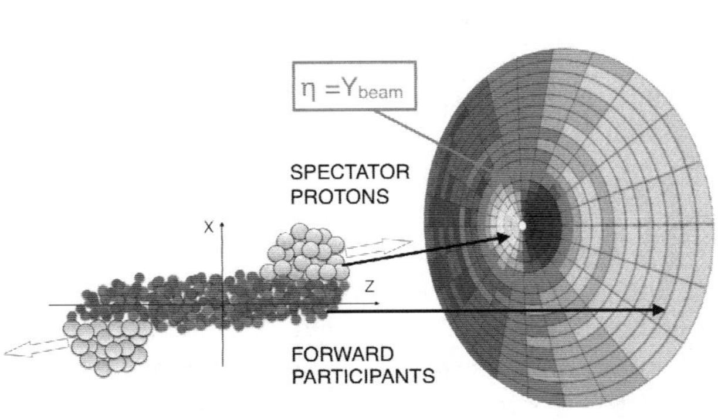

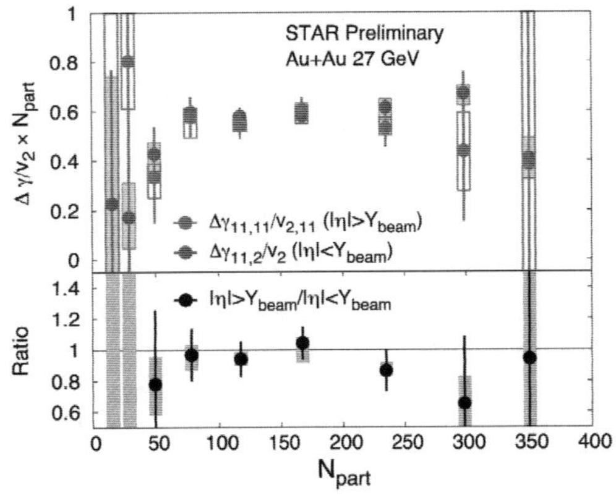

Figure 4. (Left) Figure showing EPD detector acceptance cover beam rapidity and detecting both forward participants and spectators in 27 GeV Au+Au collisions. (Right) γ-correlators scaled by v_2 across different event-planes and double ratio of spectators/participant event planes which should be unity for no-CME scenario.

an automated algorithm must be developed in this step and the related codes have to be frozen. The stability of the automated QA algorithm is tested with some of the existing data sets of Au+Au and U+U collisions.

The second step shown in Fig.3 (by blue circle) is referred to as the "isobar-blind analysis" or "unmixed-blind analysis". From this step on-wards the analysts are allowed to run their previously frozen codes. The main purpose of this step is to perform run-by-run QA of the data sample. For this the analysts are provided with files each of which contain data from a single species that is either Ru or Zr. However, there are two conditions: the files contain limited number of events that cannot lead to any statistically significant result and the species information is not revealed. Although a pseudo-run-number is used for each file, the time ordering is preserved with a unique mapping that is unknown to the analysts. It is important to maintain the time ordering to identify time-dependent changes in detectors and run conditions as a part of the run-by-run quality assurance. With this limited data sample the analysts need run the frozen automated algorithm to identify bad runs. A similar automated algorithm is also used for identifying and rejecting bad runs. After this step no more changes are allowed in terms of QA.

The final step of isobar blind analysis is shown by red circle in Fig.3 is referred to as "isobar-unblind" analysis. In this step the species information will be revealed and the physics results will be produced by the analysts using the previously frozen codes. The finding from this step will be directly be submitted for publication without any kind of alteration. If a mistake is found in the analysis code, the erroneous results will also accompany the corrected results.

11. Post-isobar era and prospects for CME search at lower collision energies

Regardless of the outcome of the measurements with the isobar program, that will be performed at the top RHIC energy, one question will remain. What happens at lower collision energy? In this context a new idea has emerged. The newly installed event-plane detector (EPD) upgrade provides a new capability at STAR towards CME search at lower collision energy and for the Beam Energy Scan phase-II program [37]. The idea is simple, at lower energies EPD acceptance ($2.1 < |\eta| < 5.1$) falls in the region of beam rapidity ($Y_{\rm beam}$) and can measure the plane of strong directed flow (Ψ_1) of spectator protons, beam fragments and stopped protons, therefore strongly correlated to the B-field direction (See fig4). The next step is to measure $\Delta\gamma$ with respect to

Ψ_1 and compare it with the measurement of $\Delta\gamma$ along Ψ_2 planes from outer regions of EPD and TPC at mid-rapidity that are weakly correlated to the B-field directions. A test of CME scenario will be to see if large difference is observed in the measurements. First preliminary measurements from STAR as shown in Fig 4 is dominated by uncertainty but seems to show a lot of prospects for the CME search at lower energies.

12. Summary

Despite several challenges experimental efforts have been continued towards disentangling the CME signals from non-CME background in the measurement of charge separation across reaction plane. The highly anticipated results from the blind analysis of isobar collisions data provides us the best opportunity to make a decisive test of the CME in heavy-ion collisions.

13. Reference

[1] Kharzeev D and Pisarski R D 2000 *Phys.Rev.* **D61** 111901 (*Preprint* hep-ph/9906401)
[2] Kharzeev D 2006 *Phys.Lett.* **B633** 260–264 (*Preprint* hep-ph/0406125)
[3] Voloshin S A 2004 *Phys. Rev.* **C70** 057901 (*Preprint* hep-ph/0406311)
[4] Abelev B I *et al.* (STAR) 2009 *Phys. Rev. Lett.* **103** 251601 (*Preprint* 0909.1739)
[5] Abelev B I *et al.* (STAR) 2010 *Phys. Rev.* **C81** 054908 (*Preprint* 0909.1717)
[6] Abelev B *et al.* (ALICE) 2013 *Phys. Rev. Lett.* **110** 012301 (*Preprint* 1207.0900)
[7] Adamczyk L *et al.* (STAR) 2013 *Phys. Rev.* **C88** 064911 (*Preprint* 1302.3802)
[8] Adamczyk L *et al.* (STAR) 2014 *Phys. Rev.* **C89** 044908 (*Preprint* 1303.0901)
[9] Adamczyk L *et al.* (STAR Collaboration) 2014 *Phys.Rev.Lett.* **113** 052302 (*Preprint* 1404.1433)
[10] Adam J *et al.* (ALICE) 2016 *Phys. Rev.* **C93** 044903 (*Preprint* 1512.05739)
[11] Khachatryan V *et al.* (CMS) 2016 *Phys. Rev. Lett* [Phys. Rev. Lett.118,122301(2017)] (*Preprint* 1610.00263)
[12] Acharya S *et al.* (ALICE) 2018 *Phys. Lett.* **B777** 151–162 (*Preprint* 1709.04723)
[13] Sirunyan A M *et al.* (CMS) 2018 *Phys. Rev.* **C97** 044912 (*Preprint* 1708.01602)
[14] Adam J *et al.* (STAR) 2019 *Phys. Lett. B* **798** 134975 (*Preprint* 1906.03373)
[15] Adam J *et al.* (STAR) 2020 (*Preprint* arXiv:2006.05035)
[16] Pratt S 2010 (*Preprint* 1002.1758)
[17] Schenke B, Shen C and Tribedy P 2019 *Phys. Rev.* **C99** 044908 (*Preprint* 1901.04378)
[18] Jiang Y, Shi S, Yin Y and Liao J 2018 *Chin. Phys. C* **42** 011001 (*Preprint* 1611.04586)
[19] Belmont R and Nagle J 2017 *Phys. Rev. C* **96** 024901 (*Preprint* 1610.07964)
[20] Schenke B, Shen C and Tribedy P 2020 *Phys. Lett. B* **803** 135322 (*Preprint* 1908.06212)
[21] Kharzeev D, Tu Z, Zhang A and Li W 2018 *Phys. Rev. C* **97** 024905 (*Preprint* 1712.02486)
[22] Magdy N, Shi S, Liao J, Ajitanand N and Lacey R A 2018 *Phys. Rev. C* **97** 061901 (*Preprint* 1710.01717)
[23] Tang A 2020 *Chin. Phys. C* **44** 054101 (*Preprint* 1903.04622)
[24] Adam J *et al.* (STAR) 2020 (*Preprint* arXiv:2006.04251)
[25] Zhao J, Li H and Wang F 2019 *Eur. Phys. J. C* **79** 168 (*Preprint* 1705.05410)
[26] Tribedy P (STAR) 2017 *Nucl. Phys. A* **967** 740–743 (*Preprint* 1704.03845)
[27] Zhao J (STAR) 2020 (*Preprint* 2002.09410)
[28] Xu H j, Zhao J, Wang X, Li H, Lin Z W, Shen C and Wang F 2018 *Chin. Phys.* **C42** 084103 (*Preprint* 1710.07265)
[29] Voloshin S A 2018 *Phys. Rev.* **C98** 054911 (*Preprint* 1805.05300)
[30] Lin Y (STAR) 2020 (*Preprint* 2002.11446)
[31] Voloshin S A 2010 *Phys. Rev. Lett.* **105** 172301 (*Preprint* 1006.1020)
[32] Deng W T, Huang X G, Ma G L and Wang G 2018 *Phys. Rev. C* **97** 044901 (*Preprint* 1802.02292)
[33] Xu H J, Wang X, Li H, Zhao J, Lin Z W, Shen C and Wang F 2018 *Phys. Rev. Lett.* **121** 022301 (*Preprint* 1710.03086)
[34] Hammelmann J, Soto-Ontoso A, Alvioli M, Elfner H and Strikman M 2019 (*Preprint* 1908.10231)
[35] STAR 2017 BUR https://drupal.star.bnl.gov/STAR/system/files/STAR_BUR_Run1718_v22_0.pdf
[36] Adam J *et al.* (STAR) 2019 (*Preprint* 1911.00596)
[37] Adams J *et al.* 2019 (*Preprint* 1912.05243)

Probing the QCD phase structure using event-by-event fluctuations

Tapan K. Nayak
CERN, CH-1211, Geneva 23, Switzerland, and
National Institute of Science Education and Research, HBNI, Bhubaneswar, India
E-mail: Tapan.Nayak@cern.ch

Abstract. Heavy-ion collisions at relativistic energies probe matter at extreme conditions of temperatures and energy densities. The study of event-by-event fluctuations of experimental observables is crucial to probe the QCD phase transition, locate the critical point, and learn about the associated critical phenomena. At the critical point, all thermodynamic quantities behave anomalously. Fluctuation measurements provide access to thermodynamic response functions. We discuss the methods for obtaining the isothermal compressibility using particle multiplicity fluctuation, and specific heat using fluctuations in mean transverse momentum, temperature, and energy. Lattice QCD calculations have predicted non-monotonic behavior in the higher-order cumulants of conserved quantities at the critical point. Fluctuations in the multiplicity of charged to neutral particles have been measured to understand the formation of domains of disoriented chiral condensates. We review the recent fluctuation results as a function of collision centrality and energy from experiments at SPS, RHIC, and LHC. In addition, we propose to map the temperature fluctuations in η-ϕ plane to probe local fluctuations of temperature and energy density.

1. Introduction

According to the theory of Quantum Chromo Dynamics (QCD), under extreme conditions of temperatures and energy densities, normal hadronic matter goes through a phase transition to a system of deconfined quarks and gluons, the quark-gluon plasma (QGP). The QCD phase structure between these two distinct states of matter span a wide range of baryon chemical potential ($\mu_{\rm B}$) and temperature (T) as shown in Fig. 1. Lattice QCD calculations indicate that at vanishing $\mu_{\rm B}$, the transition from the QGP to a hadron gas is a smooth crossover [1, 2, 3, 4], while at large $\mu_{\rm B}$, the phase transition is of first order [5, 6]. The point in T and $\mu_{\rm B}$ where the first order transition ends and instigates a crossover transition is denoted as the QCD critical point. Theoretical and experimental studies explore the rich landscape of the QCD phase diagram to understand the nature of the phase transition, locate the critical point, and to learn about the properties of the matter formed.

The experimental program to study the QCD phase structure started more than three decades ago at Bevelac, Berkeley and since then has covered four generations of experiments at the Brookhaven National Laboratory (BNL) and CERN. The collisions at the Large Hadron Collider (LHC) and at the top energy at the Relativistic Heavy Ion Collider (RHIC) probe the conditions at low $\mu_{\rm B}$. The beam energy scan (BES) program at RHIC [7, 8] is specially designed to probe the location of the critical point by varying the collision energy at close intervals in T-$\mu_{\rm B}$. In

Journal of Physics: Conference Series **1602** (2020) 012003 doi:10.1088/1742-6596/1602/1/012003

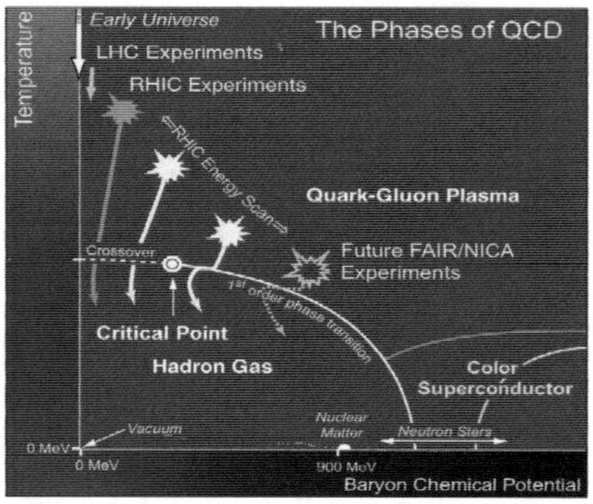

Figure 1. A schematic QCD phase diagram in the temperature (T) and baryonic chemical potential (μ_B) plane. The regions probed by different accelerator facilities are indicated.

future, this program will be complemented by upcoming facilities at Dubna, Russia, and GSI, Germany to explore large μ_B regions in the phase diagram. The regions probed by different accelerator facilities are indicated in Fig. 1.

Fluctuations play a crucial role in the study of phase transition and any associated critical phenomena. Event-by-event fluctuations in a number of observables have been predicted as signatures of the QCD phase transition and the critical point [9, 10, 11]. Several thermodynamic quantities show varying fluctuation patterns when the system goes through the phase boundary. The defining characteristics of the QCD phase transition are the abrupt changes in the physical properties of the system which can be inferred through the analysis of fluctuations of different observables. At the critical point, the fluctuations are expected to be very large. The main fluctuation signatures emanate in the form of event-by-event measurement in the number of particles, momenta of particles, as well as the spatial and energy driven patterns of multiplicity distributions. In this article, we discuss some of the fluctuation techniques, experimental results, and future prospects.

- Thermodynamic response functions: Response functions such as isothermal compressibility (k_T), specific heat (c_v), and speed of sound, are related by the equation of state (EOS), which governs the evolution of the system. Nature of phase transitions in a system can be understood by the measurement of thermodynamic response functions. These quantities can be accessed experimentally by the fluctuation of measured quantities. The heat capacity is related to the fluctuations in temperature [12, 13], whereas in the grand canonical ensemble (GCE) framework k_T is related to the fluctuation in particle multiplicity [14]. Skewness of mean transverse momentum ($\langle p_T \rangle$) fluctuations have recently been proposed as a probe of hydrodynamic behavior in nuclear collisions [15]. By measuring the event-by-event fluctuations in particle multiplicity (N), $\langle p_T \rangle$, and mean transverse energy, we can get access to the response functions.
- Fluctuations of conserved quantities: Lattice QCD calculations reveal that the higher order cumulants of conserved quantities, such as net-charge (Q), net-proton (B), and net-strangeness (S), within a limited acceptance, are proportional to the powers of the correlation length and are expected to diverge at the critical point [16, 17]. Experimentally, it is possible to measure Q, B, and S on an event-by-event basis and obtain the cumulants of these distributions. In addition, off-diagonal cumulants explore the flavor carrying susceptibilities of the system [18, 19, 20, 21]. Assuming that the signal at freeze-out survives dissipation during the evolution of the fireball from the hadronization stage, the higher cumulants can be used as one of the preferred tools for locating the critical point.
- Disoriented chiral condensates (DCC): Disoriented chiral condensates (DCC), localized in phase space, have been predicted to be formed in high energy heavy-ion collisions when

Journal of Physics: Conference Series **1602** (2020) 012003 doi:10.1088/1742-6596/1602/1/012003

the chiral symmetry is restored at high temperatures [22, 23]. Anomalous event-by-event fluctuations of the neutral to charged pions as well as neutral to charged kaons have been predicted as signatures of the formation of DCC. A fresh look at RHIC and LHC energies is needed to infer about the formation of the DCC domains.

- Fluctuation map: The physics of heavy-ion collisions at ultra-relativistic energies has often been compared to the Big Bang phenomenon of the early Universe. Observation of the cosmic microwave background radiation (CMBR) by various satellites confirms the Big Bang evolution, inflation and provide important information regarding the early Universe and its evolution with excellent accuracy. The matter produced at extreme conditions of energy density and temperature in heavy-ion collisions is a Big Bang replica on a tiny scale [24, 25, 26]. We propose to map the temperature fluctuations in η-ϕ plane to probe local fluctuations of temperature and energy density.

2. Multiplicity fluctuation and estimation of isothermal compressibility

Isothermal compressibility (k_{T}) is the measure of the relative change in volume with respect to change in pressure [14],

$$k_T|_{T,\langle N\rangle} = -\frac{1}{V}\left(\frac{\partial V}{\partial P}\right)\Big|_{T,\langle N\rangle} \tag{1}$$

where V, T, P represent volume, temperature, and pressure of the system, respectively, and $\langle N\rangle$ stands for the mean yield of the particles. In the Grand Canonical Ensemble (GCE) framework, the variance of the number of particles is directly related to k_{T}:

$$\omega_{\mathrm{ch}} = \frac{k_{\mathrm{B}}T\langle N\rangle}{V}k_{\mathrm{T}}, \tag{2}$$

where ω_{ch} is the scaled variance of multiplicity distribution. This formalism may be applied to experimental measurements of multiplicity at mid-rapidity, as energy and conserved quantum numbers are exchanged with the rest of the system.

At the chemical freeze-out surface, the inelastic collisions cease, and thus the hadron multiplicities get frozen. While the ensemble average thermodynamic properties like the temperature and volume can be extracted from the mean hadron yields, k_{T} can be accessed through the measurements of the event-by-event multiplicity fluctuations. This has been explored in Ref. [27] and the estimated values of k_{T} are shown in Fig. 2 for available experimental data and estimations from the HRG calculations. The HRG calculations reveal a sharp increase in the value of k_{T} around $\sqrt{s_{\mathrm{NN}}}\sim$20 GeV. It will be interesting to explore the behavior of k_{T} at BES-II energies and future facilities.

3. Temperature and $\langle p_{\mathrm{T}}\rangle$ fluctuations and estimation of specific heat

Heat capacity (C) is a response function which expresses how much the temperature of a system changes when heat is transferred to it. The specific heat (c_{v}) is heat capacity divided by the number of charged particles. Equivalently, for a system in thermal equilibrium to a bath at temperature, T, we can write [12, 13]:

$$C = \left(\frac{\partial E}{\partial T}\right)_V = \frac{(\langle E^2\rangle - \langle E\rangle^2)}{\langle T\rangle^2}. \tag{3}$$

where V and E are volume and energy of the system, respectively. By expressing the transverse momentum spectra of emitted particles in terms of event-by-event temperature fluctuation

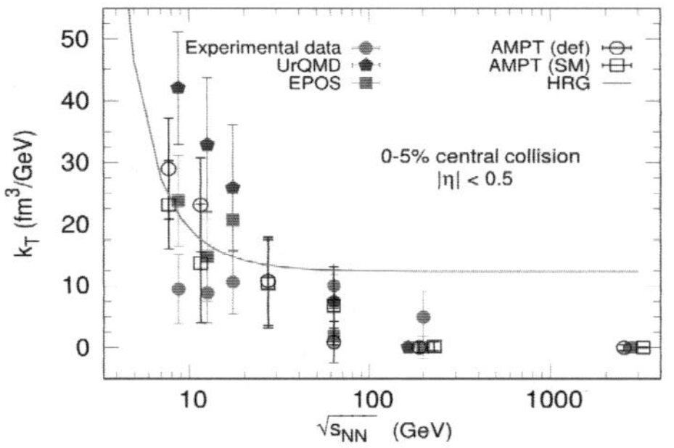

Figure 2. Isothermal compressibility, $k_{\rm T}$ as a function of collision energy for central collisions [27].

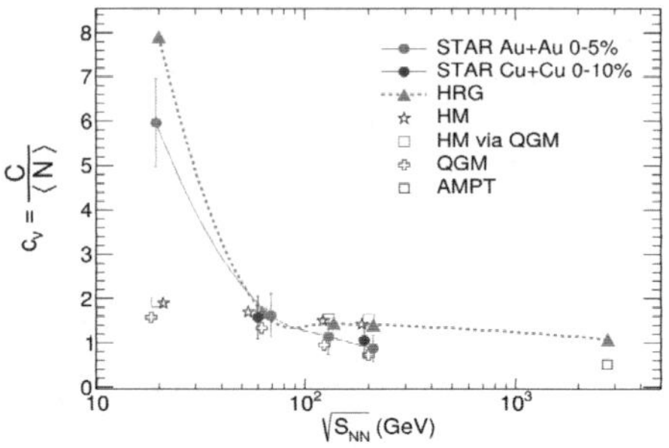

Figure 3. Specific heat, $c_{\rm v}$ as a function of collision energy for central collisions [28].

$(\Delta T = T - \langle T\rangle)$, we obtain:

$$P(T) \sim \exp[-\frac{C}{2}\frac{(\Delta T)^2}{\langle T\rangle^2}], \qquad \frac{1}{C} = \frac{(\langle T^2\rangle - \langle T\rangle^2)}{\langle T\rangle^2}. \tag{4}$$

Heat capacity thus can be estimated from the fluctuations in energy or temperature. For a system in equilibrium, the mean values of T and E are related by an equation of state. However, the fluctuations in energy and temperature have very different behavior. Energy being an extensive quantity, its fluctuation has a volume dependent component. So energy is not suited for obtaining the heat capacity. On the other hand, temperature fluctuations provide a good major for the estimation of specific heat. The temperature of the system can be obtained from the transverse momentum ($p_{\rm T}$) spectra of the emitted particles. For a range of $p_{\rm T}$ within a to b, we obtain [28]:

$$\langle p_{\rm T}\rangle = \frac{\int_a^b p_{\rm T}^2 F(p_{\rm T})dp_{\rm T}}{\int_a^b p_{\rm T} F(p_{\rm T})dp_{\rm T}} = 2T_{\rm eff} + \frac{a^2 e^{-a/T_{\rm eff}} - b^2 e^{-b/T_{\rm eff}}}{(a+T_{\rm eff})e^{-a/T_{\rm eff}} - (b+T_{\rm eff})e^{-b/T_{\rm eff}}}. \tag{5}$$

This equation links the value of event-by-event $\langle p_{\rm T}\rangle$ within a specified range of $p_{\rm T}$ to the apparent or effective temperature ($T_{\rm eff}$). The dynamical component of the $\langle p_{\rm T}\rangle$ and T distributions have been obtained for RHIC energies after subtracting the corresponding mixed event distributions. The heat capacity is calculated using the dynamical part of the fluctuation and the kinetic temperature. In Ref. [28], $\langle p_{\rm T}\rangle$ results from the STAR collaboration have been used to calculate heat capacity. The results of the specific heat as a function of collision energy are presented in Fig. 3. It shows a sharp rise in $c_{\rm v}$ below $\sqrt{s_{\rm NN}} = 62.4$ GeV. In order to probe the QCD critical point, a finer scan of beam energy at RHIC is essential.

4. Dynamical net-charge fluctuations

Net-charge fluctuations are strongly dependent on which phase the fluctuations originate. This is because, in the QGP phase, the charge carriers are quarks with fractional charges, whereas the particles in a hadron gas carry unit charge. The fluctuations in the net charge depend on the squares of the charge states present in the system, and so, the net-charge fluctuations in the QGP phase are significantly smaller compared to that of a hadron gas. The initial QGP phase is strongly gluon dominated, and so the fluctuation per entropy may further be reduced as the hadronization of gluons increases the entropy. The charge fluctuations are best studied

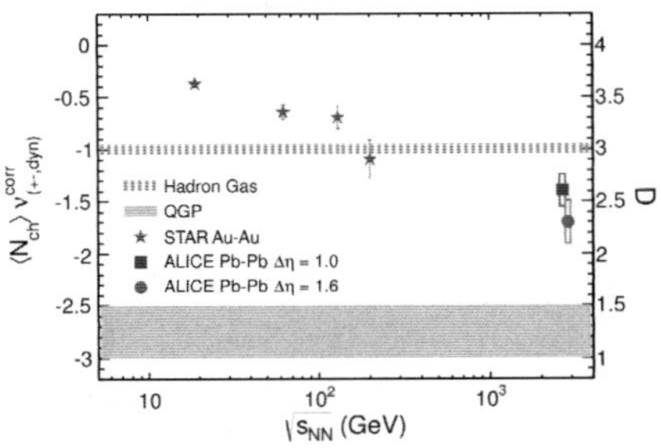

Figure 4. Collision energy dependence of the net-charge fluctuations in central collisions [29].

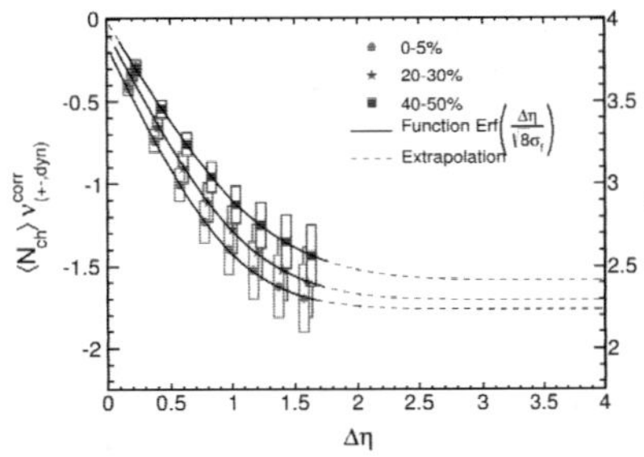

Figure 5. Growth of net charge fluctuations as a function the pseudorapidity window ($\Delta\eta$) [29].

by calculating the quantity $\nu_{(+-,\mathrm{dyn})}$, defined as:

$$\nu_{(+-,\mathrm{dyn})} = \frac{\langle N_+(N_+-1)\rangle}{\langle N_+\rangle^2} + \frac{\langle N_-(N_--1)\rangle}{\langle N_-\rangle^2} - 2\frac{\langle N_-N_+\rangle}{\langle N_-\rangle\langle N_+\rangle}, \tag{6}$$

which is a measure of the relative correlation strength of particle pairs. A negative value of $\nu_{(+-,\mathrm{dyn})}$ signifies the dominant contribution from correlations between pairs of opposite charges. The $\nu_{(+-,\mathrm{dyn})}$ has been found to be robust against random efficiency losses. It is related to the fluctuation measure (D) by:

$$\langle N_{\mathrm{ch}}\rangle\nu_{(+-,\mathrm{dyn})} \approx D - 4, \tag{7}$$

$$\text{where} \quad D = 4\frac{\langle\delta Q^2\rangle}{\langle N_{\mathrm{ch}}\rangle}. \tag{8}$$

Here $Q = N_+ - N_-$, and $N_{\mathrm{ch}} = N_+ - N_-$. Figure 4 shows the values of $\langle N_{\mathrm{ch}}\rangle\nu^{\mathrm{corr}}_{(+-,\mathrm{dyn})}$ as a function of collision energy, measured by the STAR and ALICE Collaborations [29, 30]. Here $\nu_{(+-,\mathrm{dyn})}$ is corrected for global charge conservation and finite detector acceptance. A monotonic decrease in the value of D, measured has been observed.

The measured fluctuations gets diluted during the evolution of the system from hadronization to kinetic freeze-out because of the diffusion of charged hadrons in rapidity [31, 32, 33]. This has been studied by plotting dependence of net-charge fluctuations on the width of the rapidity window as shown in Fig. 5 for Pb–Pb collisions at $\sqrt{s_{\mathrm{NN}}}$=2.76 TeV. We observe that for a given centrality bin, the D–measure shows a strong decreasing trend with the increase of $\Delta\eta$. In fact, the curvature of D has a decreasing slope with a flattening tendency at large $\Delta\eta$ windows. The data points are fitted with a functional form, $\mathrm{erf}(\Delta\eta/\sqrt{8}\sigma_{\mathrm{f}})$, which represents the diffusion in rapidity space. The value of σ_{f}, which represents the diffusion at freeze-out, turns out to be 0.41 ± 0.05 for central collisions. Taking the dissipation into account, the asymptotic value of fluctuations may represent the primordial fluctuations. It would be intriguing to make this study for a wider $\Delta\eta$ range in which most of the fluctuations are captured [34].

5. Fluctuations of conserved quantities: diagonal cumulants

Lattice QCD calculations have shown that higher order cumulants of the distributions of conserved charges are related to the corresponding higher-order thermodynamic susceptibilities and to the correlation length (ξ) of the system. These cumulants go through rapid changes near the critical point. Thus the measurement of cumulants of Q, B, and S provide a direct

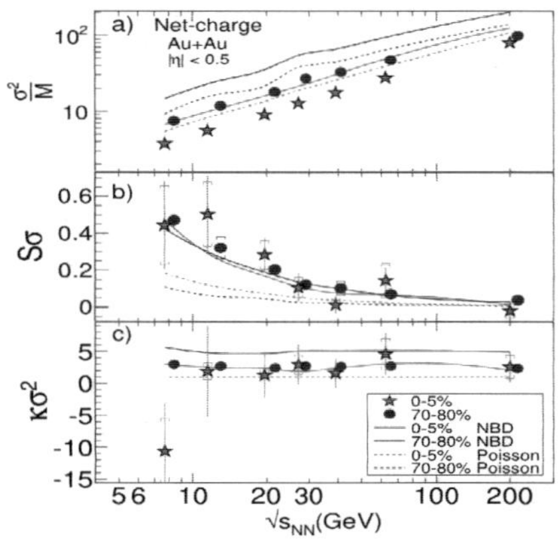

Figure 6. Collision energy dependence of the net-charge fluctuations in central collision [35].

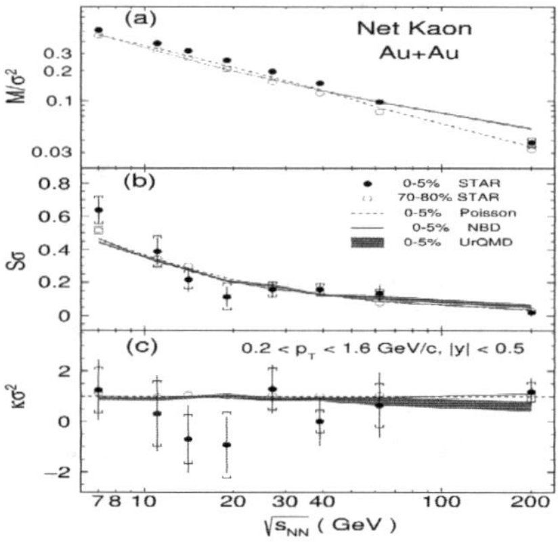

Figure 7. Collision energy dependence of the net-kaon fluctuations in central collisions [36].

correspondence to the lattice calculations and serve as an important probe for the critical point search.

The event-by-event distributions of the conserved quantities within a limited acceptance are characterized by mean (M), standard deviation (σ), skewness (S), kurtosis (κ), and other higher order cumulants. The products of the moments, such as σ^2/M, $S\sigma$ and $\kappa\sigma^2$are constructed to cancel the volume term. The net-charge multiplicity distributions directly probe the charge quantum number, whereas the net-proton and net-kaon distributions provide good proxies for the net-baryon and net-strangeness conserved quantities.

The STAR Collaboration has made extensive measurements of the cumulants of net-charge, net-kaon, and net-proton multiplicity distributions in Au–Au collisions at a wide range of energies starting from $\sqrt{s_{\rm NN}}$=7.7 GeV to 200 GeV. The collision centrality and energy dependence of the products of moments have been reported in a series of publications. Here we reproduce some of the salient features of the recent results.

The results of the cumulants of net-charge multiplicity distributions [35] for $|\eta| < 0.5$ are shown in Fig. 6. Weak centrality dependence is observed for both $S\sigma$ and $\kappa\sigma^2$ at all energies. However, within the present uncertainties, no non-monotonic behavior as a function of collision energy. The net-kaon results [36] are shown in Fig. 7. The collision energy dependence of products of cumulants is seen to be smoothly varying as a function of collision energy. No significant collision centrality dependence is observed for both M/σ^2 and $S\sigma$ at all energies. Although the collision centrality and energy dependence of $\kappa\sigma^2$ look very intriguing, no definitive statement can be made within the current experimental uncertainties.

The experimental results of the net-proton multiplicity distributions have been reported in Ref. [37, 38, 39], and summarized in Fig. 8. The collision energy dependence of $S\sigma$ shows a smooth variation as a function of energy except at very low energies. On the other hand, the nature of $\kappa\sigma^2$ variation is very different. For peripheral collisions, there is no variation with energy, whereas a non-monotonic variation (with 3.0σ significance) with beam energy is observed for $\kappa\sigma^2$. This variation could not be explained by hadron resonance gas (HRG) calculation and the UrQMD transport model simulation. This observation is most likely compatible with the theoretical predictions of the critical point. This signature has not been observed in other observables so far.

STAR collaboration has recently reported the beam energy dependence of net-Λ

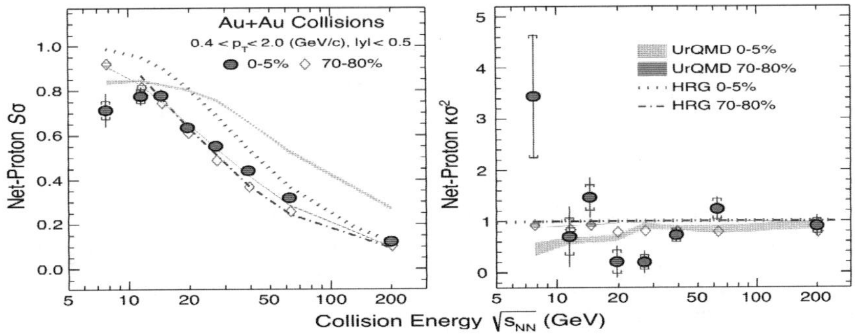

Figure 8. Collision energy dependence of the net-proton fluctuations in central and peripheral collisions [37].

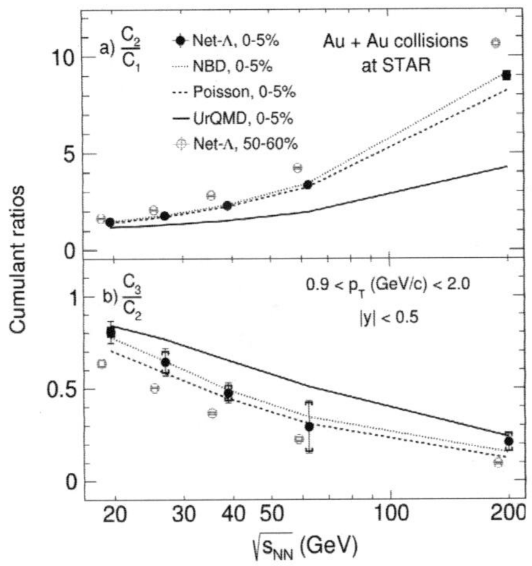

Figure 9. Beam energy dependence of net-Λ cumulant ratios, C_2/C_1, and C_3/C_2 in most central and peripheral Au–Au collisions [40].

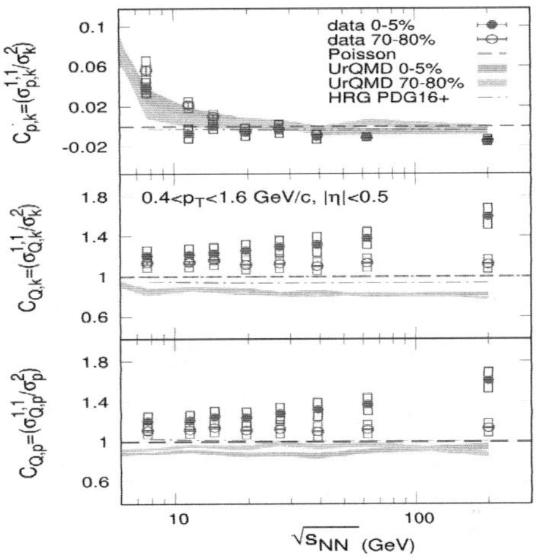

Figure 10. Beam energy dependence of cumulant ratios $C_{\mathrm{p,k}}$, $C_{\mathrm{Q,k}}$, and $C_{\mathrm{Q,p}}$, of net-proton, net-kaon and net-charge for Au–Au collisions [41].

cumulants [40], which is potentially interesting towards our comprehensive understanding of particle production mechanisms and their correlations as Λ carry both baryon and strangeness quantum number. Results of this challenging measurement in terms of the beam energy dependence of the ratios of first, second and third order cumulants are shown in Fig. 9, which show no non-monotonic behavior for the energies and cumulants studied. These results are important for the understanding of the freeze-out temperature in the context of both baryon number and strangeness conservation.

6. Off-diagonal cumulants of conserved quantities

The diagonal and off-diagonal susceptibilities of second order can be expressed in terms of second order central moments (σ):

$$\begin{pmatrix} \sigma_Q^2 & \sigma_{QB}^{1,1} & \sigma_{QS}^{1,1} \\ \sigma_{BQ}^{1,1} & \sigma_B^2 & \sigma_{BS}^{1,1} \\ \sigma_{SQ}^{1,1} & \sigma_{SB}^{1,1} & \sigma_S^2. \end{pmatrix}$$

The measurement of these observables give information on the flavor carrying susceptibilities.

The ratios of diagonal and off-diagonal elements are constructed to cancel the volume effect. In the quasiparticle picture of quarks and gluons, the ratios χ_{BS}^{11}/χ_S^2 and χ_{QS}^{11}/χ_S^2 are -1/3 and 1/3, respectively. Other ratios, like χ_{QB}^{11}/χ_B^2 has no contribution from both light and strange quarks. Thus the following ratios are constructed:

$$C_{BS} = -3\frac{\chi_{BS}^{11}}{\chi_S^2}, \qquad C_{SB} = -\frac{1}{3}\frac{\chi_{BS}^{11}}{\chi_B^2}, \tag{9}$$

$$C_{QS} = 3\frac{\chi_{QS}^{11}}{\chi_S^2}, \qquad C_{SQ} = \frac{\chi_{QS}^{11}}{\chi_Q^2}, \tag{10}$$

$$C_{QB} = \frac{\chi_{QB}^{11}}{\chi_B^2}, \qquad C_{BQ} = \frac{\chi_{QB}^{11}}{\chi_Q^2}. \tag{11}$$

Beam energy dependence of the ratios ($C_{\rm p,k}, C_{\rm Q,k}$, and $C_{\rm Q,p}$) [41] are shown in Fig. 10 for central and peripheral collisions. The values of $C_{\rm p,k}$ are negative at 200 GeV and change sign around 19.6 GeV for most central collisions. Both $C_{\rm Q,p}$ and $C_{\rm Q,k}$ show strong centrality dependence indicating the presence of a large excess correlation in central events in comparison with peripheral events. In addition, strong dependence of the cumulants is observed with the phase space window of measurements. With higher statistics datasets and improved acceptance of the STAR detector during the second phase of the BES program (BES-II) it will be possible to measure higher-order off-diagonal cumulants.

7. Exploring the formation of DCC domains

The DCC domains are expected to emit pions coherently from the collision volume, resulting in large fluctuations in the fraction of charged to neutral pions. The neutral pion fraction for DCC domains is predicted to follow a probability distribution of the form $P(f) = 1/2\sqrt{f}$, which is different from normal events. Heavy-ion experiments at the CERN SPS [42, 43, 44, 45] have put upper limits on the DCC formation as shown Fig 11, whereas anomalous fluctuations have been reported at RHIC [46]. A fresh look at RHIC and LHC energies using both pion and kaon sectors [47] is needed to infer about the formation of the DCC domains.

8. Mapping the little bang: local fluctuations

In little bangs, the produced fireball goes through a rapid evolution from an early state of partonic quark-gluon plasma (QGP) to a hadronic phase, and finally freezes out within a few tens of fm. Heavy-ion experiments are predominantly sensitive to the conditions that prevail at the later stage of the collision as majority of the particles are emitted near the freeze-out. As a result, a direct and quantitative estimation of the properties of hot and dense matter in the early stages and during each stage of the evolution has not yet been possible.

Relativistic hydrodynamic calculations reveal fluctuations of initial energy density and temperature, which may survive till the freeze-out. Initial fluctuating conditions have been found to be necessary for explaining observed elliptic flow in central collisions and substantial triangular flow of charged particles [25]. The initial state fluctuations may have their imprint on the bin to bin local fluctuations within an event. Event-by-event hydrodynamic calculations

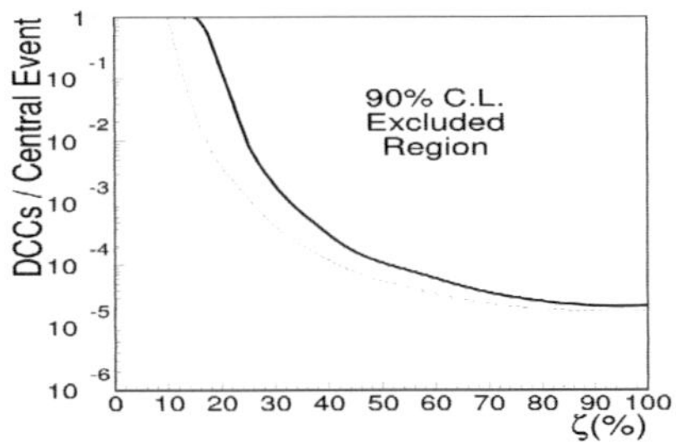

Figure 11. 90% C.L upper limit on DCC production per central event in 158 AGeV Pb–Pb Collisions as a function of the fraction of DCC pions[43].

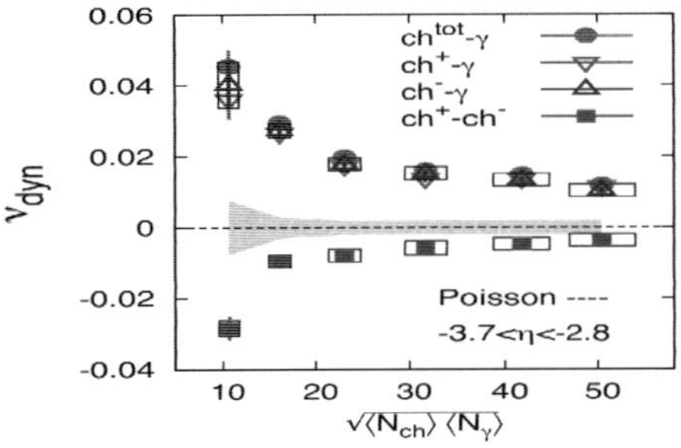

Figure 12. Dynamical correlation between positive and negative charged particles in Au–Au collisions at $\sqrt{s_{\mathrm{NN}}}$=200GeV [46].

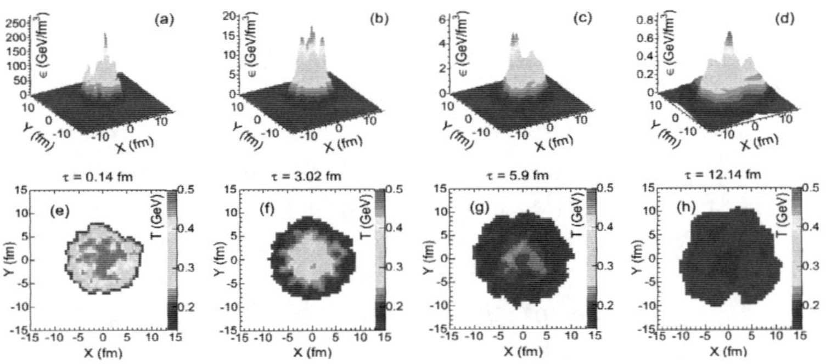

Figure 13. Hydrodynamic calculations of energy density (upper panels) and temperature (lower panels) in the transverse plane for a single Pb–Pb event at LHC energy [48].

provide a strong theoretical basis for studying the global and local fluctuations in ϵ and T [48]. The local fluctuations have been quantified throughout the evolution by simulating central Pb–Pb events at LHC energy by the use of a (2+1)-dimensional event-by-event ideal hydrodynamical framework with lattice-based EOS [25].

In Fig. 13, we present distributions of ϵ and T in the transverse plane at four proper times (τ). At early times, sharp and pronounced peaks in ϵ and hotspots in T are observed. Large bin-to-bin fluctuations observed in ϵ and T indicate that the system formed immediately after collision is quite inhomogeneous in phase space. As time elapses, the system cools, expands, and the bin-to-bin variations in ϵ and T become smooth. These observations are quantified in terms of the average over all the bins and bin to bin fluctuations in ϵ and T, plotted as a function of τ, shown in Fig. 14. We observe that ϵ decreases sharply up to $\tau = 1$ fm, and then the decrease is slow till freeze-out. The fall of T with τ is rather smooth. At early times, the fluctuations are observed to be very large, and then decrease rapidly. It is thus clear that a detailed insight to the evolution of fluctuations is possible by studying local fluctuations in ϵ and T.

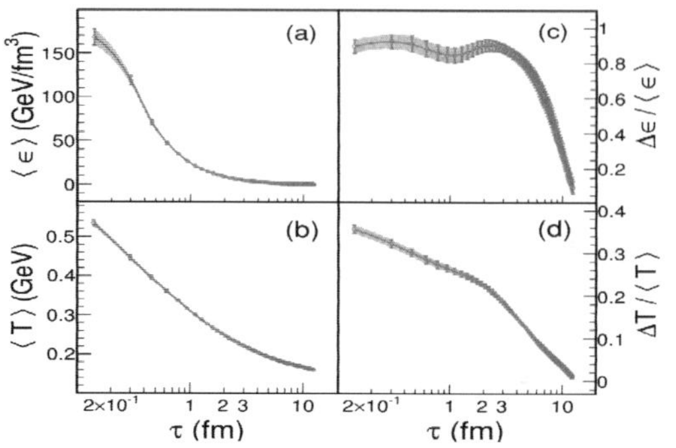

Figure 14. Temporal evolution of (a) average energy density, (b) average temperature, (c) fluctuations in energy density, and (d) fluctuations in temperature for central Pb–Pb collisions at LHC energies obtained from hydrodynamic calculations [48].

9. Summary and outlook

In this article, we have discussed a number of fluctuation techniques for understanding the nature of QCD phase transition over a wide range of baryon chemical potential as well as to locate the critical point. One of the first fluctuation studies is in terms of event-by-event fluctuations of higher order cumulants of conserved quantities, both diagonal and off-diagonal elements. A comparison with lattice and HRG model calculations provide measures of the freeze-out conditions. Assuming that the signal at freeze-out survives dissipation during the evolution of the fireball from the hadronization stage, the higher cumulants can be used as one of the preferred tools for locating the critical point. The product of the cumulants are observed to have smooth variation as a function of collision energy for net-charge, net-kaon and net-Λ distributions. But the experimental results of $\kappa\sigma^2$ at RHIC energies show a sign of non-monotonic behavior in the net-proton multiplicity distributions. We look forward to getting the confirmation of the critical point with future higher statistics data for higher order cumulants of conserved quantities as well as more differential measurements in rapidity and $p_{\rm T}$.

We have discussed fluctuations in particle multiplicity, mean transverse momentum ($\langle p_{\rm T}\rangle$) and temperature to extract isothermal compressibility, specific heat, and the speed of sound. These observables, being sensitive to the phase transition, provide important measures for the nature of the transition and to locate the critical point. Charge-neutral fluctuations in the pion and kaon sectors are discussed in terms of providing signatures of the formation of disoriented chiral condensates. Another topic in terms of fluctuation is to construct local fluctuation maps in rapidity and azimuthal bins. By making a correspondence of measured fluctuations with the time evolution of the fluctuations from theoretical calculations, it is possible to infer about the thermodynamic conditions at different stages of the QGP evolution. Most of the fluctuation studies require large coverage of the detectors and large statistics measurements. The STAR experiment with RHIC BES-II is ideal for the search of the critical point. The next-generation multipurpose detector at the LHC as a follow-up to the present ALICE experiment will have a large coverage, which will be suitable for event-by-event fluctuations [49] to probe critical fluctuations at $\mu_{\rm B}$=0.

10. References

[1] K. Fukushima, T. Hatsuda, Rept. Prog. Phys. **74**, 014001 (2011).
[2] C. Ratti, Rept. Prog. Phys. **81**, 084301 (2018).

[3] Y. Aoki, G. Endrodi, Z. Fodor, S.D. Katz and K.J. Szabo, Nature **443**, 675 (2006).
[4] M. A. Stephanov, Prog. Theo. Phys. Sup. 153, 139 (2004); Int. J. Mod. Phys. A **20**, 4387 (2005).
[5] S. Ejiri, Phys. Rev. D **78**, 074507 (2008).
[6] C. Herold, M. Nahrgang, I. Mitshustin and M. Bleicher, Nucl. Phys. A **925**, 14 (2014).
[7] M.M. Aggarwal *et al.* (STAR Collaboration), arXiv:1007.2613.
[8] A. Bzdak, S. Esumi, V. Koch, J. Liao, M. Stephanov, N. Xu, Phys. Rep. **853**, 1 (2020).
[9] M. Stephanov, K. Rajagopal, and E. Shuryak, Phys. Rev. D **60**, 114028 (1999).
[10] H. Heiselberg, Phys. Rept. **351**, 161 (2001).
[11] F. Karsch, S. Ejiri, K. Redlich, Nucl. Phys. A774, 619(2006).
[12] L.D. Landau.and E.M. Lifshitz, Statistical Physics, v.5, 3rd, rev. and enlarged ed.) Pergamon Press (1980).
[13] L. Stodolsky, Phys. Rev. Lett. **75**, 1044 (1995).
[14] S. Mrowczynski, Phys. Lett. B 430 (1998) 9.
[15] G. Giacalone, F.G. Gardim, J. Noronha-Hostler, and J-Y Ollitrault, arXiv:2004.09799 [nucl-th].
[16] C. Athanasiou, K. Rajagopal,and M. Stephanov, Phys. Rev. D **82**, 074008 (2010).
[17] M.A. Stephanov, Phys. Rev. Lett. **102**, 032301 (2009).
[18] V. Koch, A. Majumder and J. Randrup, Phys. Rev. Lett. 95 182301 (2005).
[19] A. Majumder and B Muller, Phys. Rev. C 74, 054901 (2006).
[20] R. V. Gavai, Sourendu Gupta, Phys. Rev. D 73, 014004 (2006).
[21] A. Chatterjee, S. Chatterjee, T.K. Nayak, N.R. Sahoo, Jour. Phys. G **43**, 125103 (2016).
[22] J. D. Bjorken, Acta Phys. Polon. B **28**, 2773 (1997).
[23] K. Rajagopal and F. Wilczek, Nucl. Phys. B **399**, 395 (1993).
[24] U. Heinz, J. Phys.: Conf. Ser. **455**, 012044 (2013).
[25] H. Holopainen, H. Niemi, and K. Eskola, Phys. Rev. C **83**, 034901 (2011).
[26] S. Floerchinger, U.A. Wiedemann, Phys. Rev. **C 88**, 044906 (2013).
[27] M. Mukherjee, S. Basu, A. Chatterjee, S. Chatterjee, S. Adhya, S. Thakur, T. Nayak, Phys. Lett. B **784**, 1 (2018).
[28] S. Basu, S. Chatterjee, R. Chatterjee, B. Nandi, and T.K. Nayak, Phys. Rev. C **94**, 044901 (2016).
[29] B. I. Abelev *et al.* (ALICE Collaboration), Phys. Rev. Lett. **110**, 152301 (2013).
[30] B. I. Abelev *et al.* (STAR Collaboration), Phys. Rev. C **79**, 024906 (2009).
[31] S. Pratt and C. Plumberg, arXiv:1904.11459 [nucl-th].
[32] E.V.Shuryak,M.A.Stephanov, Phys. Rev. C **63**, 064903 (2001).
[33] M. A. Aziz, S. Gavin, Phys. Rev. C **70**, 034905 (2004).
[34] B. Sharma, M.M. Aggarwal, N.R. Sahoo, T.K. Nayak, Phys. Rev. C. **91**, 024909 (2015).
[35] L. Adamczyk *et al.* (STAR Collaboration), Phys. Rev. Lett. **113**, 092301 (2014).
[36] L. Adamczyk *et al.* (STAR Collaboration), Phys. Lett. B **785**, 551 (2018).
[37] J. Adam *et al.* (STAR Collaboration), arXiv:2001.02852 [nucl-ex].
[38] M. M. Aggarwal *et al.* (STAR Collaboration), Phys. Rev. Lett. **105**, 022302 (2010).
[39] L. Adamczyk *et al.* (STAR Collaboration), Phys. Rev. Lett. **112**, 032302 (2014).
[40] J. Adam *et al.* (STAR Collaboration), arXiv:2001.06419 [nucl-ex].
[41] J. Adam *et al.* (STAR Collaboration), Phys. Rev.C **100**, 014902 (2019).
[42] T.K. Nayak *et al.* (WA98 Collaboration), Nucl. Phys. A**638**, 249c (1998).
[43] M.M. Aggarwal *et al.* (WA98 Collaboration), Phys. Lett. B 420, 169 (1998).
[44] M.M. Aggarwal *et al.* (WA98 Collaboration), Phys. Rev. C64, 011901 (2001).
[45] M.M. Aggarwal *et al.* (WA98 Collaboration), Phys. Rev. C67, 044901(2003).
[46] L. Adamczyk *et al.* (STAR Collaboration), Phys. Rev. C91,034905 (2015).
[47] R. Nayak, S. Dash, B. Nandi, and C. Pruneau, Phys. Rev. C **101**, 054904 (2020)
[48] S. Basu, R. Chatterjee, B.K. Nandi, T.K. Nayak, arXiv:1504.04502 [nucl-ex].
[49] D. Adamova *et al.* arXiv:1902.01211 [physics.ins-det].

Journal of Physics: Conference Series **1602** (2020) 012004 doi:10.1088/1742-6596/1602/1/012004

On feasibility of azimuthal flow studies with Principal Component Analysis

Igor Altsybeev

Saint-Petersburg State University, Universitetskaya nab. 7/9, St. Petersburg, 199034, Russia

E-mail: i.altsybeev@spbu.ru

Abstract.
It is shown that the Principal Component Analysis applied to azimuthal single-particle distributions allows to perform flow analysis in ways that are analogous to the traditional approaches based on multi-particle correlations. In particular, symmetric cumulants are considered. It is demonstrated also that statistical fluctuations due to a finite number of particles per event practically do not play a role for higher order PCA-based cumulants.

1. Introduction

Principal Component Analysis (PCA) is a method for decorrelation of multivariate data. PCA finds the most optimal basis for a given problem and thus reduces its dimensionality. Recently, it was suggested to apply PCA to heavy-ion collisions data on two-particle azimuthal correlations [1], in order to reveal hidden patterns of the collective behaviour of the hadronic medium. In [2] it was shown that PCA applied directly to single-particle azimuthal (φ) distributions in A–A collisions reveals Fourier harmonics as the natural and the most optimal basis.

In the latter approach, technically, one needs to take distributions of particles in M bins in each out of N events, normalize them and subtract the mean, and then apply PCA to the obtained $N \times M$ matrix (see more details in [2, 3]). As an output from PCA, we have a set of orthonormal eigenvectors ($\mathbf{e}_i, i = 1, ..., M$), and also a set of coefficients $\alpha_i^k (k = 1, ..., N)$ of PCA decomposition, such that the particle distribution in k-th event (denoted as $\mathbf{x}^{(k)}$ that is a vector with M elements) can be written as

$$\mathbf{x}^{(k)} = \sum_{i=1}^{M} \alpha_i^{(k)} \mathbf{e}_i \,. \tag{1}$$

By construction, the first K components ($K < M$) contain the most of the total variance of a dataset.

The azimuthal flow in heavy-ion collisions is typically studied using expansion of particle azimuthal probability density in a series:

$$f(\varphi) = \frac{1}{2\pi}\big[1 + 2\sum_{n=1}^{\infty} v_n \cos\big(n(\varphi - \Psi_n)\big)\big], \tag{2}$$

where v_n are the flow coefficients. As mentioned above, PCA applied to event-by-event azimuthal single-particle distributions reveals the Fourier basis, and thus coefficients of the PCA decomposition gain a definite meaning. If the decomposition (2) is applied event-by-event, values of $\hat{v}_n$ observed in the k-th event are related to the PCA coefficients (assuming that the elliptic flow dominates, and the next is the triangular flow) as follows: $\hat{v}_2^{(k)} = \sqrt{\frac{M}{2}}\sqrt{\alpha_1^{k\,2} + \alpha_2^{k\,2}}$, $\hat{v}_3^{(k)} = \sqrt{\frac{M}{2}}\sqrt{\alpha_3^{k\,2} + \alpha_4^{k\,2}}$, and so on.

In this paper, it is demonstrated that the coefficients of the PCA decomposition can be combined into expressions that are equivalent to the multi-particle cumulants used in the flow studies. The first observable considered in Section 2 is the flow amplitude calculated via two- and four-particle correlations. So-called symmetric cumulants that measure correlations between amplitudes of flow harmonics of different orders are studied in Section 3. It is estimated also how the statistical fluctuations contribute to the values extracted using PCA.

2. Higher-order cumulants from PCA

In flow studies, the simplest way to get an estimate for the amplitudes v_n is to use the two-particle cumulants $c_n\{2\}$:

$$v_n\{2\} = \sqrt{c_n\{2\}} = \sqrt{\langle v_n^2 \rangle}. \tag{3}$$

It is well-known that this quantity suffers from the so called non-flow contributions coming from e.g. resonance decays and jets. Suppression of the non-flow is typically done by utilizing the multi-particle cumulants. For example, the amplitude of the n-th harmonic can be estimated from the fourth-order cumulant $c_n\{4\}$ [4] as

$$v_n\{4\} = \sqrt[4]{-c_n\{4\}} = \sqrt[4]{2\langle v_n^2 \rangle^2 - \langle v_n^4 \rangle}. \tag{4}$$

We may try to adopt (3) and (4) for the flow studies with the PCA. When one deals with event-by-event particle distributions (like in the present application of the PCA), it is essential to investigate the influence of the statistical fluctuations due to a limited number of particles per event. Following the approach used, for instance, in [5,6], we denote a *true* amplitude of the n-th Fourier harmonic in a given event as v_n and an amplitude of the statistical noise as a_n. If $\hat{v}_n$ is the *observed* amplitude, extracted by PCA in a given event, projections of the corresponding flow vector on x and y axes in the transverse plane are

$$\hat{v}_{n,x} = v_{n,x} + a_{n,x}, \qquad \hat{v}_{n,y} = v_{n,y} + a_{n,y}. \tag{5}$$

The squares of the v_n, a_n and $\hat{v}_n$ are

$$v_n^2 = v_{n,x}^2 + v_{n,y}^2, \tag{6}$$

$$a_n^2 = a_{n,x}^2 + a_{n,y}^2, \tag{7}$$

$$\hat{v}_n^2 = \hat{v}_{n,x}^2 + \hat{v}_{n,y}^2 = v_n^2 + a_n^2 + 2(v_{n,x}a_{n,x} + v_{n,y}a_{n,y}). \tag{8}$$

After averaging (8) over events, we get

$$\langle \hat{v}_n^2 \rangle = \langle v_n^2 \rangle + \langle a_n^2 \rangle + 2\langle v_{n,x}a_{n,x} \rangle + 2\langle v_{n,y}a_{n,y} \rangle \tag{9}$$

If we assume that signal and the statistical noise are uncorrelated, the last two terms factorize: $\langle v_{n,x}a_{n,x} \rangle = \langle v_{n,x} \rangle \langle a_{n,x} \rangle$ and $\langle v_{n,y}a_{n,y} \rangle = \langle v_{n,y} \rangle \langle a_{n,y} \rangle$. Since event-averaged values of the x- and y-components of v_n and a_n are zero, (9) becomes

$$\langle \hat{v}_n^2 \rangle = \langle v_n^2 \rangle + \langle a_n^2 \rangle, \tag{10}$$

and the true value v_n is found by inverting (10):

$$\langle v_n^2 \rangle = \langle \hat{v}_n^2 \rangle - \langle a_n^2 \rangle. \tag{11}$$

This result was obtained in [2] and can be used to get an estimation of the Fourier amplitudes $v_n\{2\}$ using (3). Values $\langle a_n^2 \rangle$ measure the statistical fluctuations. They can be calculated by applying PCA to the same events, but with randomized φ-angles.

The effect from the statistical noise correction on the v_2 is shown in Figure 1 for Pb-Pb events simulated in AMPT event generator. Uncorrected raw $\hat{v}_2$ values (upper gray diamonds), extracted directly from PCA, after the correction become blue open circles. These circles are on top of the values obtained with the traditional two-particle cumulant method (full circles). It can be seen that effect from the correction is more pronounced for the peripheral events, where a number of particles per event is lower.

For the fourth power of the observed magnitude, we can use (8) again:

$$\hat{v}_n^4 = \left[v_n^2 + a_n^2 + 2(v_{n,x}a_{n,x} + v_{n,y}a_{n,y})\right]^2. \tag{12}$$

Averaging (12) over events and taking into account that x and y components of the noise are independent and their variances are equal, $\langle a_{n,x}a_{n,y} \rangle = \langle a_{n,x} \rangle \langle a_{n,y} \rangle$ and $\langle a_{n,x}^2 \rangle = \langle a_{n,y}^2 \rangle$, we get

$$\langle \hat{v}_n^4 \rangle = \langle v_n^4 \rangle + \langle a_n^4 \rangle + 4\langle v_n^2 \rangle \langle a_n^2 \rangle. \tag{13}$$

From (11) and (13),

$$2\langle \hat{v}_n^2 \rangle^2 - \langle \hat{v}_n^4 \rangle = 2\langle v_n^2 \rangle^2 - \langle v_n^4 \rangle + 2\langle a_n^2 \rangle^2 - \langle a_n^4 \rangle, \tag{14}$$

and, inverting (14) and substituting into (4), we get the following estimation for the v_n:

$$v_n\{4\} = \sqrt[4]{2\langle \hat{v}_n^2 \rangle^2 - \langle \hat{v}_n^4 \rangle - \left(2\langle a_n^2 \rangle^2 - \langle a_n^4 \rangle\right)}. \tag{15}$$

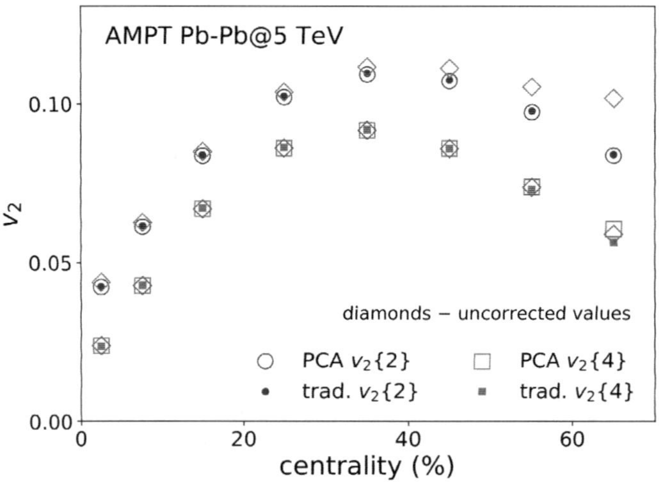

Figure 1. Amplitudes v_2 of the second Fourier harmonic as a function of centrality in Pb-Pb collisions at 5 TeV in AMPT. Values extracted by PCA are shown by open markers: circles – $v_2\{2\}$, squares – $v_2\{4\}$. Small full markers show calculations using standard 2- and 4-particle cumulant methods. Gray diamonds denote PCA values before the correction on the statistical noise. Analysis is performed for charged particles with pseudorapidity $|\eta| < 0.8$ within transverse momentum (p_T) range is 0.2–5 GeV/c. Number of φ bins used for PCA is $M = 48$.

From this expression, one may note that when $v_n \gg a_n$, values of $v_n\{4\}$ are remarkably insensitive to the statistical noise. Indeed, for a hypothetical case of constant magnitudes of the flow and the statistical noise, $v_n\{4\} = \sqrt[4]{\hat{v}_n^4 - a_n^4}$, while $v_n\{2\} = \sqrt{\hat{v}_n^2 - a_n^2}$.

In Figure 1, open squares correspond to estimations of $v_n\{4\}$ based on PCA coefficients, formula (15). They match small closed squares that stand for $v_n\{4\}$ calculated with the traditional approach using four-particle correlations. At the same time, it can be seen that the raw values $\hat{v}_n\{4\}$ (diamonds), calculated by (15) without taking into account the statistical term, are almost on top of the squares even for peripheral events. This indicates that the correction for statistical fluctuations is almost irrelevant for the $v_n\{4\}$ as soon as the number of particles in events is large enough. Toy studies showed that this is the case when the flow magnitude is $v_n \sim 0.1$ and a number of particles is $\gtrsim$100.

3. Symmetric cumulants with PCA

Following the same strategy, we can investigate feasibility of the studies of the so-called symmetric cumulants [7] with the PCA. This observable measures correlations between the amplitudes of the n-th and m-th Fourier harmonics and is defined as

$$\mathrm{SC}(n,m) = \langle v_n^2 v_m^2 \rangle - \langle v_n^2 \rangle \langle v_m^2 \rangle. \tag{16}$$

Previous attempt to study the symmetric cumulants with the PCA was done in a paper [3]. Since the PCA basis obtained in [3] was somehow distorted (i.e. not identical to the Fourier harmonics), extracted SC in this paper values did not match the "truth" values.

We start with the single-event raw quantity:

$$\begin{aligned}\hat{v}_n^2 \hat{v}_m^2 &= (\hat{v}_{n,x}^2 + \hat{v}_{n,y}^2)(\hat{v}_{n,x}^2 + \hat{v}_{n,y}^2) = \\ &= \left[(v_{n,x} + a_{n,x})^2 + (v_{n,y} + a_{n,y})^2\right]\left[(v_{m,x} + a_{m,x})^2 + (v_{m,y} + a_{m,y})^2\right] = \\ &= \left[v_n^2 + a_n^2 + 2(v_{n,x}a_{n,x} + v_{n,y}a_{n,y})\right]\left[v_m^2 + a_m^2 + 2(v_{m,x}a_{m,x} + v_{m,y}a_{m,y})\right].\end{aligned} \tag{17}$$

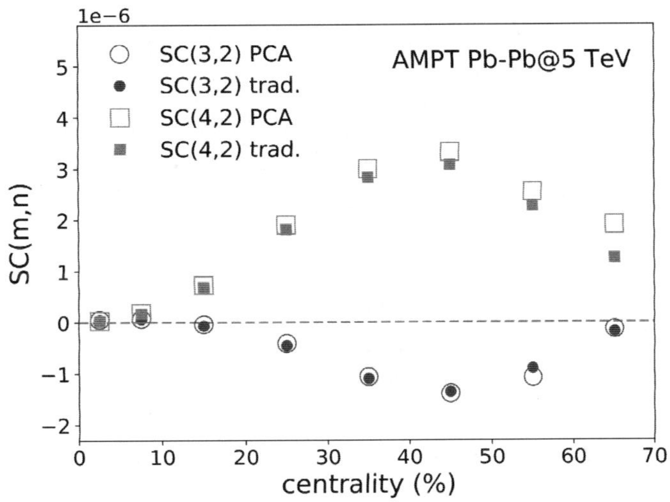

Figure 2. Centrality dependence of the symmetric cumulants SC(3,2) and SC(4,2) in Pb-Pb collisions from AMPT. Open markers – values are extracted from PCA, closed markers – by traditional method of multi-particle correlations. Particles within $|\eta| < 2.4$, p_T range 0.2–5 GeV/c.

Averaging over events and taking into account $\langle v_{n,x}\rangle = \langle v_{n,y}\rangle = \langle a_{n,x}\rangle = \langle a_{n,y}\rangle = 0$ (the same for the m-th harmonic), and also the factorization of the noise harmonics $\langle a_n^2 a_m^2\rangle = \langle a_n^2\rangle\langle a_m^2\rangle$ as well as the noise and the signal $\langle v_n^2 a_m^2\rangle = \langle v_n^2\rangle\langle a_m^2\rangle$, we obtain

$$\langle \hat{v}_n^2 \hat{v}_m^2\rangle = \langle v_n^2 v_m^2\rangle + \langle v_n^2\rangle\langle a_m^2\rangle + \langle v_m^2\rangle\langle a_n^2\rangle + \langle a_n^2\rangle\langle a_m^2\rangle. \tag{18}$$

Using (11) and (18), the desired term $\langle v_n^2 v_m^2\rangle$ can be expressed via "measurable" quantities as

$$\langle v_n^2 v_m^2\rangle = \langle \hat{v}_n^2 \hat{v}_m^2\rangle - \langle \hat{v}_n^2\rangle\langle a_m^2\rangle - \langle \hat{v}_m^2\rangle\langle a_n^2\rangle + \langle a_n^2\rangle\langle a_m^2\rangle. \tag{19}$$

The final expression for the $\mathrm{SC}(n,m)$ is thus

$$\mathrm{SC}(n,m) = \langle v_n^2 v_m^2\rangle - \langle v_n^2\rangle\langle v_m^2\rangle = \langle \hat{v}_n^2 \hat{v}_m^2\rangle - \langle \hat{v}_n^2\rangle\langle \hat{v}_m^2\rangle. \tag{20}$$

It is remarkable that all terms related to the statistical noise are canceled.

Figure 2 shows centrality dependence of the symmetric cumulants SC(3,2) and SC(4,2) in Pb-Pb collisions from AMPT. It can be seen that values extracted from PCA (open markers) match with calculations using multi-particle correlations (closed markers). For SC(4,2), there are slight deviations in peripheral centrality classes, a possible reason is the event-plane correlation of these two harmonics. Detailed investigation of this is out of the scope of this article.

4. Summary

It was shown that the Principal Component Analysis applied to event-by-event azimuthal single-particle distributions allows to perform flow analyses that are analogous to the traditional approaches based on multi-particle correlations. As the first example, flow amplitudes based on the fourth-order cumulant were considered. As the second case, correlations between flow amplitudes in terms of symmetric cumulants were calculated. Using realistic events from the AMPT generator, PCA results were directly compared to calculations using the traditional techniques, a good correspondence was obtained. It was demonstrated also that a contribution from statistical fluctuations due to a finite number of particles per event to the higher-order PCA-based cumulants is small.

Acknowledgements
This study is supported by Russian Science Foundation, grant 17-72-20045.

References

[1] R.S. Bhalerao, J.Y. Ollitrault, S. Pal, and D. Teaney, PRL 114, (2015) 152301.
[2] I. Altsybeev, Phys.Part.Nucl. 51 (2020) 3, 314-318; arXiv:1909.03979 [nucl-th].
[3] Z. Liu, W. Zhao, H. Song, Eur.Phys.J.C 79 (2019) 10, 870.
[4] N. Borghini, P. M. Dinh and J. Y. Ollitrault, Phys. Rev. C 64 (2001) 05.
[5] R. He, J. Qian, L. Huo, arXiv:1702.03137 [nucl-th] (2017).
[6] J. Jia and P. Huo, Phys. Rev. C 90, 034905 (2014).
[7] A. Bilandzic et al., Phys. Rev. C 89, 064904 (2014).

Journal of Physics: Conference Series **1602** (2020) 012005 doi:10.1088/1742-6596/1602/1/012005

SM antideuteron background to indirect dark matter signals in galactic cosmic rays

D.M. Gomez Coral[1,2] and Arturo Menchaca-Rocha[1]

[1] Instituto de Fisica, UNAM, A.P. 20-364, CP 10000, CDMX, Mexico

[2] Department of Physics and Astronomy, University of Hawaii at Manoa, 2505 Correa Rd, Honolulu, HI 96822

E-mail: dgomezco@hawaii.edu.

Abstract. Antideuteron production cross-sections estimated using EPOS-LHC with a coalescence afterburner, tuned to reproduce published experimental data over a wide range of energy were used here as input to the galactic propagator code GALPROP, validated with comparing to existing proton, helium fluxes as well as boron-to-carbon ratio data. The resulting near-Earth antideuteron flux, including solar modulation, is compared to previous estimates. An overall factor of two increment in the antideuteron flux is predicted, the origin of which is also discussed. However, this standard model source of antideuteron background still lies well below the AMS-02, and the expected GAPS, sensitivities, as well as the fluxes predicted by several dark matter models.

1. Introduction

Within the Standard Model (SM), cosmic deuteron formation is understood as the result of nuclear interactions of primary CRs, mainly protons, with the interstellar matter (ISM). This cosmic deuteron source, known as secondary production, is dominated by two contributions: fragmentation of ISM nuclei, in which deuterons are direct products; and the resonant reaction $p+p \rightarrow \pi+d$. Thus, cosmic deuteron flux provides important information about CR propagation in the Galaxy, such as the mean amount of ISM that primary CR's encounter as they travel from their sources to the Earth. Accelerator experiments allowed the study of a third deuteron production mechanism, explained within the framework of the so-called coalescence model. This applies to free nucleons resulting from CR-ISM interactions, in which residual protons and neutrons lie sufficiently close enough in phase space recombine (coalesce) to form deuterons. Such free nucleons may be the result of p-nuclei fragmentation interactions. But, at sufficiently high energies, p-p and p-nuclei interactions can also create multiple nucleon-antinucleon pairs, generating conditions for the formation of deuterons through the coalescence mechanism. Note that, of the three deuteron-producing mechanisms described, this is the only one that also allows the formation of secondary antideuterons. The result is an antideuteron flux having an energy distribution with a maximum at ≈ 4 GeV. Lower energy antideuterons, with maximum at ≈ 1 GeV and a similarly energy distribution, can also be produced in the interaction of primary protons with secondary antiprotons, with smaller probability. Yet, in that energy range, when all possible mechanisms are taken into account, as illustrated by T. Aramaki et al. [1] and reproduced in Fig. 1, the predicted abundance of antideuterons, estimated using simplified coalescence model calculations and geometrical cross sections, turns out to be at least one order of magnitude smaller than what several models predict for antideuteron production from dark matter particle-antiparticle annihilation.

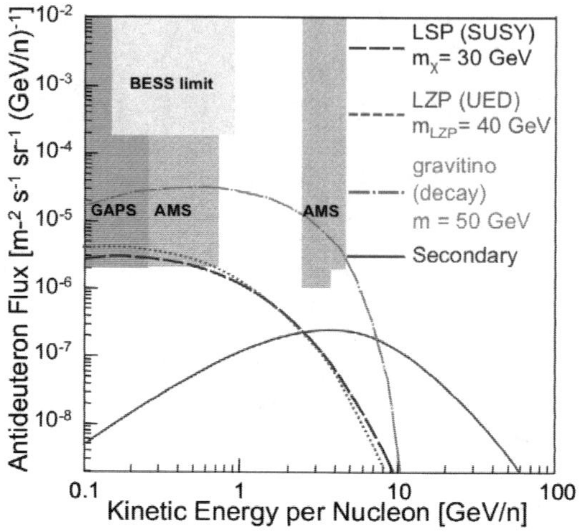

Figure 1. Expected antideuteron flux from three different dark matter candidates [2, 3, 4] compared to expected secondary production predicted by Ibarra et al. [6] and the corresponding detector sensitivity limits.

The predicted low energy secondary SM antideuteron-free window (E_{kin} = 0.1 GeV to 1 GeV) shown in Fig. 1 generated great interest in dark matter research. Given the continuous improvement of Monte Carlo particle-interaction simulators, as well as the development by our group of an afterburner scheme to include in them the deuteron/antideuteron coalescence mechanism [7], in Ref. [8] we estimated secondary antideuteron cross sections predictions (see Fig. 2), based on available accelerator data over a wide range of energies.

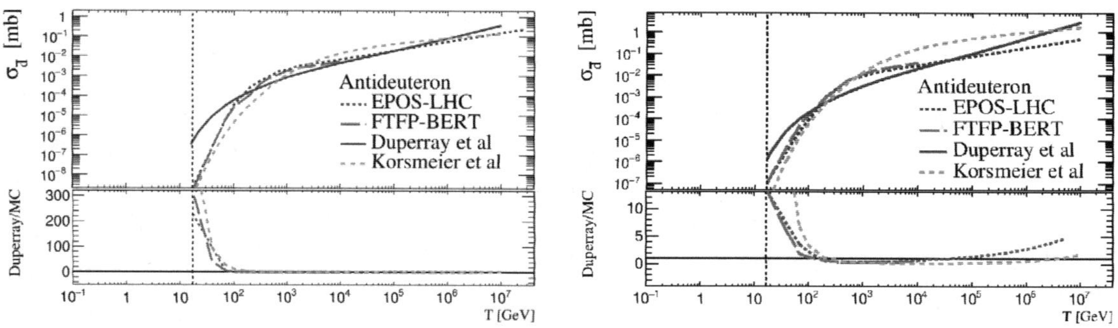

Figure 2. Antideuteron total production cross-section [8] in p+p collisions (left), and in p+He collisions (right). The expected antideuteron cross-section from Duperray's parametrization [5] has been added. In the lower panels Duperray and MC predictions for antideuteron are compared. Vertical broken lines represent the antideuteron production threshold.

In Fig. 2, as expected, simplified cross section estimations work best at high energies but show important deviations from our HEP simulators in the low energy region, where the best possible SM background knowledge is required.

Here, we present preliminary results [9] of using the cross sections shown in Fig. 2 as input to the galactic-transport code GALPROP [10], to calculate the near-Earth antideuteron flux. This code incorporates as much realistic astrophysical input as possible, together with the latest theoretical developments to simulate the propagation of cosmic-ray nuclei, antiprotons, electrons, and positrons, and computes diffuse γ-rays and synchrotron emission in the same framework.

2. Transport

GALPROP [10] solves the transport equation considering a source distribution and boundary conditions for all cosmic-ray species. The equation includes Galactic wind (convection), diffusive re-acceleration in the interstellar medium, energy losses, nuclear fragmentation, and nuclear decay. The numerical solution of the transport equation is based on a second-order implicit scheme. The spatial boundary conditions assume free particle escape. The transport equation is written as

$$\frac{\partial f(p,\vec{r},t)}{\partial t} = \vec{\nabla}\cdot(D_{xx}(p,\vec{r})\vec{\nabla}f - \vec{V}f) + \frac{\partial}{\partial p}p^2 D_{pp}\frac{\partial}{\partial p}\frac{1}{p^2}f - \frac{\partial}{\partial p}\left[\dot{p}f - \frac{p}{3}(\vec{\nabla}\cdot\vec{V})f\right] - \frac{1}{\tau_f}f - \frac{1}{\tau_r}f + q(p,\vec{r},t), \quad (1)$$

where $f(p;\vec{r};t)$ is the density of particles in the ISM per unit of total momentum. D_{xx} is the spatial diffusion coefficient, $\vec{V}$ is the convection velocity, D_{pp} is the diffusion coefficient in momentum space describing re-acceleration, $\dot{p}$ is the momentum loss rate, τ_f is the timescale for fragmentation, and τ_r is the timescale for the radioactive decay. The last term of Eq. 1 represents the source, integrated of all possible secondary sources contributing to a particular isotope species. For antideuterons, the secondary source terms obtained with the corresponding EPOS-LHC cross sections are shown on the left of Fig. 3. From this, it is concluded the p+p channel represents around 48% of the total flux, while the antiproton channels account for $\approx 5.2\%$ of the antideuteron flux. Note, however, that the latter contribution is dominant below 1 GeV. The contribution from He+He has also been included. On the right side of Fig. 3, the EPOS-LHC source term ($q_{\bar{d}}$) is compared to results from previous works [5, 6]. These calculations predict a higher antideuteron production in CRs collisions, reaching around twice the value from Ibarra et al. [6] and ≈ 1.2 times the result from Duperray et al. [5]. Furthermore, the maximum of the source distribution is slightly shifted in the antideuteron kinetic energy in comparison to parametrizations, being located at ≈ 5GeV/n. This shift is a consequence of competing factors folded into the secondary source term, i.e., the rising production cross-section that dominates at low energies (see Fig. 2), and the rapidly decreasing incident CR flux that governs the high-energy region. Thus, in contrast with previous calculations, the production cross-section from EPOS-LHC is more significant in the high energy range, causing an increase observed in $q_{\bar{d}}$ and, despite the decreasing incident flux. It also produces a broader distribution with a flattened shape on the top of the curve, and the shifting of its maximum.

3. Validation

The first step to calculate the propagation of deuterons and antideuteron in the Galaxy correctly is to be sure the transport model is correctly set up. Special care has to be taken concerning the proton and helium fluxes, since they are the primary incident particles. Additionally, the precise measurement of the Boron-to-Carbon ratio by AMS-02 [11] helps restricting the diffusion coefficient value, hence, improving the predictability of the model. In Fig. 4 the fluxes obtained using GALPROP with solar modulation effects taken from [12, 13], are compared to AMS-02 data. On the left-top side is the proton flux, where the solid line represents the GALPROP prediction using a Fisk potential of 570 MeV in the solar modulation model. As a reference, in the same figure the dashed line is the same GALPROP result but using the solar modulation

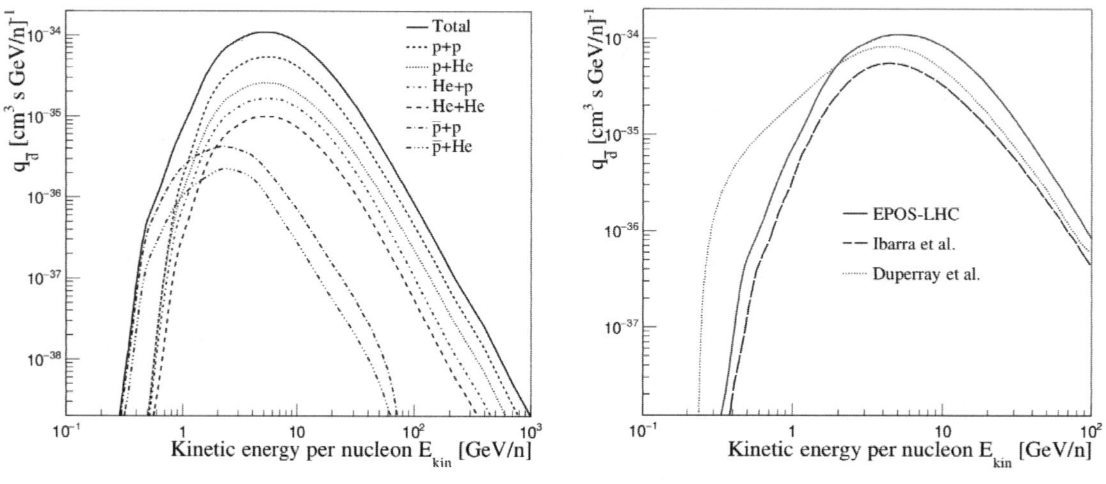

Figure 3. Secondary source term for antideuterons as function of kinetic energy per nucleon. On the left side the "Total" contribution from the most important collisions is shown. On the right side, the result is compared to previous works [5, 6].

model HelMod [14]. In this theoretical approach, the entire CRs transport equation, from the supernova termination shock, to the Earth orbit is Monte Carlo-simulated. On the right-top side of Fig. 4, the corresponding GALPROP predictions for the Boron-to-Carbon ratio using HelMod, are compared to the AMS-02 [11] data. In both cases, the relative difference between simulation and data is less than 20%, and the solar modulation model used is the force field approximation, again with a Fisk potential of 570 MeV. As shown at the bottom of this figure, Helium flux is compared as well, the results differ by a maximum of 10% from AMS-02 data.

As already mentioned, in addition to the most abundant CRs species simulated, and validated in Fig. 4, antiprotons play an essential role in the antideuteron production, and for that reason, their generation in GALPROP has to be analyzed and compared with recent AMS-02 [11, 15] experimental data.

3.1. Antiprotons

The next step is to determine the antiproton flux using EPOS-LHC and GALPROP to compare with parametrizations. As in the case of antideuterons, to obtain the antiproton flux, the secondary and tertiary source terms have to be calculated [9]. The dominant channels for the production of the secondary antiprotons are: $p+p$ at roughly $50-60\%$ of the total contribution, and $p+He$, $He+p$, and $He+He$ at $10-20\%$ each; while the channels involving heavier incoming CRs contribute only up to a few percent. Thus, here the contributions from the reactions $p+p$, $p+He$, $He+p$, and $He+He$ were included, and antiprotons created in heavier CRs collisions were ignored. The antiproton multiplicity produced in CRs includes antiprotons from CRs collisions and from hyperon and antineutron decay. This is shown on the left side of Fig. 5, where the total antiproton source is presented. The distribution shows a maximum at around 2 GeV/n, although ours is slightly broader than previous studies [5, 6, 16], also shown.

Our "Total" antiproton calculation also predicts a larger flux antiproton production than Duperray [5], Korsmeier [16], and Winkler [17]. An estimate of the source term using EPOS-LHC, but considering the antiproton contribution from antineutron decay as in Korsmeier [16] shows a lower antiproton production, while conserving the shape of the original EPOS-LHC calculation (Total). This indicates antiproton overproduction generated by EPOS-LHC is related to an

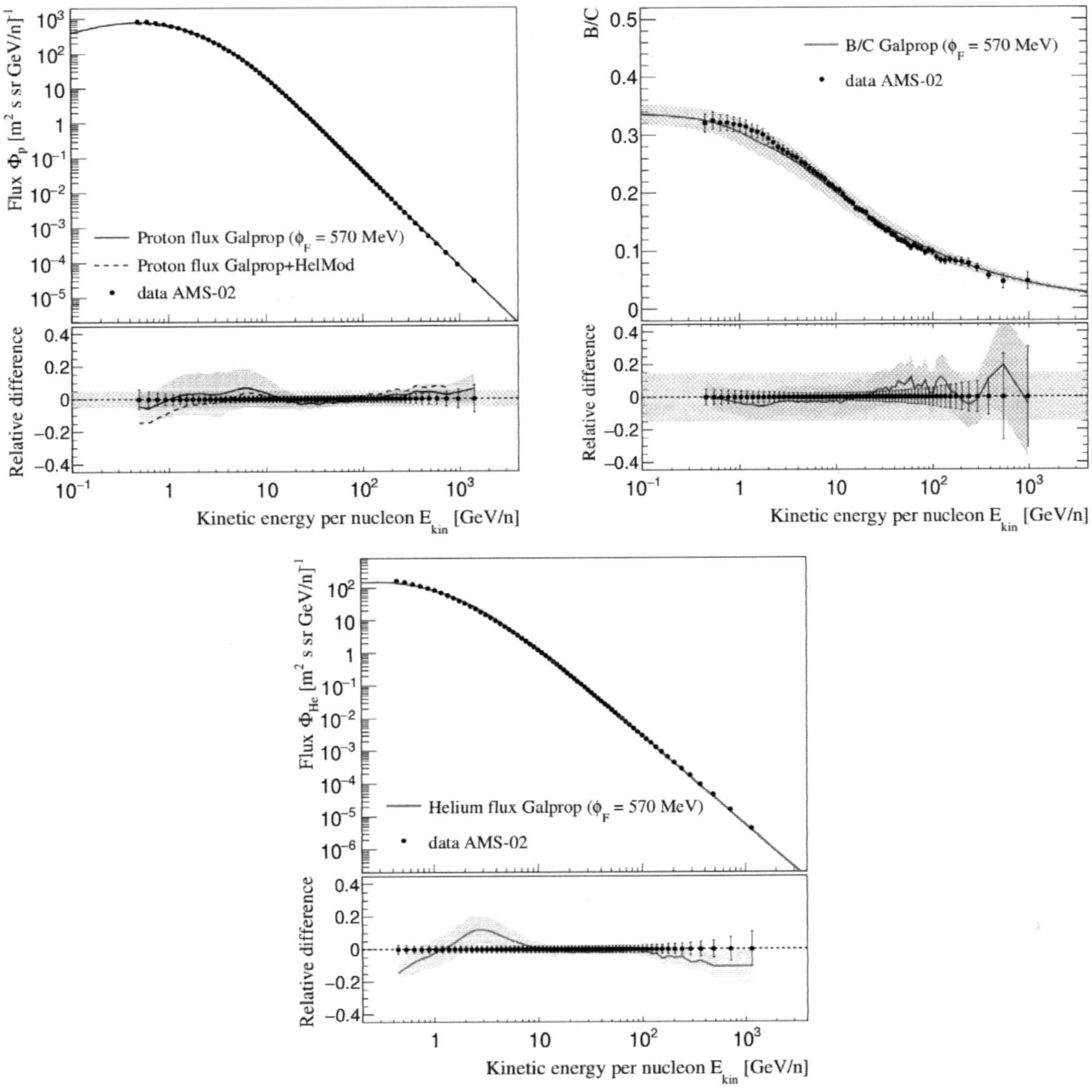

Figure 4. Results obtained with GALPROP and the parameters for proton, Boron-to-Carbon ratio and Helium compared to the most recent data from AMS-02 [11] experiment.

excess in the contribution from antineutron decays. On the right side of Fig. 5 the antiproton flux is compared to the one obtained by Winkler et al. [17] and the data from AMS-02 [15]. Our EPOS-LHC simulation predicts a larger antiproton flux below 10 GeV/n, reaching a magnitude 1.5 times larger than Winkler calculation. At high energies the antiproton flux calculated with EPOS-LHC shows similar results as the flux from Winkler.

3.2. Deuterons and Antideuterons

Once the propagation parameters have been defined and tested, it is safe to proceed with the transport of deuterons and antideuterons, as shown in Fig. 6. On the left side of the figure, the deuteron flux times $E_{kin}^{2.7}$ is plotted as a function of the deuteron kinetic energy. Besides the deuteron result, the helium flux is included in the same plot to show the good description of

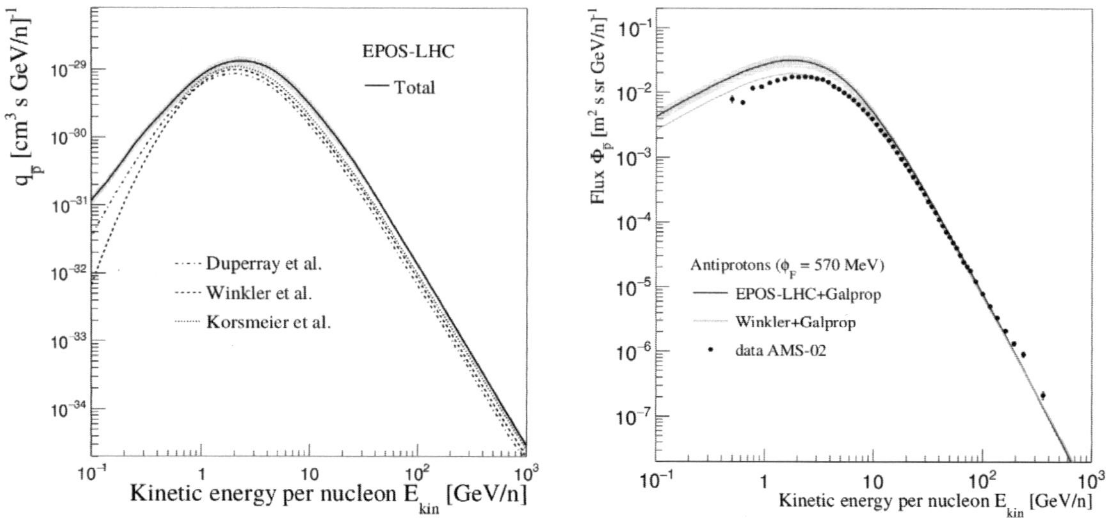

Figure 5. (Left) Antiproton secondary source term for EPOS-LHC. (Right) Antiproton flux obtained with EPOS-LHC and GALPROP compared to AMS02 [15] data and Winkler [17] prediction.

data by GALPROP. Data from AMS-1 [18] and AMS-02 [15] are presented in the figure together with data from CAPRICE [19], which is the only deuteron measurement at high energies. The deuteron flux considering coalescence [8] is represented by the thin solid line. The dot-dashed line is the result without coalescence.

As can be observed in Fig. 6, the deuteron flux including coalescence production of deuterons at high energies shows an improved description of data. On the right side of Fig. 6, the deuteron-to-helium ratio results are presented and compared to data from CAPRICE [19] and SOKOL [20]. The CAPRICE [19] ratio was reevaluated with the helium data from AMS-02 [11]. Again, as in the left side figure, it is observed deuteron simulation containing coalescence production describes data satisfactorily; however, this simulation is not enough to account for the excess predicted by SOKOL [20].

4. Result

The end result of the antideuteron flux calculated with EPOS-LHC as event generator to simulate CR collisions, simultaneously with the afterburner to produce antideuterons through the coalescence model, and the propagation of these particles in the Galaxy by means of GALPROP is presented in Fig. 7. The first notorious feature on the left side of this figure is that the antideuteron flux evaluated here (solid line) is the largest one compared with previous works. Indeed, the predicted antideuteron flux is around 30% larger than the flux calculated by Duperray et al. [5] and nearly to 4 times the one inferred by Ibarra et al. [6]. The last result is not surprising, since from the source term in Fig. 3 it was evident EPOS-LHC+afterburner produce a higher number of antideuterons as a consequence of the larger cross-section. The predicted antideuteron flux obtained with EPOS-LHC, also shows a wider distribution compared with the mentioned studies, due to a broader high energy contribution originated in a high production cross-section. An effect of the energy-dependent coalescence momentum and the cross-section derived from that analysis [8] is the slight shift of the maximum in the antideuteron flux that now is around $4-5$ GeV/n, when other calculations place it around $3-4$ GeV/n.

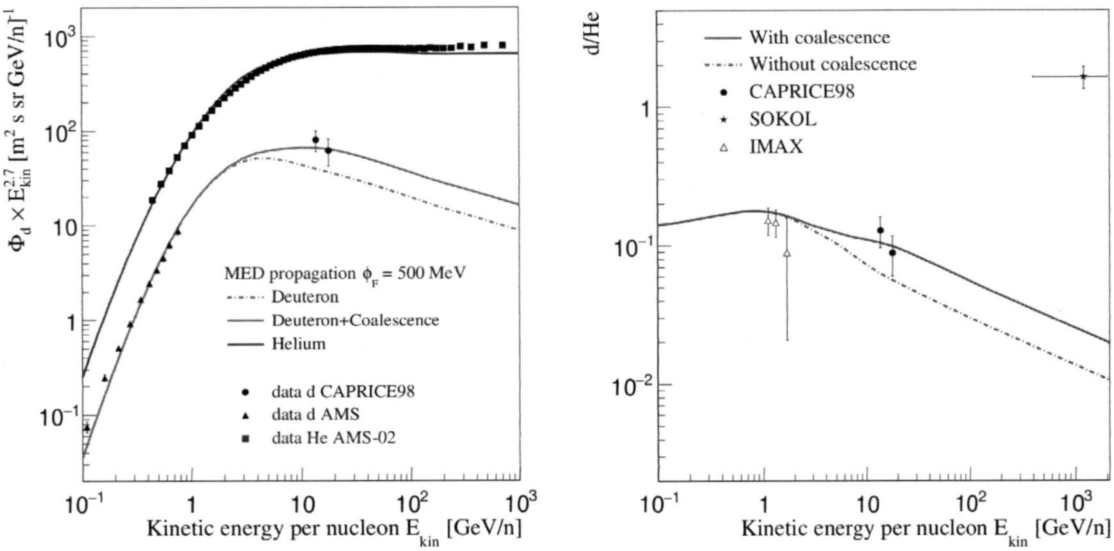

Figure 6. (Left) Deuteron flux produced using QGSJET-II-04 with coalescence (thin solid line) and without coalescence (dot-dashed line) propagated with GALPROP and compared to data from AMS [11, 18], CAPRICE [19]. (Right) Deuteron-to-Helium ratio compared to data from CAPRICE [19], SOKOL [20], and IMAX [21].

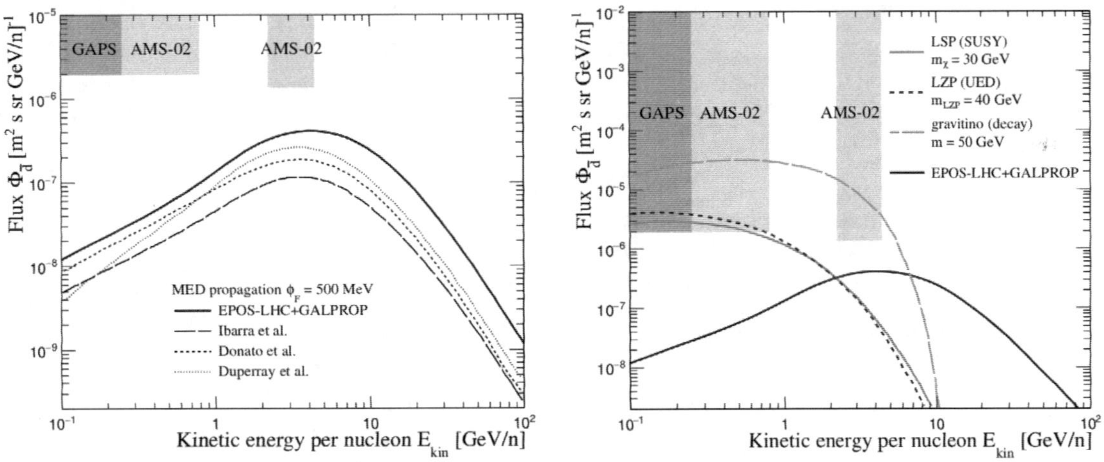

Figure 7. Antideuteron flux at the top of the atmosphere obtained with EPOS-LHC and GALPROP. (Left) The result is compared to previous works from Ibarra [6], Donato [22] and Duperray [5]. (Right) The result is compared to antideuteron fluxes expected from dark matter annihilation or decay [2, 3, 4]. Sensitivity limits for AMS-02 and GAPS are also included.

If the coalescence momentum would have been considered constant at all energies, then the shape of the distribution would be the same as Duperray et al. [5], or Ibarra et al. [6]. But since the suppression of the coalescence momentum was taken into account near the energy threshold, the low energy region of the flux (< 1 GeV) is also suppressed, and therefore the shape of the distribution is modified. On the right side of Fig. 7, the resulting antideuteron flux is plotted along with three different expected antideuteron fluxes from dark matter annihilation or decay. These candidates, already shown in Fig. 1, continue having a considerable larger flux below 1 GeV/n compared to the secondary, despite its increase using EPOS-LHC.

5. Conclusion

In conclusion, the antideuteron flux obtained here shows a larger magnitude compared to other studies, and a slight difference in the shape distribution, as a consequence of the energy dependence of the coalescence parameter (Fig. 7). Although the resulting flux is 1.3 times larger than the predicted by Duperray et al. [5], and four times than Ibarra et al. [6], it is still below the expected sensitivities of AMS-02 and GAPS. These findings strengthen the idea that if these experiments detect cosmic antideuterons, they are probably from exotic sources.

6. Acknowledgements

The authors would like to thank J. Martinez Mendoza, E. Lopez Pineda and S. Aguilar Salazar for technical aid, as well as CONACyT and PAPIIT-DGAPA: IN109617 for financial support.

References

[1] Aramaki, T., et al., *Phys. Rept.* **618**, 1 (2016), arXiv:1505.07785 [hep-ph].
[2] Donato, F., Fornengo, N., and Maurin, D., *Phys. Rev. D*, **78**:043506, Aug 2008.
[3] Baer, W., and Profumo, S., *Journal of Cosmology and Astroparticle Physics*, **2005** (12):008, 2005.
[4] Dal, L.A., and Raklev, A.R., *Phys. Rev. D*, **89**:103504, May 2014.
[5] Duperray, R., et al. *Physical Review D*, **71**(8):083013, April 2005.
[6] Ibarra, A. and Wild, S., *Phys. Rev. D*, **88**:023014, Jul 2013.
[7] Serradilla, E. *PhD thesis*, Universidad Complutense de Madrid, Madrid, España, November 2014; Adam, J., et al. . Rev. C, **93**:024917, Feb 2016.
[8] Gomez-Coral, D.M., Menchaca-Rocha, A., Grabski, V., Datta, A.,von Doetinchem, Ph. and Shukla, A., . Rev. D, **98**:023012, Jul 2018.
[9] Gómez-Coral, D.M., *PhD thesis*, Universidad Nacional Autónoma de México, Ciudad de México, Mexico, 2019.
[10] Strong, A.W., et al. *GALPROP Version 54: Explanatory Supplement. Users Guide*, Sep 2011. GALPROP website, 2018.
[11] Aguilar, M., et al., *Phys. Rev. Lett.*, **123**:181102, 2019.
[12] Gleeson, L.J. and Axford, W.I., *The Astrophysical Journal*, **154**:1011, December 1968.
[13] Perko, J.S., *Astronomy and Astrophysics*, **184**:119:121, October 1987.
[14] Boschini, M.J., et al., *The Astrophysical Journal*, **840**(2):115, 2017.
[15] Aguilar, M., et al., *Phys. Rev. Lett.*, **117**:091103, 2016.
[16] Korsmeier, M., Donato, F., and Di Mauro, M., *Phys. Rev. D*, **97**:103019, May 2018.
[17] Winkler, M.W., *Journal of Cosmology and Astroparticle Physics*, **2017**(02):048, 2017.
[18] Aguilar, M., et al., *Physics Reports*, **366**(6):331-405, 2002.
[19] Papini, P., et al., *The Astrophysical Journal*, **615**(1):259, 2004.
[20] Turundaevskiy, A., and Podorozhnyi, D., *Advances in Space Research*, **59**(1):496-501, 2017.
[21] de Nolfo, G.A., et al. *AIP Conference Proceedings*, **528**(1):425-428, 2000.
[22] Donato, F., Fornengo, N., and Salati, P., *Phys. Rev. D*, **62**:043003, Jul 2000.

Journal of Physics: Conference Series **1602** (2020) 012006 doi:10.1088/1742-6596/1602/1/012006

Real-Time Event Reconstruction and Analysis in CBM and STAR Experiments

Ivan Kisel
(for CBM and STAR Collaborations)

[1] Goethe-University Frankfurt, Theodor-W.-Adorno-Platz 1, 60323 Frankfurt am Main, Germany
[2] FIAS Frankfurt Institute for Advanced Studies, Ruth-Moufang-Str. 1, 60438 Frankfurt am Main, Germany
[3] Helmholtz Research Academy Hesse for FAIR, Max-von-Laue-Str. 12, 60438 Frankfurt am Main, Germany
[4] GSI Helmholtz Centre for Heavy Ion Research, Planckstr. 1, 64291 Darmstadt, Germany

E-mail: I.Kisel@compeng.uni-frankfurt.de

Abstract. Within the FAIR Phase 0 program, the algorithms of the FLES (First-Level Event Selection) package developed for the CBM experiment (FAIR/GSI, Germany) are adapted for the STAR experiment (BNL, USA). Use of the same algorithms creates a bridge between online and offline, which makes it possible to combine online and offline resources for data processing. In this way, an express data production chain was created on the basis of the STAR HLT farm, that extends the functionality of HLT in real time up to the analysis of physics. It is important, that the express analysis chain does not interfere with the standard analysis chain. A particular advantage of express analysis is that it allows calibration, production and analysis of the data to begin immediately after they are collected. Therefore, the use of express analysis is beneficial for BES II data production and helps to speed up scientific discovery by helping to obtain results within one year after the end of data acquisition. The specific features of express data production are presented and discussed as well as the results of online production and analysis, such as real-time reconstruction of short-lived particles in the BES-II STAR environment.

1. Introduction

Within the framework of the Facility for Antiproton and Ion Research (FAIR) project, a large international centre is being constructed to study the structure and fundamental properties of matter. It will be a new generation accelerator complex that will provide unique opportunities for detailed investigations in the most interesting areas of modern science: nuclear, hadron and particle physics, atomic and anti-matter physics, high density plasma physics, and applications in condensed matter physics, biology and bio-medical sciences [1].

In the Compressed Baryonic Matter (CBM) [2] experiment with heavy ions, the highest baryon densities will be created, and the properties of super-dense nuclear matter will be investigated in various extreme states that are similar to, for example, the conditions of matter in the center of neutron stars, where matter is at the final stage of evolution before transition to the black hole. The CBM experiment will thus complement the experimental heavy-ion program at the LHC accelerator complex at CERN where the properties of the hot matter similar to that after the Big Bang are investigated.

The scientific program of the CBM experiment includes:

- explore properties of super-dense nuclear matter;
- search for in-medium modifications of hadrons;
- search for the transition from dense hadronic matter to quark-gluon matter;
- search for the critical endpoint in the phase diagram of strongly interacting matter;
- investigate the structure of neutron stars and the dynamics of core-collapse supernovae.

The experiment will measure rare and penetrating probes such as dilepton pairs from light vector mesons and charmonium, open charm, multistrange hyperons, together with collective hadron flow and fluctuations in heavy-ion collisions at rates up to 10^7 collisions per second.

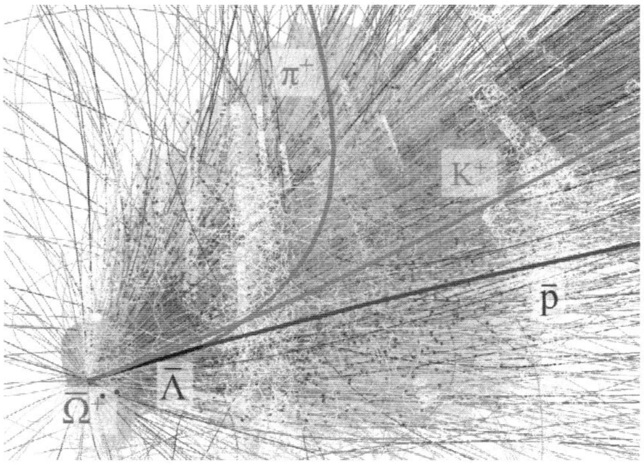

Figure 1. A simulated central Au-Au collision at 25 *A*GeV energy with about 1000 charged particles in the CBM experiment.

CBM is characterized by high collision rates, large amount of produced particles, non-homogeneous magnetic fields and a very complex detector system. Event reconstruction is the most complicated and time consuming task of the data analysis in modern high-energy physics experiments. It is a key part of success in the CBM experiment with up to thousand particles per central collision (Fig. 1). An additional complication in CBM is its continuous data stream represented in form of time slices. This makes the reconstruction of such 4-dimensional data with time stamps and the search for interesting physics extremely difficult. All of the above mentioned makes necessary to develop fast and efficient algorithms for data analysis and to optimize them for running on a modern high-performance computer cluster [3].

2. First Level Event Selection

The First Level Event Selection (FLES) package [4, 5] of the CBM experiment is intended to reconstruct online the full event topology including tracks of charged particles and short-lived particles. The FLES package consists of several modules: Cellular Automaton (CA) track finder, kalman Filter (KF) track fitter, KF Particle Finder and physics selection. In addition, a quality check module is implemented, that allows to monitor and control the reconstruction process at all stages. The FLES package is platform and operating system independent. The package is portable to different many-core CPU architectures, vectorized using SIMD (Single Instruction, Multiple Data) instructions and parallelized between CPU cores. All algorithms are optimized with respect to the memory usage and the speed.

2.1. Cellular Automaton (CA) track finder
The 4-dimensional (4D, space and time) Cellular Automaton (CA) track finder [5, 6] takes as input hit measurements from the tracking detector in the form of a time-slice, which includes time and spacial measurements. The track finding procedure starts with combining the hits into triplets, combination of three hits on adjacent stations. The triplet structure was chosen, since it allows to estimate the momentum of a particle, which could produce it. The triplets with two common hits are combined into track candidates. The track candidates should survive a dedicated selection based on the track length and calculated χ^2-value to be accepted to the reconstructed tracks.

Input time information is used in the algorithm to the same extent and in similar manner as it is done with the spacial coordinates. The same logic is used while constructing triplets: the hits in the triplet should belong to the same particle, therefore they should correlate not only in space, but also in time. The resulting track reconstruction efficiencies for the cases of event-by-event analysis (so-called 3D analysis) as well as for the 4D case (with included time measurement, as well as 3-dimensional spacial information) while reconstructing time-slices are similar thus there is no efficiency degradation in the much more complicated case of time slices. The same is valid for the speed of the 4D CA tracks finder with respect to the 3D case.

2.2. Kalman Filter (KF) track fit
High precision of the parameters of particle trajectories (tracks) and their covariance matrices is a prerequisite for finding rare signal events among hundreds of thousands of background events. Such high precision is usually obtained by using the estimation algorithms based on the Kalman filter (KF) method. High speed of the reconstruction algorithms on modern many-core computer architectures can be accomplished by: optimizing with respect to the computer memory, in particular declaring all variables in single precision, vectorizing in order to use the SIMD instruction set and parallelizing between cores within a compute node.

2.3. 4D event builder

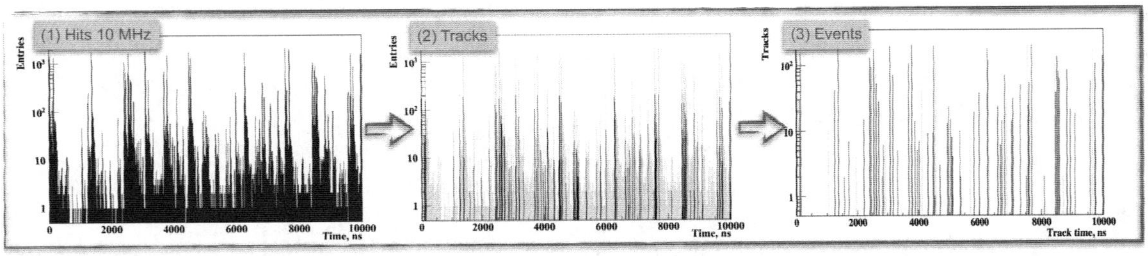

Figure 2. Reconstructed tracks in time slices clearly represent groups, which correspond to the original events with 85% of single events, no splitted events, and final event building is done at the vertexing stage using TOF information.

After all tracks are found and their parameters are reconstructed, the tracks are grouped into events. This is done by clustering tracks based on their time parameters in the area of the target. The left distribution of Fig. 2 shows hits within a time slice for 10^7 interaction rate. One can see that the traditional grouping of hits into events at this stage is impossible. The track distribution at the middle shown against the same hits displays grouping of tracks belonging to the same event. The right distribution with different colors shows different clusters of tracks, close in time in the target area. One can see that already at this stage it is possible to build events with efficiency more than 85%. The task of event building is finalized at the stage of searching the primary vertex, where it is possible to additionally use the proximity of tracks in space, as well as more accurate time measurements of the TOF detector.

2.4. KF Particle Finder — a package for reconstruction of short-lived particles

Today the most interesting physics is hidden in the properties of short-lived particles, which are not registered, but can be reconstructed only from their decay products. A fast and efficient KF Particle Finder package [4, 7], based on the Kalman filter (hence KF) method, for reconstruction and selection of short-lived particles is developed to solve this task. A search of more than 100 decay channels has been currently implemented (Fig. 3).

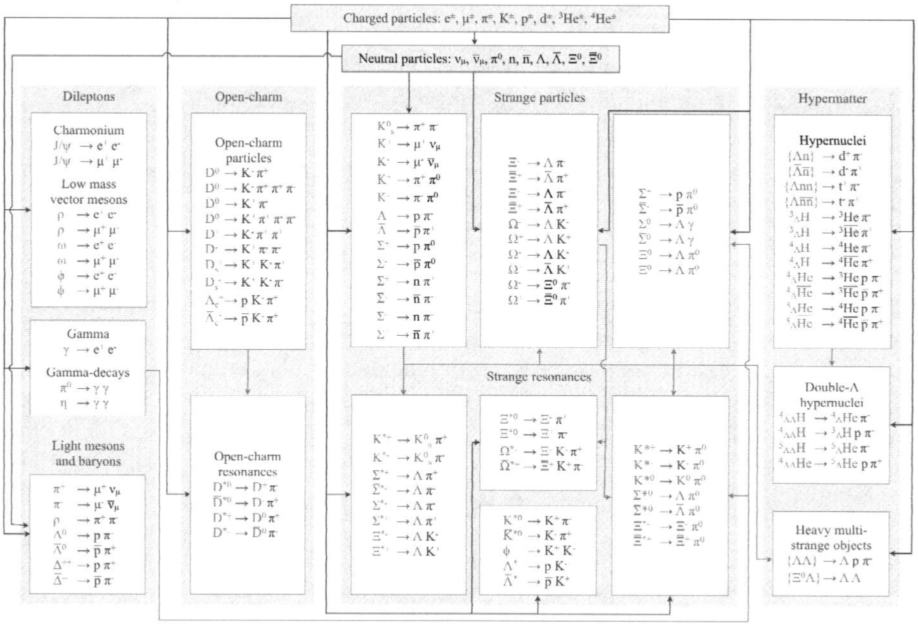

Figure 3. Block diagram of the KF Particle Finder package. The particle parameters, such as decay point, momentum, energy, mass, decay length and lifetime, together with their errors are estimated using the Kalman filter method.

In the package all registered particle trajectories are divided into groups of secondary and primary tracks for further processing. Primary tracks are those, which are produced directly in the collision point. Tracks from decays of resonances (strange, multi-strange and charmed resonances, light vector mesons, charmonium) are also considered as primaries, since they are produced directly at the point of the primary collision. Secondary tracks are produced by the short-lived particles, which decay not in the point of the primary collision and can be clearly separated. These particles include strange particles (K^0_s and Λ), multi-strange hyperons (Ξ and Ω) and charmed particles (D_0, $D^{\pm}$, $D^{\pm}_s$ and Λ_c). After that tracks are combined according to the block diagram in Fig. 3. The package estimates the particle parameters, such as decay point, momentum, energy, mass, decay length and lifetime, together with their errors. The package has a rich functionality, including particle transport, calculation of a distance to a point or another particle, calculation of a deviation from a point or another particle, constraints on mass, decay length and production point. All particles produced in the collision are reconstructed at once, that makes the algorithm local with respect to the data and therefore extremely fast.

In addition, simultaneous reconstruction in the KF Particle Finder of different decay channels of the same particle, including also decays with a neutral particle in the final state, makes it possible to calculate the efficiency of the reconstruction of rare particles and reliably estimate their systematic errors.

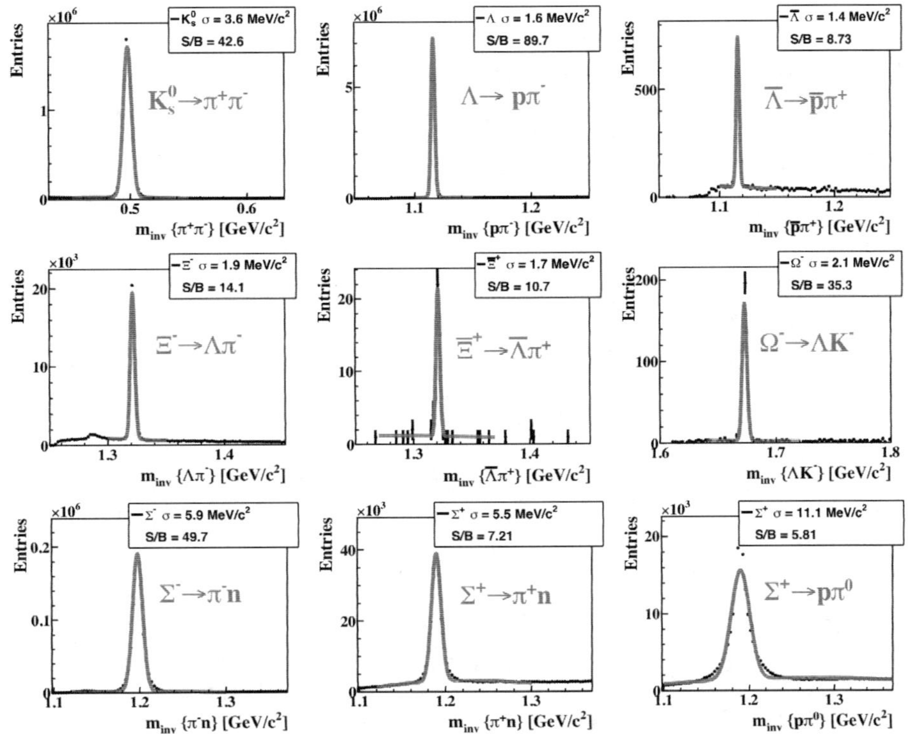

Figure 4. The FLES package provides clean probes of various stages of the collision: results of the search for short-lived particles are shown for 5M central AuAu UrQMD events at 10 AGeV with realistic PID.

The use of the Kalman filter at all stages of particle reconstruction allows in many cases to get rid almost completely of the combinatorial background and to obtain clean sets of particles, which can serve as probes of various stages of the collision (Fig. 4).

2.5. Deep learning for quark-gluon plasma detection

In addition to the *macroscopic* inverse approach [8] we investigate the *microscopic* inverse approach by using artificial neural networks to classify processes in heavy ion collisions. We have created two types of neural networks: fully connected (FC) and deep convolutional (CNN) neural networks. These networks were then used to identify quark-gluon plasma simulated within the Parton-Hadron-String Dynamics (PHSD) microscopic off-shell transport approach for central Au+Au collision at a fixed energy.

For FC networks we use a 64-neuron fully-connected hidden layer with batch normalization, Leaky Rectified Linear Unit (LReLU) activation and dropout. The number of neurons is chosen empirically and is fixed to allow comparison of FC neural networks with one, two and three layers. Batch normalisation and dropout are used to reduce overfitting and therefore improve overall performance. LReLU is used as it performs similarly to the most commonly used Rectified Linear Unit (ReLU) activation function but avoids dead neuron issues.

The CNN consists of two three-dimensional convolutional layers, each followed by a max pooling layer, and two sequential fully-connected layers.

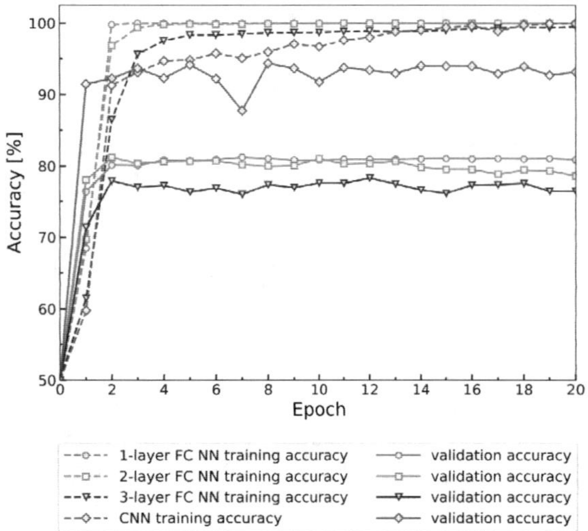

Figure 5. Training and validation accuracy for the FC networks and the CNN.

The Fig. 5 shows that the accuracy on the validation set rapidly increases for all four network architectures up to the fifth epoch, when the rise slows down and the curves level off. At the same time, the precision on the training set continues to go up until it reaches 100%, which suggests that overfitting occurs after the fifth epoch. Nevertheless, the fully-connected networks reach 80% precision while the convolutional neural network attains the best performance of more than 90% accuracy.

3. Express reconstruction and analysis in STAR

The STAR (Solenoidal Tracker At RHIC) experiment [9] at the RHIC (Relativistic Heavy Ion Collider) facility of the Brookhaven National Laboratory (BNL, USA) is designed to study nuclear matter under extreme conditions of relativistic heavy ion collisions, including hadron production and search for signs of quark-gluon plasma formation and its properties.

Very important for RHIC is the possibility to collide ions, covering the range of baryon chemical potential (μ_B) from 20 to 420 MeV, which corresponds to a wide range of energies (so called Beam Energy Scan, BES). The BES results once again confirmed evidence of QGP discovery in the upper RHIC energy $\sqrt{s_{NN}} = 200$ GeV. The results of the search for the critical point and the first-order phase boundary have narrowed the region of interest to collision energies below $\sqrt{s_{NN}} = 20$ GeV. Thus, the phase II of the Beam Energy Scan (BES-II), scheduled for 2019-2020 [10], covers the energy range $\sqrt{s_{NN}}$ <20 GeV. Current lattice QCD calculations suggest that key features of the phase diagram like the critical point and the first-order phase transition lie within μ_B of the BES Phase-II program.

There are three main improvements to the STAR detector for the BES-II phase: inner TPC (iTPC), end-cap Time-Of-Flight (eTOF), and event plane detector (EPD). These new detector systems cover pseudo-rapidity in the range of $1 < |\eta| < 5.4$. While iTPC and EPD cover both sides of STAR, eTOF is only installed on one side. The upgrade of iTPC and eTOF allows better identification and acceptance of particles. The EPD provides event plane reconstruction and centrality detection. All three detector systems improve significantly the statistical accuracy of STAR measurements in the low energy range close to the CBM energy range.

The FLES package of the CBM experiment is ported to the STAR experiment. Since 2013 (online) and 2016 (offline) the CA track finder is the standard STAR track finder for data production. Use of CA provides 25% more D_0 and 20% more W. The KF particle finder provides a factor 2 more signal particles than the standard approach in STAR. The package is used in HLT and its integration into the official STAR repository for use in the standard physics analysis is currently in progress.

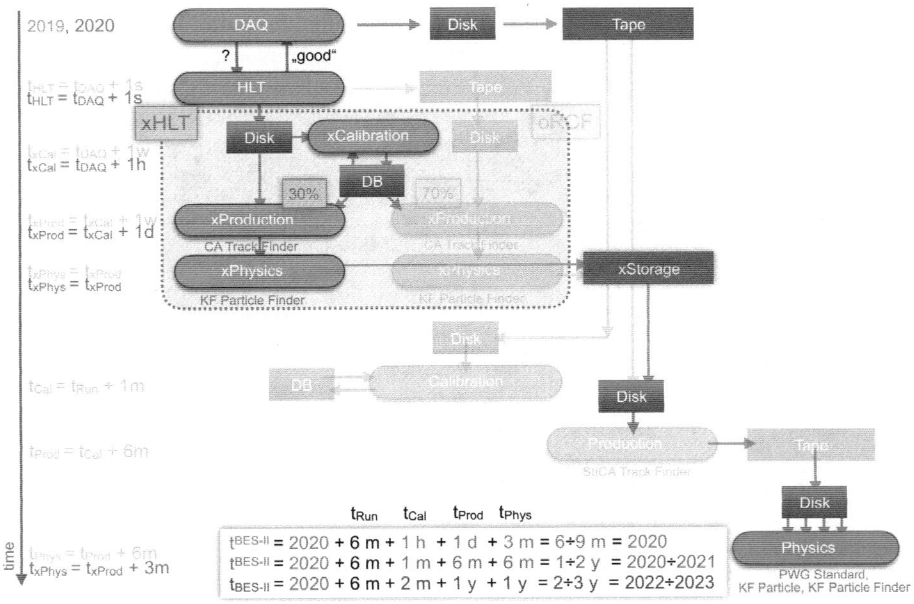

Figure 6. The HLT express and the standard data production and analysis workflows.

The use of the CA track finder and the KF particle finder in online extends significantly the functionality of HLT (Fig. 6). The standard calibration, production and analysis remain unchanged. HLT starts the calibration procedure as soon as data become available. The express chain makes possible physics analysis of the data as soon as the calibration is reasonable. It unifies approaches in extended (x)HLT and online (o)RCF to speed up the express workflow, and combines high competence of xHLT and oRCF experts involved in online operation. In addition, it provides physics working groups with instant and uncomplicated access to the data, like picoDST etc.

With the express calibration and alignment one can reconstruct hyperons with high significance and low level of background, as it is shown in Fig. 7. Hyperons are clearly seen at all BES-II energies: 3, 3.2, 3.9, 7.7, 9.1, 14.5, 19.6, 27 GeV. In addition, high significance allows extraction of spectra.

4. Conclusion

The CBM experiment with 10^7 input rate will require the full event reconstruction and physics analysis of the experimental data online. As the same HPC farm will be used for offline and online processing of experimental data, the main reconstruction and analysis algorithms will work both offline and online. Errors and insufficient accuracy in online data processing, physics analysis or selection of interesting collisions by the reconstruction algorithms will lead to complete loss of all experimental data, since only the incorrectly selected data will be stored in this case. Therefore only immediate comparison of the results of online analysis with the predictions of theoretical models using ANNs can guarantee the proper operation of the whole experiment. It has been demonstrated, that the core algorithms of the FLES package, the Cellular Automaton for searching for particle trajectories (100 μs/core/track) and the Kalman Filter to estimate their parameters (0.5 μs/core/track), have a very high level of intrinsic parallelism for their fast

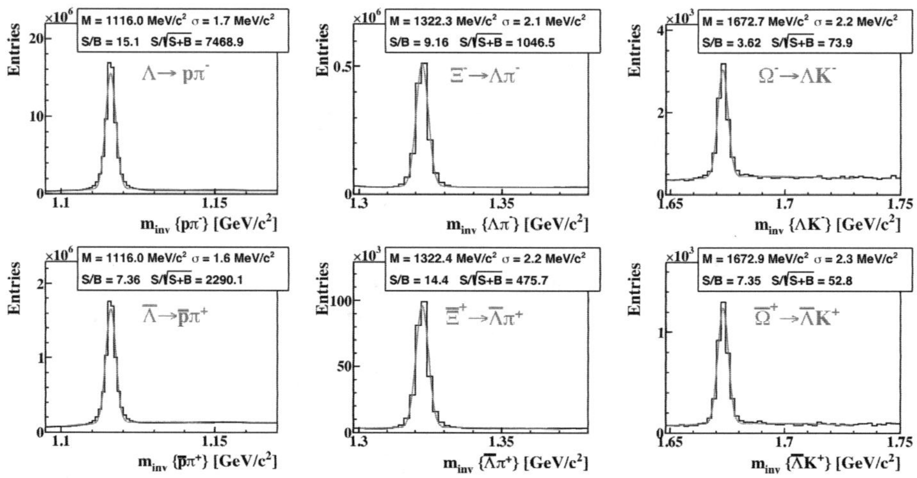

Figure 7. Online search for hyperons on 200M AuAu events at 14.5 GeV (2019 BES-II express production).

and efficient implementation on many-core CPU/GPU architectures. The KF particle finder package with more than 150 decay channels implemented (100 μs/core/decay) is a common platform for offline physics analysis and for real-time express analysis at 10^7 interaction rate in CBM.

Adaptation of the FLES algorithms within the FAIR Phase-0 program to the STAR experiment with its excellent detector performance, high quality experimental data and a well established reconstruction chain is the first and successful step in preparing the FLES algorithms for reconstruction and analysis of CBM real data at Day-1. Use of the CA track finder and KF particle finder developed in the CBM experiment can be beneficial for other experiments as the experimental heavy ion physics becomes more and more challenging.

References

[1] FAIR — Facility for Antiproton and Ion Research. Green Paper. The Modularized Start Version. GSI. October 2009.
[2] CBM Collaboration, Compressed Baryonic Matter Experiment, Tech. Stat. Rep., GSI, Darmstadt, 2005; 2006 update.
[3] I. Kisel, Event reconstruction in the CBM experiment, Nucl. Instr. and Meth. A566 (2006) 85-88.
[4] I. Kisel, I. Kulakov and M. Zyzak, Standalone first level event selection package for the CBM experiment, IEEE Trans. Nucl. Sci. 60 (5) (2013) 3703-3708.
[5] V. Akishina and I. Kisel, Online 4-dimensional reconstruction of time-slices in the CBM experiment, IEEE Trans. Nucl. Sci. 62 (6) (2015) 3172-3176.
[6] V. Akishina, "4D Event Reconstruction in the CBM Experiment", Dissertation thesis, Goethe university, Frankfurt am Main (2017).
[7] M. Zyzak, "Online Selection of Short-Lived Particles on Many-Core Computer Architectures in the CBM Experiment at FAIR", Dissertation thesis, Goethe university, Frankfurt am Main (2016).
[8] I. Kisel, Event Topology Reconstruction in the CBM Experiment, J. Phys. Conf. Ser. 1070 (2018), 97.
[9] K.H. Ackermann et al. (STAR Collaboration), "STAR detector overview", Nucl. Instr. Meth. A499 (2003) 624.
[10] STAR Collaboration, "Studying the Phase Diagram of QCD Matter at RHIC", 01 June 2014. https://drupal.star.bnl.gov/STAR/files/BES_WPII_ver6.9_Cover.pdf

Journal of Physics: Conference Series **1602** (2020) 012007 doi:10.1088/1742-6596/1602/1/012007

Probing strangeness production in small systems through new multi-differential measurements with ALICE at the LHC

Adrian Nassirpour, on behalf of The ALICE Collaboration

E-mail: adrian.nassirpour@hep.lu.se

Abstract. ALICE has observed that the relative fraction of strange hadrons grows strongly with multiplicity in small collision systems (proton-proton and proton-lead collisions) at LHC energies, in particular for multi-strange baryons. This implies that a proton-proton collision cannot be described as an incoherent sum of parton-parton collisions, an idea that has been central in most proton-proton generators, for example PYTHIA.
To accommodate the new ALICE results, models have to introduce significant final-state interactions. These final-state effects can have very different phenomenological origin. To be able to eventually discriminate experimentally between different final-state models, new experimental tools are required and in this paper different multi-differential observables will be tested with this goal in mind. Transverse Spherocity (S_O) is an observable that allows a topological selection of events that are "isotropic" (dominated by multiple soft processes) and "jetty" (where a single hard process is responsible for a significant part of the multiplicity). The underlying event activity is another observable that can be used to vary the amount of soft processes. It can be estimated by measuring the charged-particle multiplicity in the Transverse region (R_{T}).
Using these new observables, ALICE has obtained results for π, K, ϕ, p, and Ξ production at mid-rapidity ($|\eta| < 0.8$) as a function of event shape and underlying event activity in proton-proton collisions at $\sqrt{(s_{\mathrm{NN}})} = 13$ TeV. Finally, this contribution will report on how these new multi-differential measurements compare to predictions from PYTHIA and EPOS-LHC.

1. Introduction

Measurements in high-multiplicity proton-proton (pp) and proton-lead (p-Pb) collisions have revealed that small collision systems (i.e pp and p-Pb) exhibit some signatures that were previously considered unique features of heavy-ion collisions. Some of these signatures, such as an enhanced production of strange hadrons, or a collective expansion of the system, could be explained by the formation of a strongly interacting medium. However, it is difficult to argue for the presence of a strongly interacting medium in small-systems within current theoretical frameworks, because of the limits to the achievable lifetime and volume of the produced medium, as well as temperature/baryon density, see e.g [1] for discussions.
The observation of strangeness enhancement in small-systems at the Large Hadron Collider (LHC) indicates that pp collisions at these energies can no longer be seen as incoherent sums of parton-parton collisions, an idea that is central to most general-purpose, Quantum-Chromo Dynamics (QCD)-inspired Monte-Carlo event generators, such as PYTHIA [2]. EPOS-LHC is an event generator which forms a two-phase state consisting of a dense core of QGP, and a

dilute corona (core-corona). The strangeness enhancement in EPOS-LHC is due to a change in the relative contribution of the corona (low strangeness production) and core (high strangeness production) with multiplicity. Strange-particle production constitutes a good probe to test if models such as PYTHIA or EPOS-LHC are better at describing hadronization in high-multiplicity small-system collisions.
This contribution presents new results from A Large Ion Collider Experiment (ALICE) on strangeness production in small systems. The data analyzed are from pp collisions at $\sqrt{(s_{\mathrm{NN}})}$ = 13 TeV. The observables relate to the event topology and the activity of the underlying event; Transverse Spherocity $S_{\mathrm{O}}^{p_{\mathrm{T}}=1}$and the self-normalized charged-particle density in the transverse region, R_{T}. New results from a third type of measurement, Ξ-h correlations, can be found here [3].

2. π/K/p/ϕ/Ξ Production as a Function of Unweighted Transverse Spherocity $S_{\mathrm{O}}^{p_{\mathrm{T}}=1}$

2.1. Method
The premise of the analysis, discussed in the following [4], is that the event topology in the azimuthal plane will reflect the main mode of particle production. Events dominated by a single hard scattering will have pronounced back-to-back jet structures, while events dominated by multiple soft scatterings will result in an isotropic topology, see Fig. 1 for an illustration. The unweighted transverse spherocity $S_{\mathrm{O}}^{p_{\mathrm{T}}=1}$is used as an estimator to disentangle these two topological limits:

$$S_{\mathrm{O}}^{p_{\mathrm{T}}=1} = \frac{\pi^2}{4} \min_{\hat{n}} \left(\frac{\Sigma_i |p_{\mathrm{T},i} \times \hat{n}|}{N_{\mathrm{trks}}} \right). \tag{1}$$

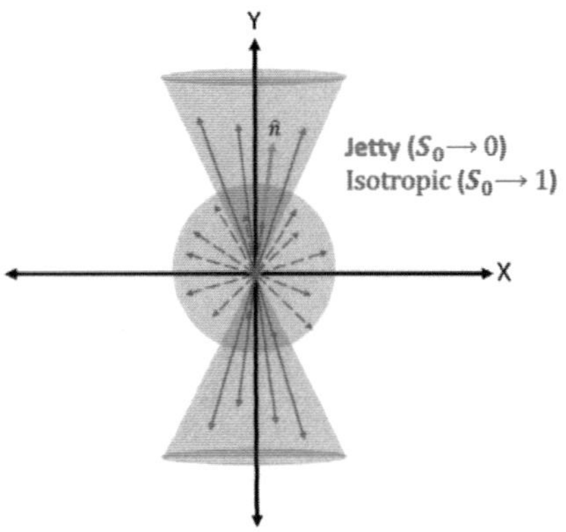

Figure 1: Topological illustration of "jetty" and "isotropic" event classes [5].

$S_{\mathrm{O}}^{p_{\mathrm{T}}=1}$is calculated using charged particle tracks that have $p_{\mathrm{T}} > 0.15$ GeV. Unlike the estimator discussed in [4], the transverse momentum (p_{T}) of each track is normalized to 1 ($p_{\mathrm{T}} = 1$). This modification is required to minimize biases which affect neutral particle yields. N_{trks} is the sum of charged particles in a given event, and $\hat{n}$ is a unit vector that minimizes the function. The charged particle tracks are reconstructed using the ALICE Time-Projection Chamber (TPC), which is a gas detector that reconstructs charged particles, with a high efficiency and good momentum resolution, within the pseudorapidity interval $|\eta| < 0.8$.
$S_{\mathrm{O}}^{p_{\mathrm{T}}=1}$will vary by construction between values of 0 and 1, where the two limits will correspond to the two different topological limits; Events where $S_{\mathrm{O}}^{p_{\mathrm{T}}=1} \rightarrow 0$ will be dominated by a single back-to-back jet, whereas events where $S_{\mathrm{O}}^{p_{\mathrm{T}}=1} \rightarrow 1$ will be dominated by more isotropic particle production (absence of a preferred direction). The events in the bottom 20% quantile of the $S_{\mathrm{O}}^{p_{\mathrm{T}}=1}$distribution will henceforth be referred to as "jetty" events (not in the strict sense of a jet-finding algorithm, but rather that the azimuthal topology is pencil-like), whereas the top 20% quantile is referred to as "isotropic" events.
In addition to the $S_{\mathrm{O}}^{p_{\mathrm{T}}=1}$selection, a top 10% high-multiplicity (multiplicity classes I-III [4]) requirement is also imposed for event selection in order to maintain sensitivity to the event structure. Strangeness enhancement is a phenomenon observed in high-multiplicity pp collisions,

making this a region of specific interest. Furthermore, minimum-bias pp collisions mainly consist of collisions that produce few tracks, and it is not clear what a topology means in this case. For this reason, in addition to the top 10% multiplicity selection, events are also required to have at least $N_{trks} \geq 10$ to be considered (which is already the case for most of the top 10% events). The multiplicity is estimated using two different sub-detector systems in ALICE. The V0M scintillators measure the charged particle multiplicity at forward rapidity, in ranges of $2.8 < \eta < 5.1$ and $-3.7 < \eta < -1.7$. The multiplicity is also estimated at mid-rapidity ($|\eta| < 0.8$), using the ALICE Inner Tracking System (ITS), by measuring the number of short-track segments N_{SPD}, referred to as the CL1 multiplicity in the following, (as it is related to the CL1 trigger). The main difference between the two multiplicity estimators is that for V0M-based selections, the rapidity regions for multiplicity estimation and for the measurement of the particle spectra are different, whereas for CL1-based selections, the multiplicity and particle spectra are measured in the same rapidity region. Results from the CL1 is thus more sensitive to local fluctuations (auto-correlations, such as jets).
Particle Identification (PID) is used in order to measure the spectra of identified particle species in the different multiplicity and $S_{\mathrm{O}}^{p_{\mathrm{T}}=1}$regions. Pions, kaons and protons are identified from the specific energy loss $\mathrm{d}E/\mathrm{d}x$ measured by the TPC, together with the time-of-flight (TOF) information using the ALICE TOF detector. ϕ and Ξ yields are extracted from invariant mass distributions of their identified decay daughters (after topological selections for the Ξ).

2.2. Results

Figure 2 presents the pion $\langle p_{\mathrm{T}} \rangle$ as a function of the average pion yield in different multiplicity and spherocity bins. The figure illustrates how the multiplicity and spherocity selection biases the event samples. The V0M selected measurements maintain a similar $\langle p_{\mathrm{T}} \rangle$ between the spherocity classes, but has large differences in $\langle \mathrm{d}N_{\pi}/\mathrm{d}y \rangle$. In contrast, the $S_{\mathrm{O}}^{p_{\mathrm{T}}=1}$classes for the CL1 selected events are characterized by the $\langle p_{\mathrm{T}} \rangle$ of the events, suggesting that $S_{\mathrm{O}}^{p_{\mathrm{T}}=1}$might be able to categorize events according to their hardness. The implicit multiplicity dependence of $S_{\mathrm{O}}^{p_{\mathrm{T}}=1}$is minimized when using a mid-rapidity multiplicity estimator, which is in agreement with findings from [4].
A similar effect as observed in Fig. 2 can also be seen in Fig. 3. The V0M ratios in the lower panel display a flat p_{T}-dependence for both spherocity classes, whereas a p_{T} dependence can be observed for the CL1 triggered events. The particle ratios of K/p/ϕ/Ξ-to-π, compared to equivalent PYTHIA and EPOS-LHC ratios, are shown in Figs. 4 to 7. It is observed that neither event generator can accurately predict the "single-ratios" in the upper panels. EPOS/PYTHIA tend to overestimate/underestimate the trends respectively.

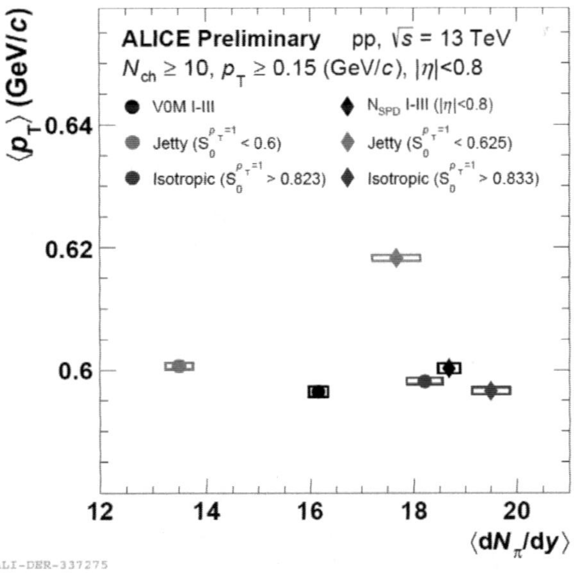

Figure 2: The $\langle p_{\mathrm{T}} \rangle$ and π meson yield for different multiplicity estimators and $S_{\mathrm{O}}^{p_{\mathrm{T}}=1}$selections.

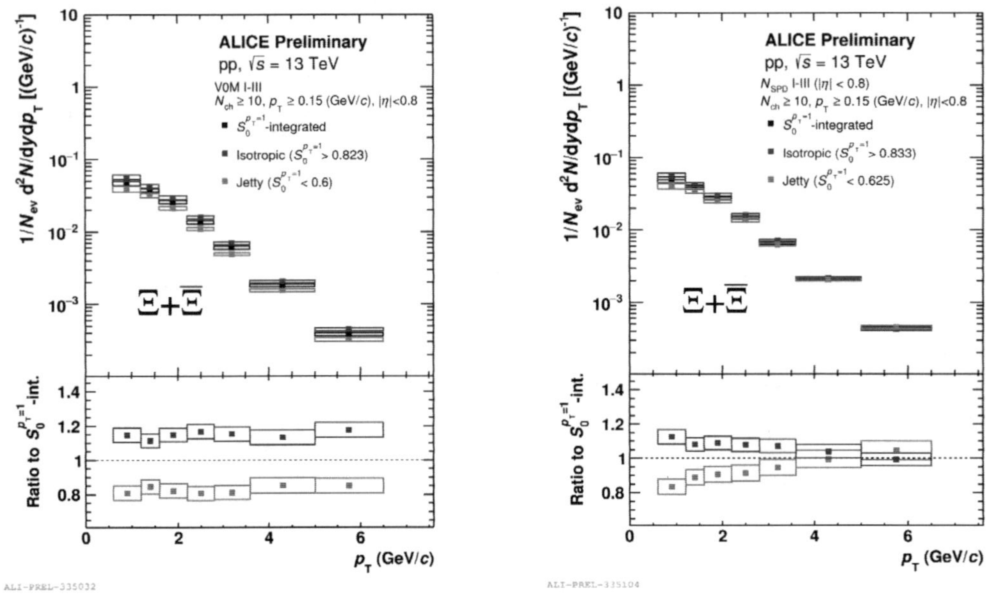

Figure 3: The Ξ spectra as a function of multiplicity and $S_{\mathrm{O}}^{p_{\mathrm{T}}=1}$. The left (right) panel displays events where the multiplicity selection was done using the V0M (CL1) multiplicity estimator. Lower panels show the ratios to $S_{\mathrm{O}}^{p_{\mathrm{T}}=1}$-integrated.

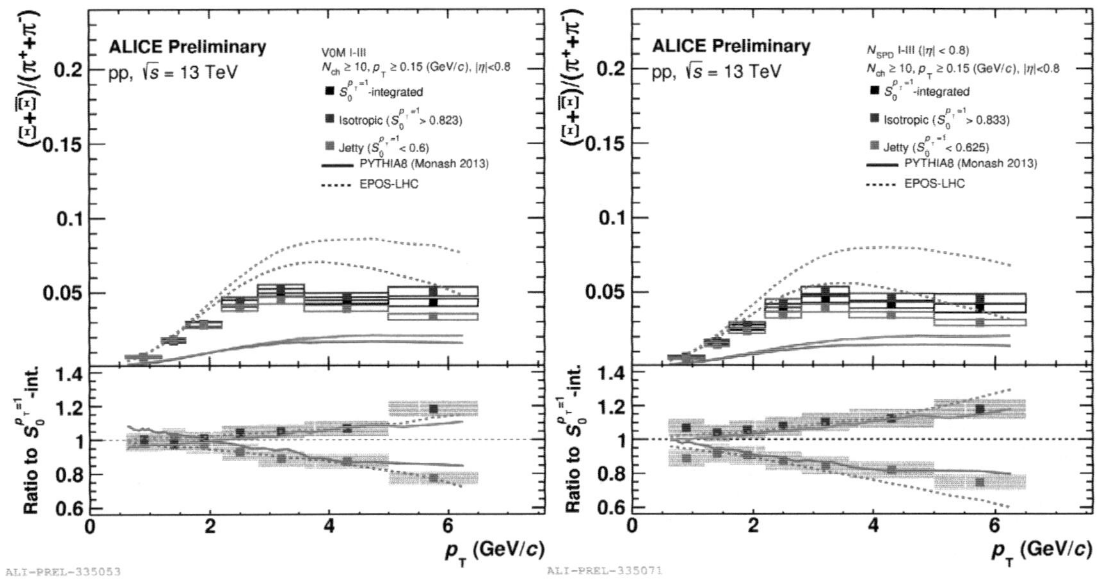

Figure 4: Ξ-to-π ratio as a function of p_{T} in bins of $S_{\mathrm{O}}^{p_{\mathrm{T}}=1}$. The left (right) panel displays events where the multiplicity selection was done using the V0M (CL1) multiplicity estimator. Lower panels show the ratios to $S_{\mathrm{O}}^{p_{\mathrm{T}}=1}$-integrated.

Both event generators are able to qualitatively describe the double-ratio (lower panels), which suggests that the simulations are able to describe the $S_{\mathrm{O}}^{p_{\mathrm{T}}=1}$selection but overestimate/underestimate the general enhancement of K/p/ϕ/Ξ production in high-multiplicity

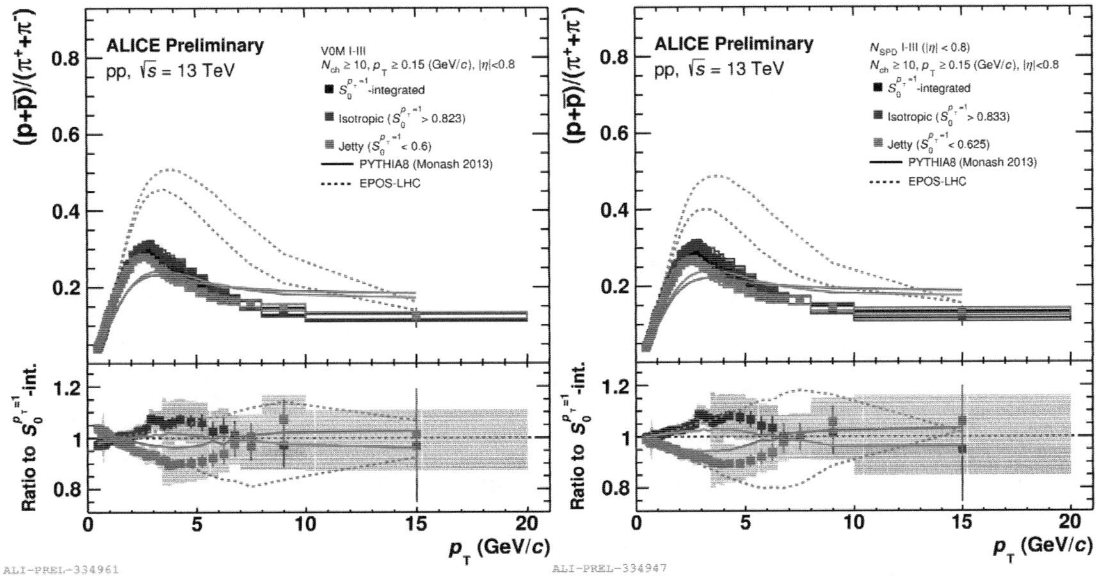

Figure 5: p-to-π ratio as a function of p_{T} in bins of $S_{\mathrm{O}}^{p_{\mathrm{T}}=1}$. The left (right) panel displays events where the multiplicity selection was done using the V0M (CL1) multiplicity estimator. Lower panels show the ratios to $S_{\mathrm{O}}^{p_{\mathrm{T}}=1}$-integrated.

events. The double-ratios highlight an overall relative enhancement of K and Ξ (Figs. 4 and 6) hadrons compared to pions in "isotropic" events, and a suppression in "jetty" events. This suggests that the enhancement of strange hadrons in high-multiplicity pp collisions is driven by soft processes rather than hard processes.

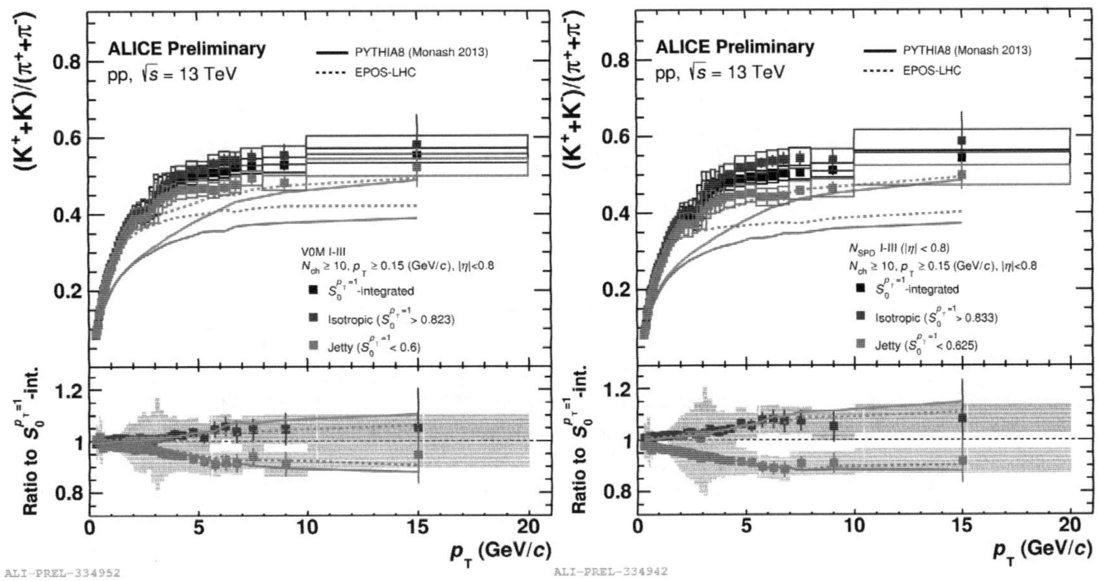

Figure 6: K-to-π ratio as a function of p_{T} in bins of $S_{\mathrm{O}}^{p_{\mathrm{T}}=1}$. The left (right) panel displays events where the multiplicity selection was done using the V0M (CL1) multiplicity estimator. Lower panels show the ratios to $S_{\mathrm{O}}^{p_{\mathrm{T}}=1}$-integrated.

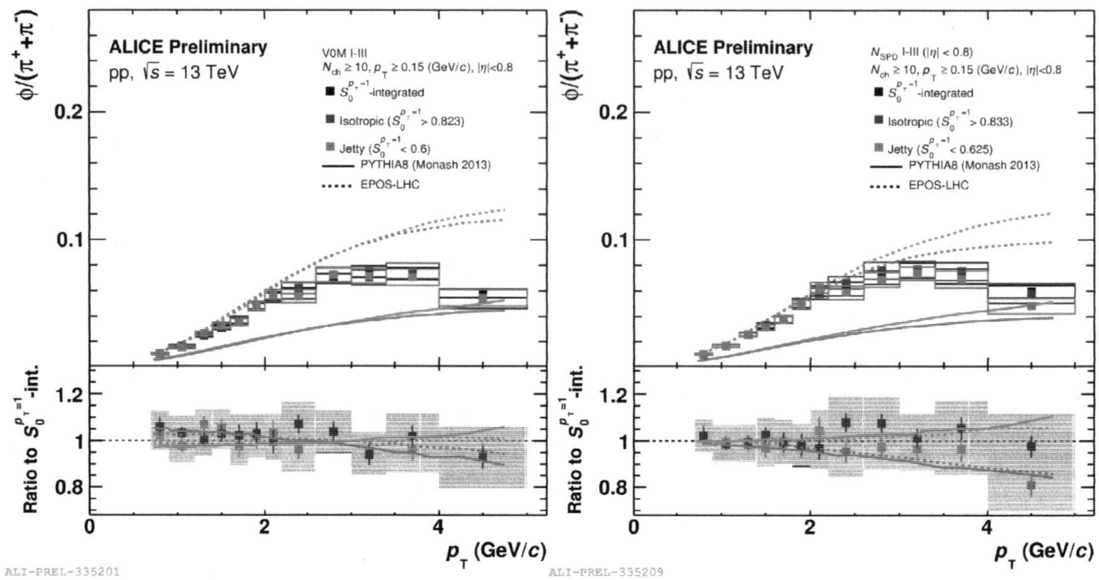

Figure 7: ϕ-to-π ratio as a function of $p_{\rm T}$ in bins of $S_{\rm O}^{p_{\rm T}=1}$. The left (right) panel displays events where the multiplicity selection was done using the V0M (CL1) multiplicity estimator. Lower panels show the ratios to $S_{\rm O}^{p_{\rm T}=1}$-integrated.

The p-to-π V0M ratios in Fig. 5 highlight that the baryon-to-meson enhancement at intermediate $p_{\rm T}$ (usually attributed to radial flow [6]) is larger in isotropic events, which suggests that these events indeed have more flow, as we expect if a strongly-interacting medium is present. In contrast, the CL1 p-to-π ratios seem to suggest a strict enhancement/suppression of proton production in the different event types, similar to effects observed in measurements of the jet-$p_{\rm T}$ evolution [7].
The ϕ-to-π ratio (Fig. 7 indicates that ϕ particle production has no significant dependence between "jetty" or "isotropic" events. Furthermore, the dependence on V0M and CL1 multiplicity slicing appears to be the same.

3. π/K/p/ϕ/Ξ Production as a Function of $R_{\rm T}$

3.1. Method
The purpose of this measurement is to estimate the production of hard probes relative to the underlying event. Analyzed events are required to have a leading trigger particle of $p_{\rm T} \geq 5.0$ GeV, to ensure that at least one hard scattering took place in the event. The event is sectioned into three different azimuthal regions, relative to the leading track, as seen in the left panel of Fig 8. Letting ϕ_L define the azimuthal angle of the leading particle, the three regions are defined as containing charged particle tracks within relative differences of azimuthal angle:

- Toward: $|\phi - \phi_L| \leq \frac{\pi}{3}$
- Away: $\frac{2\pi}{3} \leq |\phi - \phi_L| \leq 2\pi$
- Transverse: $\frac{\pi}{3} < |\phi - \phi_L| < \frac{2\pi}{3}$

Particle production in the Toward region will be dominated by jet fragmentation, as well as soft collisions due to the underlying event. The Away-side region (often) captures the back-scattered jet, and particle production in this region is also dominated by a combination of jet

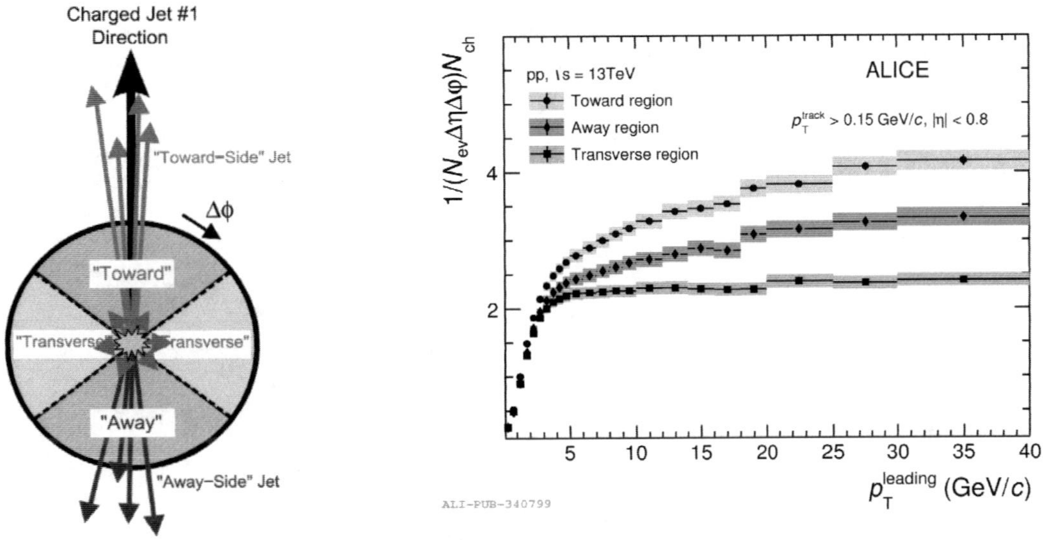

Figure 8: Left panel illustrates the three different azimuthal regions relative to the leading track [9]. Right panel showcases the average charged-particle production in the different regions as a function of the p_T of the leading track [10].

fragmentation and the underlying event. The Transverse region is perpendicular to the leading track, and by definition, is almost a pure sample of the underlying event (although with possible contamination from initial-and final-state radiation). This can also be observed in the right panel of Fig. 8. For trigger particles with $p_T \geq 5.0$ GeV, the charged-particle production in the Transverse region is largely independent of the trigger particle p_T, indicating that the underlying event has the same properties so that one can integrate over the trigger p_T.
The particle yield of $\pi/\mathrm{K}/\mathrm{p}/\phi/\Xi$ is measured in the azimuthal regions as a function of the self-normalized transverse charged particle density R_T. If N_T is defined as the number of charged particles reconstructed in the Transverse region, R_T is defined as shown in Eq. 2. R_T is used to control the underlying event relative to the hard scattering [8], where events in the limit of $R_T \to 0$ approaches "e^+e^-" physics (only consisting of a single hard scattering), and $R_T \to \infty$ approaches "AA"-like physics, where events are dominated by multiple soft collisions.

$$R_T = \frac{N_T}{\langle N_T \rangle} \tag{2}$$

3.2. Results

The particle ratios of $\mathrm{K}/\mathrm{p}/\phi/\Xi$-to-$\pi$ as a function of R_T in the Toward and Transverse regions are presented in Figs. 9 to 12. The p-to-π ratio in Fig. 9 displays a similar dependence on the event activity (R_T) as observed in multiplicity-dependent pp studies [6]. This effect is enhanced in the Transverse region. Notably, the high-R_T p-to-π ratio in the Toward region approaches the p-to-π values in the Transverse region. This effect is also observed in the Ξ-to-π ratio in Fig. 10. Furthermore, there is a large difference between the two R_T bins that is only present in the Toward region. This indicates that the relative Ξ-to-π production is proportional to R_T. The difference seen in the Toward region between the two R_T limits indicate that the relative Ξ-to-π rate is different in jet-fragmentation and underlying-event production, suggesting that the enhancement of strange-hadron production at increased multiplicity originates from the underlying event.

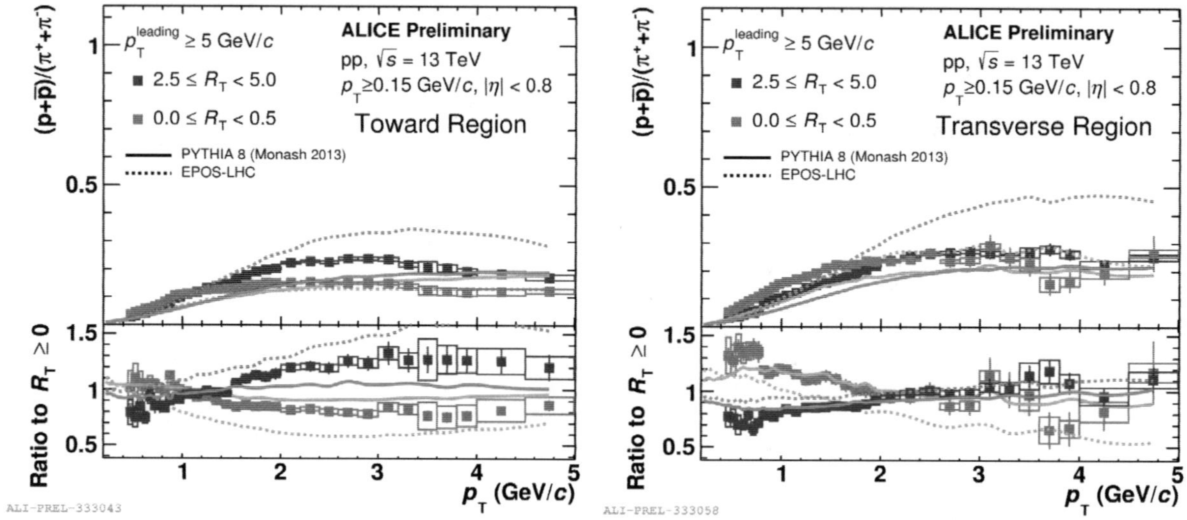

Figure 9: p-to-π ratio as a function of p_T in bins of R_T. Left panel displays the p-to-π ratio of particles extracted from the Toward region. Right panel displays the p-to-π ratio extracted from the Transverse region. Lines (dashed and solid) represent equivalent EPOS-LHC and PYTHIA comparisons respectively. Lower panels show the ratios to the integrated R_T selection.

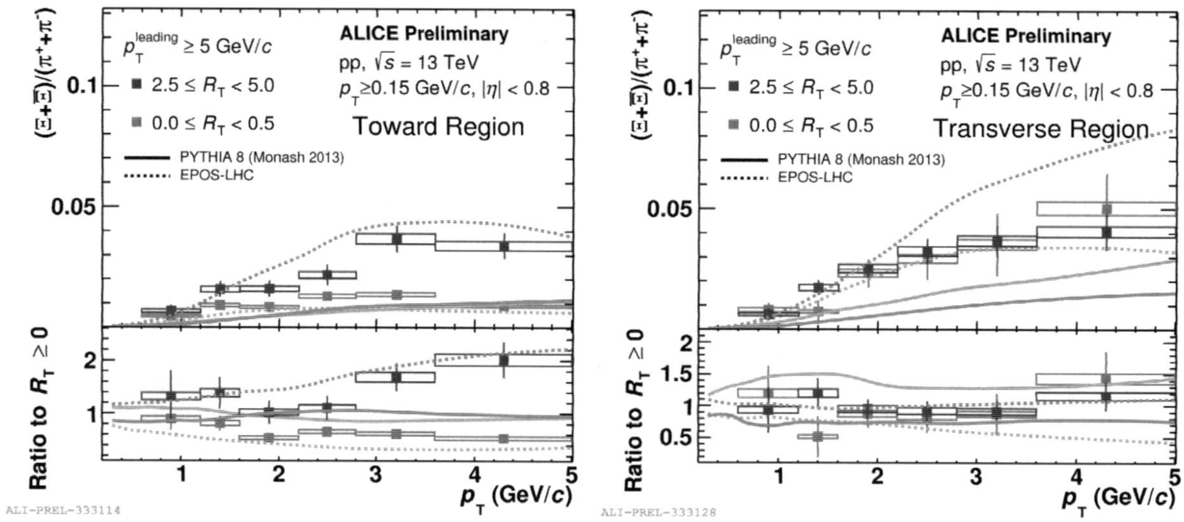

Figure 10: Ξ-to-π ratio as a function of p_T in bins of R_T. Left panel displays the p-to-π ratio of particles extracted from the Toward region. Right panel displays the p-to-π ratio extracted from the Transverse region. Lines (dashed and solid) represent equivalent EPOS-LHC and PYTHIA comparisons respectively. Lower panels show the ratios to the integrated R_T selection.

The ϕ-to-π and K-to-π ratios displayed in Figs. 11 to 12 largely resemble the same kind of phenomena observed in the previously shown baryon ratios. Notably, there is a very large modification of ϕ meson yield in the Toward region for events with large underlying event activity. Together with the Ξ-to-π results, this observation strengthens the suggestion that

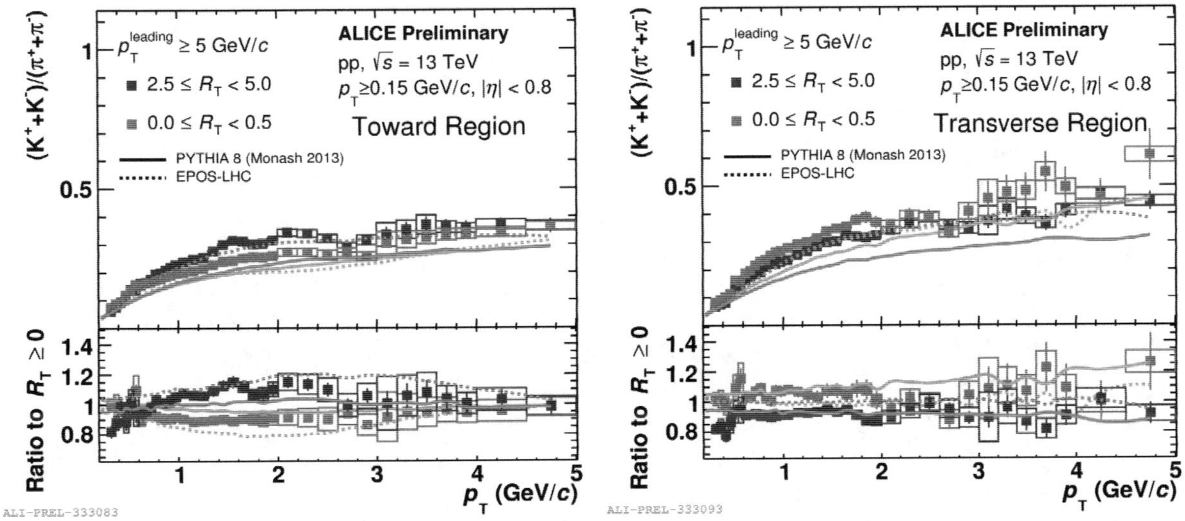

Figure 11: K-to-π ratio as a function of p_T in bins of R_T. Left panel displays the K-to-π ratio of particles extracted from the Toward region. Right panel displays the K-to-π ratio extracted from the Transverse region. Lines (dashed and solid) represent equivalent EPOS-LHC and PYTHIA comparisons respectively. Lower panels show the ratios to the integrated R_T selection.

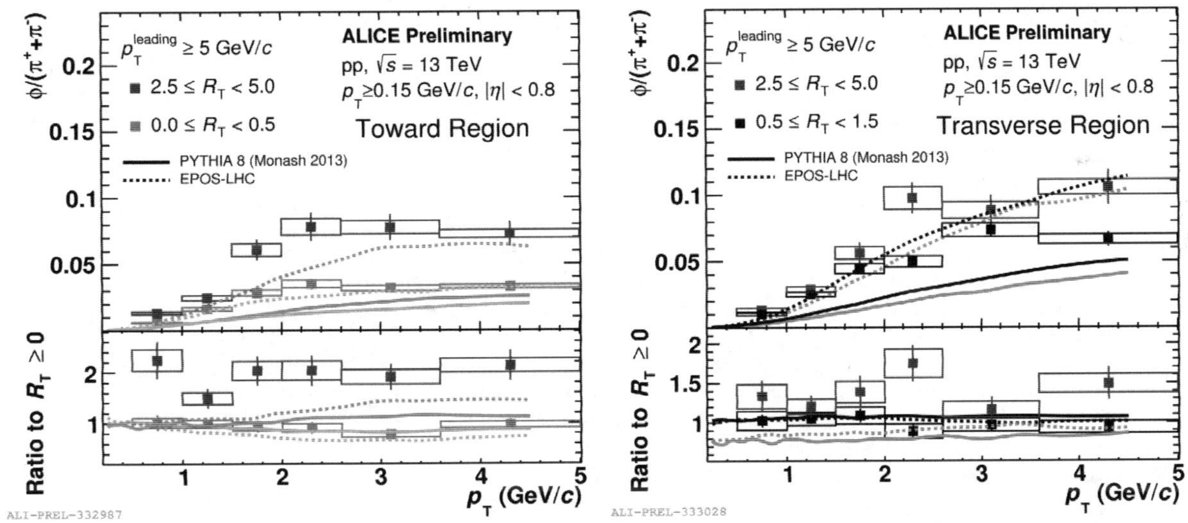

Figure 12: ϕ-to-π ratio as a function of p_T in bins of R_T. Left panel displays the ϕ-to-π ratio of particles extracted from the Toward region. Right panel displays the ϕ-to-π ratio extracted from the Transverse region. Lines (dashed and solid) represent equivalent EPOS-LHC and PYTHIA comparisons respectively. Lower panels show the ratios to the integrated R_T selection. Note that the lower R_T-limit is different for this particle specie, arising from complications in the yield extraction for low-R_T events.

strange particle production is enhanced in the underlying event.
Furthermore, while both of the models are able to qualitatively predict the relative production

of kaons and protons, they are not able to accurately account for the $R_{\rm T}$-dependent phenomena observed for ϕ and Ξ hadrons. While PYTHIA manages to describe low-$R_{\rm T}$ trends in the Toward region, it is currently not capable of describing the increase of relative ϕ and Ξ production observed as the underlying event increases, both in the single and double ratios. However, EPOS-LHC is able to model the qualitative increase of Ξ and ϕ production as a function of $R_{\rm T}$ (although slightly overestimating/underestimating the increase of the single-ratio for Ξ and ϕ production respectively).

4. Conclusions

This work has showcased two new measurements from ALICE that aim to achieve a better understanding of light-flavor hadronization in small systems. It has been demonstrated that $S_{\rm O}^{p_{\rm T}=1}$is able to disentangle events based on their azimuthal topology and overall hardness. Therefore, $S_{\rm O}^{p_{\rm T}=1}$can be used as a proxy to probe particle production in events that are either dominated by hard or soft processes. It has also been shown that $S_{\rm O}^{p_{\rm T}=1}$can be used as a tool to disentangle biases from local fluctuations that arise when measurements and multiplicity estimations occur in the same rapidity region. Furthermore, the observations presented in Figs. 4 - 7 suggest that the enhancement of K and Ξ hadrons seen in high-multiplicity small-system collisions are primarily driven by dynamics that arise from events with isotropic topologies. The ϕ meson displays no modifications dependent on $S_{\rm O}^{p_{\rm T}=1}$.

The measurements of relative p/K/ϕ/Ξ production relative to π as a function of $R_{\rm T}$ also support some of the indications seen in the $S_{\rm O}^{p_{\rm T}=1}$measurements. The two estimators are indirectly linked, as the underlying event is produced through multiple soft isotropic interactions. Events with a large underlying event will thus on average tend to have an isotropic azimuthal event topology. As seen in Fig. 9 and Fig. 12, the relative production of strange hadrons in the Toward region is greatly enhanced alongside the underlying event activity. Furthermore, the relative production of all Ξ and ϕ hadrons to π in the Toward region, in high-$R_{\rm T}$ events, tends to saturate toward the equivalent distributions for the Transverse region. This further suggests that the enhancement of strange hadron production with multiplicity is mainly driven by the underlying event.

References

[1] Kurkela, Aleksas Mazeliauskas, *Phys.Rev.Lett.* **122**, 142301 (2019)
[2] T.Sjöstrand et al, *Computer Physics Communications* **191**, 159–177 (2015)
[3] J. Adolfsssons, "Studying particle production in small systems through correlation measurements in ALICE", arXiv:2005.14675v1 **hep-ex** (2020)
[4] ALICE Collaboration, *Eur. Phys. J,* **C79**, 10, 857 (2019)
[5] A. Khuntia, S. Tripathy, A. Bisht, R. Sahoo, "Event Shape Engineering and Multiplicity dependent Study of Identified Particle Production proton+proton Collisions at $\sqrt{s}$ = 13 TeV using PYTHIA", arXiv:1811.04213v1 **hep-ph** (2018)
[6] ALICE Collaboration, *Phys.Rev.* **C99**, 024906 (2019)
[7] B.A.Hess, "Particle Identification in Jets and High-Multiplicity pp Events with the ALICE TPC", **CERN-THESIS-2015-158** (2015)
[8] T. Martin, P.Skands, S.Farrington, *Eur.Phys.J,* **C76**, 5, 299 (2016)
[9] ALICE Collaboration, *Eur. Phys. J,* **C62**, 237 (2009)
[10] ALICE Collaboration, *JHEP*, **04**, 192 (2020)

Production of hadronic resonances measured with ALICE at the LHC

Jihye Song (for the ALICE Collaboration)
Department of Physics, University of Houston, Houston, Texas, USA
E-mail: Jihye.Song@cern.ch

Abstract. Measurements of short-lived hadronic resonances are used to probe the properties of the hadronic phase in ultra-relativistic heavy-ion collisions. Since these resonances have lifetimes comparable to that of the produced fireball, they are sensitive to the competitive rescattering and regeneration effects in the hadronic gas, which modify the observed particle momentum distributions and yields after hadronization. Having different masses, quantum numbers, and quark content, hadronic resonances can provide insight into processes that determine the shapes of particle momentum spectra, strangeness production, and the possible onset of collective effects in small systems.

We here present the latest results on $\rho(770)^0$, K*(892), $\phi(1020)$, $\Sigma(1385)^{\pm}$, $\Lambda(1520)$, $\Xi(1530)^0$ and $\Xi(1820)$ production in pp, p–Pb, Pb–Pb and Xe–Xe collisions at different LHC energies. Results include system-size and collision-energy evolution of transverse momentum spectra, integrated yields, mean transverse momenta, particle ratios, and nuclear modification factors.

1. Introduction

Hadronic resonances with various lifetimes are valuable probes to study the properties of the hadronic medium formed in ultra-relativistic heavy-ion collisions, since the yield ratios of resonances to stable hadrons provide information about the re-scattering and regeneration effects in the hadronic medium. If the loss of resonances due to elastic or pseudo-elastic scattering of their decay products (re-scattering) is dominant over regeneration, the resonance yield after kinetic freeze-out will be smaller than the one originally produced at the chemical freeze out. Considering the expected lifetime of the hadronic phase ($\sim$ 10 fm/c), the measurement of the production of a comprehensive set of resonances with different lifetimes can be used to study the interplay of particle re-scattering and regeneration.

Studies of strangeness production play an important role in understanding the matter produced in heavy-ion collisions. Measurements of strange and non-strange particle yields can be described by grand-canonical thermal models in heavy-ion collisions [1], while canonical suppression is expected in small systems [2, 3]. Enhancement of strangeness production has been observed in high energy nucleus–nucleus (A–A) collisions with respect to pp collisions at RHIC (Relativistic Heavy-Ion Collider) and LHC (Large Hadron Collider) energies [4]. The study of strangeness production as a function of the charged particle multiplicity produced in the collision systems from pp to A–A, therefore, allows one to investigate the origin of the enhancement.

Journal of Physics: Conference Series **1602** (2020) 012008 doi:10.1088/1742-6596/1602/1/012008

Particles with open strangeness may be subject to canonical suppression in small collision systems with respect to large systems, whereas the ϕ, a particle with hidden strangeness, is not expected to be canonically suppressed. Comparisons of particles with open and hidden strangeness can help to understand strangeness production.

2. Experiment

The ALICE experiment [5] is particularly well suited to study the production of both identified and unidentified charged particles thanks to its excellent tracking performance coupled with extensive particle identification (PID) capabilities over a wide range of transverse momentum. In the central barrel, the Inner Tracking System (ITS) and the Time Projection Chamber (TPC) are used for charged-particle tracking and primary collision vertex reconstruction. The ITS consists of three sub-detectors of two layers each, covering a central pseudorapidity range $|\eta| < 0.9$: Silicon Pixel Detector (SPD), Silicon Drift Detector (SDD) and Silicon Strip Detector (SSD). The TPC is the main charged particle tracking detector, and has full azimuthal coverage in the pseudorapidity range $|\eta| <0.9$. Along with track reconstruction, it also provides a measurement of the momentum and excellent particle identification. The TPC provides the measured specific energy loss ($\mathrm{d}E/\mathrm{d}x$) to identify the particles, especially in low momentum range ($p < 1$ GeV/c) where the $\mathrm{d}E/\mathrm{d}x$ of particles are well separated. To extend the particle identification to higher p_{T}, the Time of Flight (TOF) detector is used in addition to the TPC information. It covers a pseudorapidity range $|\eta| < 0.9$ and provides excellent PID capabilities in the intermediate p_{T} range by exploiting the time-of-flight information. The forward V0 detector, a scintillator detector with a timing resolution less than 1 ns, is used for centrality selection, triggering and beam-induced background rejection. The V0 consists of two sub-detectors, V0A and V0C, placed at asymmetric positions, one on each side of the interaction point with full azimuthal acceptance and cover the pseudorapidity ranges $2.8 < \eta < 5.1$ and $-3.7 < \eta < -1.7$, respectively. The year and different collision energies for accumulated data in different collision systems are summarized in Table 1. The list of resonances that we report is provided in Table 2 with their quark content, decay modes exploited for the measurements, and branching ratios.

Table 1. Year for accumulated data sample and center of mass energy for different collision systems.

	Pb–Pb	Xe–Xe	p–Pb	pp
Year	2010-2011, 2015, 2018	2017	2013, 2016	2009-2013, 2015-2018
$\sqrt{s_{\mathrm{NN}}}$ [TeV]	2.76, 5.02	5.44	5.02, 8.16	0.9, 2.76, 7, 8, 5.02, 13

3. Results and discussion

3.1. Transverse momentum spectra

The p_{T} spectra for K^{*0} and ϕ in the various multiplicity classes in pp collisions at $\sqrt{s} = 13$ TeV are shown in Figure 1 with the ratios of these spectra to the inclusive INEL>0 spectrum [7]. Note that, the INEL>0 event class is defined as the set of inelastic collisions with at least one charged particle in the range $|\eta| < 1$. For $p_{\mathrm{T}} \leq 5$ GeV/c the evolution of spectral shape from low to high multiplicity has been observed. For higher p_{T}, the spectra in different multiplicity classes all have similar shape, indicating that the processes that change the shape of the p_{T} spectra in different multiplicity classes are dominant at low p_{T}.

Table 2. Lifetime of the measured resonances with their quark content, decay modes exploited for the measurements presented here and branching ratios [6].

	$c\tau$ (fm)	quark content	Decay modes	BR [%]
$\rho(770)^0$	1.3	$(u\overline{u}+d\overline{d})/\sqrt{2}$	$\pi^{+}+\pi^{-}$	100
$\mathrm{K}^*(892)^0$	4.2	$d\overline{s}$	$K^{+}+\pi^{-}$	66.6
$\Sigma(1385)^+$	5.5	uus	$\Lambda\pi^+ \to (p\pi^-)\pi^+$	87.0
$\Sigma(1385)^-$	5.0	dds	$\Lambda\pi^- \to (p\pi^-)\pi^-$	87.0
$\Xi(1820)^-$	8.1	dss	$\Lambda\mathrm{K} \to (p\pi^-)\mathrm{K}$	unknown
$\Lambda(1520)$	12.6	uds	$\mathrm{p} + \mathrm{K}^-$	22.5
$\Xi(1530)^0$	21.7	uss	$\Xi^-\pi^+ \to (\Lambda\pi^-)\pi^+ \to ((p\pi^-)\pi^-)\pi^+$	66.7
$\phi(1020)^0$	44	$s\overline{s}$	$\mathrm{K}^{+}+\mathrm{K}^{-}$	48.9

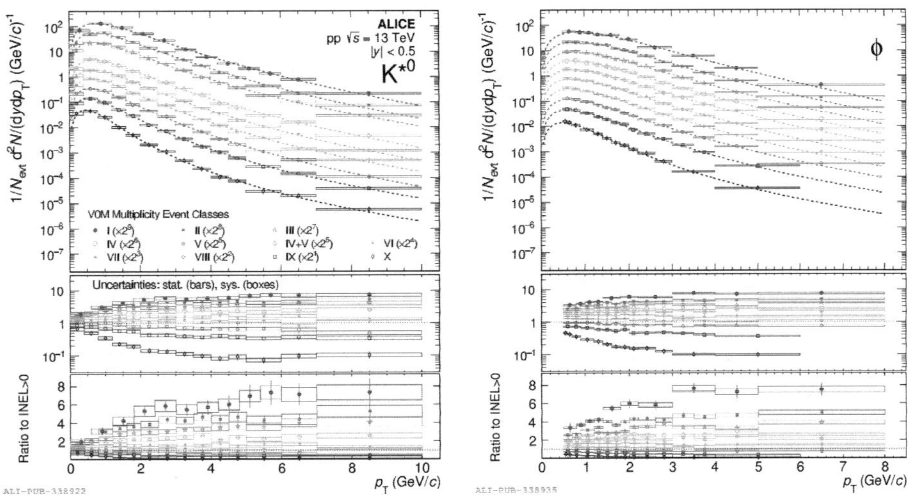

Figure 1. p_{T} spectra of K^{*0} and ϕ in pp collisions at $\sqrt{s}$ = 13 TeV for different multiplicity classes, scaled by factors as indicated [7]. The lower panels show the ratios of the multiplicity-dependent p_{T} spectra to the multiplicity-integrated INEL>0 spectra (with both linear and logarithmic vertical scales)

3.2. p_{T} integrated yields dN/dy and resonance particle ratios

The p_{T}-integrated yields $\mathrm{d}N/\mathrm{d}y$ of K^{*0} and ϕ are extracted from the p_{T} spectra in different multiplicity classes and presented in Figure 2 as functions of $\langle \mathrm{d}N_{\mathrm{ch}}/\mathrm{d}\eta_{\mathrm{lab}}\rangle_{|\eta|<0.5}$. For both particles, $\mathrm{d}N/\mathrm{d}y$ exhibits an approximately linear increase with increasing $\langle \mathrm{d}N_{\mathrm{ch}}/\mathrm{d}\eta_{\mathrm{lab}}\rangle_{|\eta|<0.5}$. Results for pp collisions at $\sqrt{s}$ = 7 and 13 TeV and for p–Pb collisions at $\sqrt{s_{\mathrm{NN}}}$ = 5.02 TeV follow approximately the same trends. This indicates that, for a given multiplicity, K^{*0} and ϕ production does not depend on the collision system or energy. The measured $\mathrm{d}N/\mathrm{d}y$ values are compared with five different model calculations: PYTHIA6 (Perugia 2011 tune) [15], PYTHIA8 (Monash 2013 tune, both with and without color reconnection) [16], EPOS-LHC [17], and DIPSY [18]. The EPOS-LHC and PYTHIA8 without color reconnection give the best description for the the K^{*0} while the other PYTHIA calculations exhibit fair agreement with the measured data, and DIPSY overestimates the K^{*0} yields. The ϕ yields tend to be slightly overestimated

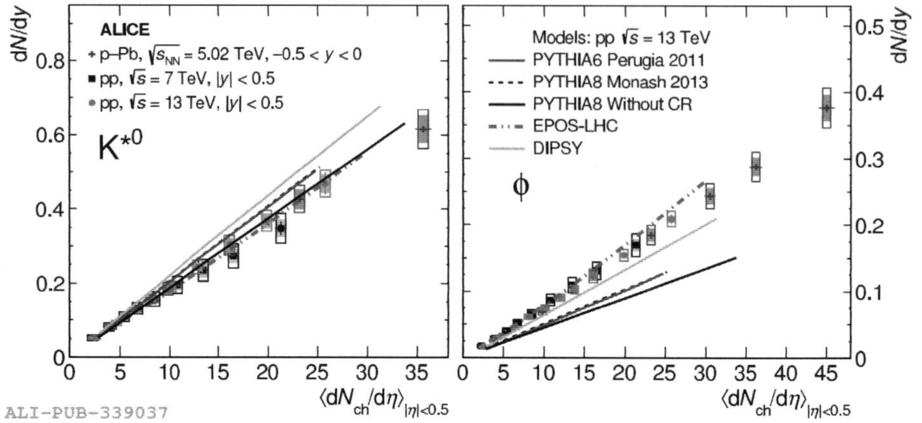

Figure 2. p_T-integrated yields dN/dy of K^{*0} (average of the particle and antiparticle) and ϕ as functions of $\langle dN_{ch}/d\eta_{lab}\rangle_{|\eta|<0.5}$. Results are shown for pp collisions at s = 13 and 7 TeV [7, 3], as well as for p–Pb collisions at $\sqrt{s_{NN}}$ = 5.02 TeV [8]. The measurements in pp collisions at s = 13 TeV are also compared with values from common event generators [18, 17, 15, 16]. Bars represent statistical uncertainties, open boxes represent total systematic uncertainties, and shaded boxes show the systematic uncertainties that are uncorrelated between multiplicity classes.

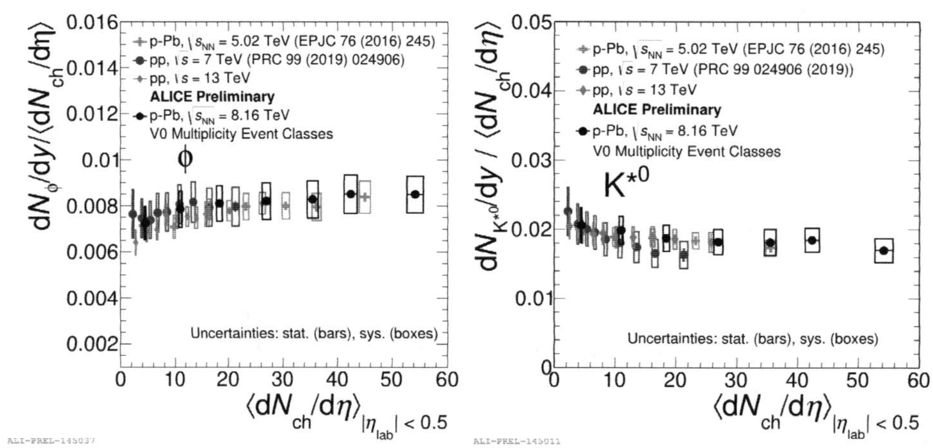

Figure 3. Integrated yields of K^{*0} (left panel) and ϕ (right panel) normalized to $\langle dN_{ch}/d\eta\rangle_{|\eta_{lab}|<0.5}$ in pp collisions at $\sqrt{s}$ = 7 and 13 TeV and p-Pb collisions at $\sqrt{s_{NN}}$ = 5.02 and 8.16 TeV for different multiplicity classes. The bars and the boxes represent the statistical and systematic errors, respectively.

by EPOS-LHC and underestimated by DIPSY and PYTHIA calculations.

The yields normalized to the $\langle dN_{ch}/d\eta\rangle$ of K^{*0} and ϕ in pp collisions at $\sqrt{s}$ = 7 and 13 TeV, p–Pb collisions at $\sqrt{s_{NN}}$ = 5.02 and 8.16 TeV are shown as a function of multiplicity ($\langle dN_{ch}/d\eta\rangle_{|\eta_{lab}|<0.5}$) in Figure 3. The scaled integrated yield is flat with multiplicity and independent of collision energy and systems for pp and p–Pb collisions within the uncertainties. This confirms that the particle production is mainly driven by charged particle multiplicity irrespectively of collision systems and energies for pp and p–Pb collisions.

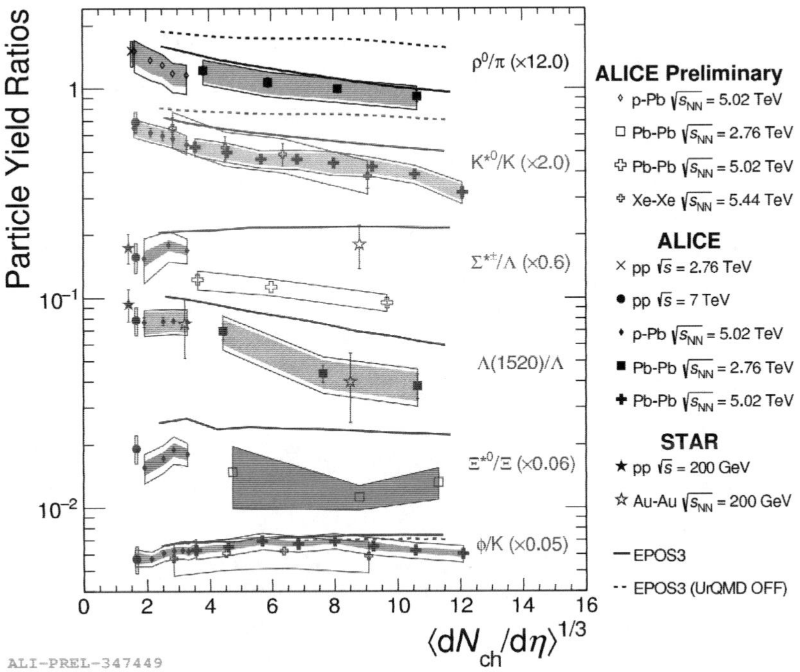

Figure 4. Ratios of resonance and stable counterpart yields are reported as a function of multiplicity for the hadronic resonances, ρ^0/π, K^{*0}/K, $\Sigma^{*\pm}/\Lambda$, $\Lambda(1520)/\Lambda$, Ξ^{*0}/Ξ and ϕ/K from inelastic pp, p–Pb, Pb–Pb and Xe–Xe collisions. The error bars show the statistical uncertainty, while the emtpy and dark-shaded boxes show the total systematic uncertainty and the uncorrelated contribution across multiplicity bins, respectively. A comparison is also shown to the ratio predicted by the EPOS3 model for Pb–Pb at $\sqrt{s_{NN}} = 2.76$ TeV, where the late hadronic cascade is simulated with UrQMD. STAR data are also shown for the $\Sigma^{*\pm}/\Lambda$ and $\Lambda(1520)/\Lambda$ ratios.

The ratios of p_T-integrated yields as a function of multiplicity are shown in Figure 4 for pp, p–Pb, Pb–Pb and Xe–Xe collisions, for ρ^0/π, K^{*0}/K, $\Sigma^{*\pm}/\Lambda$, $\Lambda(1520)/\Lambda$, Ξ^{*0}/Ξ and ϕ/K. Short-lived resonances show a sizable dependence on the multiplicity. A clear suppression is observed for ρ^0/π going from pp to Pb–Pb collisions at $\sqrt{s_{NN}} = 2.76$ TeV [9]. The EPOS3 event generator [10], which includes UrQMD [11] for the late stage hadronic cascade is able to describe the evolution with multiplicity. The same behavior is observed in the K^{*0}/K ratio, where the suppression can be observed in a wider multiplicity range with data from pp at $\sqrt{s} = 7$ TeV, Pb–Pb at $\sqrt{s_{NN}} = 5.02$ TeV and Xe–Xe at $\sqrt{s_{NN}} = 5.44$ TeV.

The ratios of $\Sigma^{*\pm}/\Lambda$ and $\Lambda(1520)/\Lambda$ show flat behavior in small systems from pp and p–Pb collisions, and the results in Pb–Pb collisions show that $\Sigma^{*\pm}$ and $\Lambda(1520)$ are suppressed with respect to the Λ production [13]. Finally, there is no significant centrality dependence of the Ξ^{*0}/Ξ and ϕ/K ratios with multiplicity for all measured systems. This is expected in the context of re-scattering, considering that the Ξ^{*0} baryon and the ϕ meson live longer than the expected fireball lifetime and therefore their decay daughters will not undergo re-scattering.

Further, to quantify the p_T-dependence of the rescattering effect observed in Pb–Pb collisions, p_T-differential yield ratios were studied as shown in Figure 5 [14]. At low p_T, the K^{*0}/K ratio for central collisions is lower than in peripheral (pp) collisions whereas the ϕ/K ratio is comparable within the uncertainties. This observation is consistent with the suppression of K^{*0} yields due

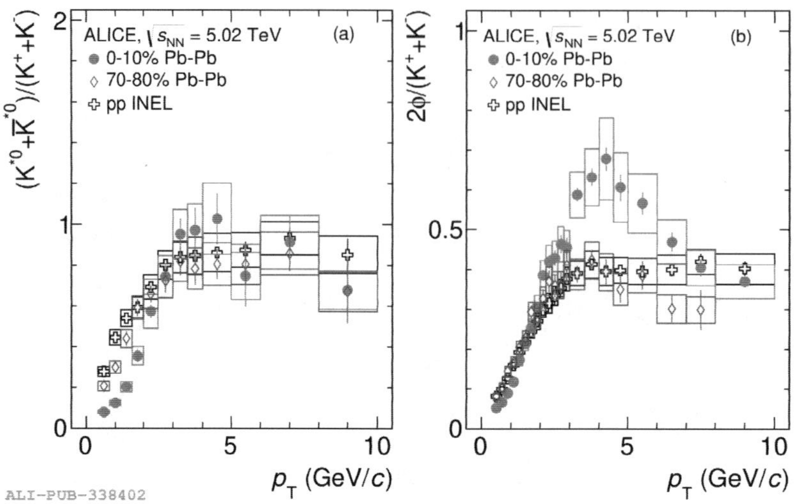

Figure 5. Particle yield ratios $(K^{*0} + \bar{K}^{*0})/(K^{+} + K^{-})$ in panel (a) and $(2\phi)/(K^{+} + K^{-})$ in panel (b), both as a function of p_T for centrality classes 0-10% and 70-80% in Pb–Pb collisions at $\sqrt{s_{NN}} = 5.02$ TeV [14]. For comparison, the corresponding ratios are also shown for inelastic pp collisions at $\sqrt{s} = 5.02$ TeV. The statistical uncertainties are shown as bars and systematic uncertainties are shown as boxes.

to rescattering in the hadronic phase. It indicates that the rescattering affects low momentum particles. At intermediate p_T, both ratios show enhancement for central Pb–Pb collisions with respect to peripheral and pp collisions. This is in agreement with the presence of a larger radial flow in central Pb–Pb collisions relative to peripheral and pp collisions. For $p_T > 6$ GeV/c, the K^{*0}/K and ϕ/K yield ratios in central collisions are similar to peripheral and pp collisions. It demonstrates that fragmentation is the dominant hadron production mechanism in this p_T region.

3.3. Mean transverse momentum

The mean transverse momentum ($\langle p_T \rangle$) values for each multiplicity event class are obtained by integrating the p_T-spectra in the measured range and by using a fit function to extrapolate the yields in the unmeasured p_T region. Figure 6 shows the $\langle p_T \rangle$ of proton, K^{*0} and ϕ as a function of the average charged particle multiplicity density ($\langle dN_{ch}/d\eta \rangle$) measured at mid-rapidity ($|\eta| < 0.5$) in pp collisions at $\sqrt{s} = 7$ and 13 TeV and compared with the results obtained in p–Pb at $\sqrt{s_{NN}} = 5.02$ TeV and Pb–Pb collisions at $\sqrt{s_{NN}} = 5.02$ and 2.76 TeV.

For all the particles, an increase in $\langle p_T \rangle$ from low to high multiplicity classes is observed. The same increasing trend of the $\langle p_T \rangle$ as a function of the multiplicity is observed in pp collisions at $\sqrt{s} = 7$ TeV and 13 TeV and a mass ordering of $\langle p_T \rangle$ is found to be followed in central and semi-central Pb–Pb collisions as expected from the hydrodynamic expansion of the system [22]. However, this breaks down for smaller systems. The increase in $\langle p_T \rangle$ is steeper for smaller systems.

3.4. Strangeness production

To understand the strangeness production, the particle yields are measured in various collision systems at different energies. Figure 5 presents yield-ratios of various strange and non-strange hadrons to pions as a function of the mean charged-particle multiplicity $\langle dN_{ch}/d\eta \rangle_{|\eta|<0.5}$ for

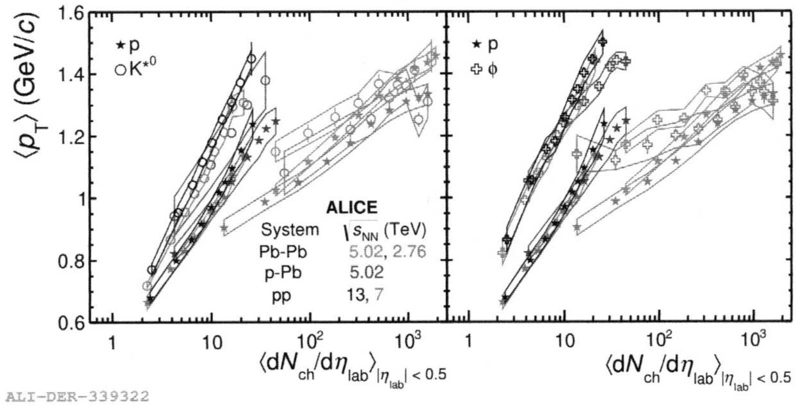

Figure 6. Mean transverse momentum (⟨p_T⟩) of K^{*0}, ϕ and p in pp at $\sqrt{s} = 7$ and 13 TeV, p–Pb at $\sqrt{s_{NN}} = 5.02$ TeV and Pb–Pb at $\sqrt{s_{NN}} = 2.76$ and 5.02 TeV collisions as function of multiplicity. The bars represent the statistical error and the lines represent the systematic error.

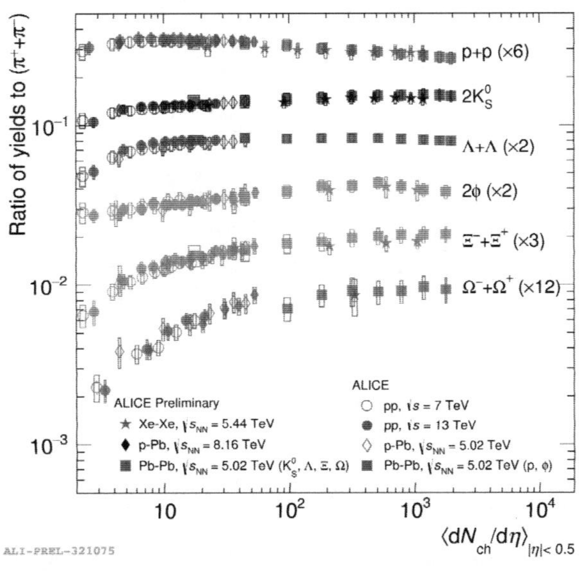

Figure 7. p/π, K^0_S/π, Λ/π, Ξ/π, Ω/π and ϕ/π ratios as a function of $\langle dN_{ch}/d\eta\rangle_{|\eta|<0.5}$ in different collision systems, pp at $\sqrt{s} = 7$ [19] and 13 TeV, p–Pb at $\sqrt{s_{NN}} = 5.02$ TeV [20, 21], Pb–Pb at $\sqrt{s_{NN}} = 5$ TeV and Xe–Xe at $\sqrt{s_{NN}} = 5.44$ TeV.

different collision systems and different energies. The ratios of all strange particle yields to pions increase with $\langle dN_{ch}/d\eta\rangle_{|\eta|<0.5}$ and saturate in larger collision systems such as Pb–Pb and Xe–Xe collisions. The magnitude of the increase depends on strangeness content. It is observed that the ratios are consistent at similar multiplicities in all collision systems at all LHC energies.

In order to investigate the behavior of hidden strangeness, p_T-integrated yield ratios of ϕ ($|S|$=0) to K ($|S|$=1) and Ξ ($|S|$=2) to ϕ ($|S|$=0) are compared as shown in Figure 8. The ϕ/K ratio shows flat or slightly increasing trend at lower multiplicities that suggests ϕ behaves like a $S \geq 1$ particle. The ratio of Ξ/ϕ increases with increasing $\langle dN_{ch}/d\eta_{lab}\rangle_{|\eta|<0.5}$ for low multiplicity

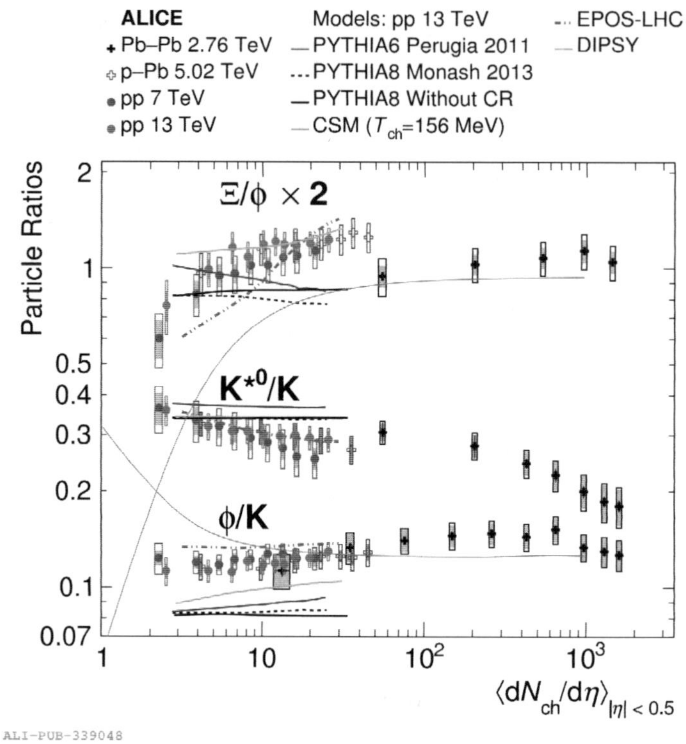

Figure 8. Ratios of p_T-integrated particle yields K^{*0}/K, ϕ/K, and Ξ/ϕ in pp collisions at $\sqrt{s}$ = 13 TeV as functions of $\langle dN_{ch}/d\eta_{lab}\rangle_{|\eta|<0.5}$ [24]. These measurements are compared with data from p–Pb collisions at $\sqrt{s_{NN}}$ = 5.02 TeV [9, 10] and Pb–Pb collisions at $\sqrt{s_{NN}}$ = 2.76 TeV [43, 44], as well as results from common event generators [28, 33, 58, 59] and a Canonical Statistical Model calculation [23].

collisions and is then fairly constant for a wide range of multiplicities from high-multiplicity pp and p–Pb collisions to central Pb–Pb collisions. The multiplicity evolution of ϕ/K and Ξ/ϕ ratios suggests that the ϕ has effective strangeness between 1 and 2 units.

3.5. Nuclear modification factors

The nuclear modification factor R_{AA} is used to study medium-induced effects in heavy-ion collisions. The R_{AA} is the ratio of the yield of a particle in nucleus-nucleus collisions to its yield in pp collisions. This ratio is scaled by the number of binary nucleon-nucleon collisions in each centrality class, which is estimated from Glauber model calculations. For each p_T bin,

$$R_{AA} = \frac{1}{\langle N_{coll}\rangle} \cdot \frac{dN_{AA}/dp_T}{dN_{pp}/dp_T} \quad (1)$$

The nuclear modification factors measured in 0-10% central Pb–Pb collisions at $\sqrt{s_{NN}}$ = 5.02 TeV for for charged pions, charged kaons, (anti)protons, K^{*0} and ϕ are presented on the left panel in Figure 9. And the R_{AA} of ρ^0 meson [9] are reported with charged pions, charged kaons, (anti)protons in 0-20% (middle panel) and 60-80% (right panel) in Figure 9. At high transverse momenta (p_T > 8 GeV/c), the production of all hadrons is suppressed by a similar amount and there is no dependence of the suppression on particle mass or quark content within uncertainties.

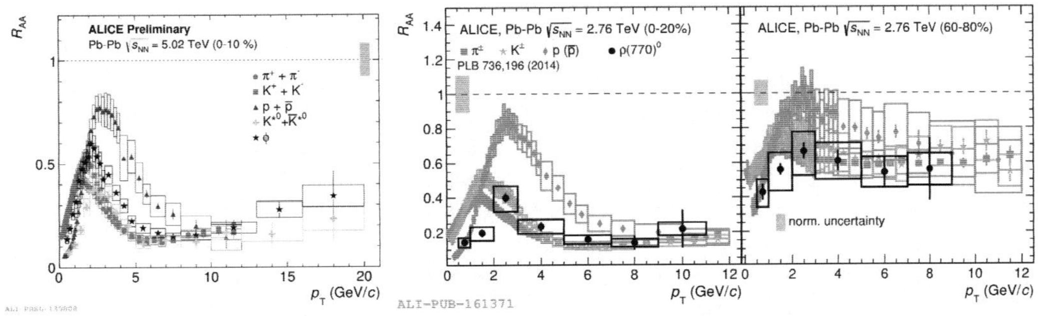

Figure 9. Nuclear modification factors R_{AA} measured for charged pions, charged kaons, (anti)protons and, K^{*0} and ϕ meson (left pannel) in 0-10% Pb–Pb collisions at $\sqrt{s_{NN}} = 5.02$ TeV. The R_{AA} of ρ^0 mesons in 0–20% (middle panel) and 60–80% (right panel) central Pb–Pb collisions at $\sqrt{s_{NN}} = 2.76$ TeV with charged pions, charged kaons, (anti)protons. The statistical and systematic uncertainties are shown as bars and boxes, respectively.

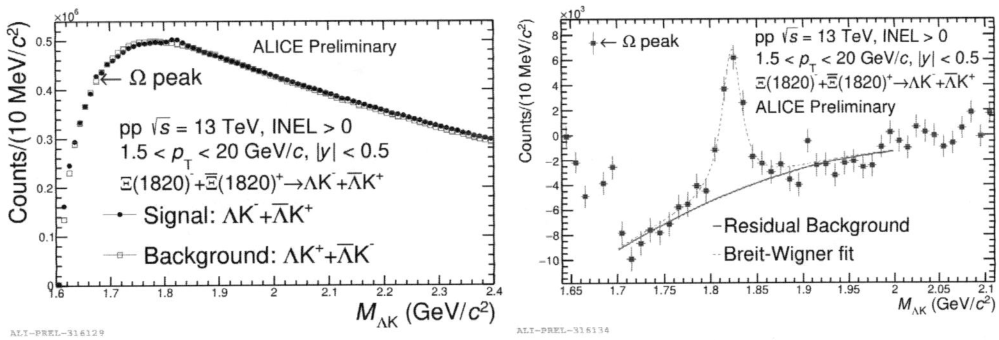

Figure 10. (Left) The ΛK Invariant mass distribution in p_T bins 1.5 – 20 GeV/c and for the INEL > 0 event class. (Right) The invariant mass distribution after subtraction of the background. The dashed curve represents the combined fit, while the solid line describes the residual background.

There is a species dependence of R_{AA} at intermediate transverse momentum (2< p_T <8 GeV/c), which is likely to be a result of an interplay between different effects such as radial flow, low-p_T suppression, species dependent p_T shapes of the pp reference spectra.

3.6. Reconstruction of Ξ(1820)

The first measurement of Ξ(1820) from collider experiments is shown in Figure 10. The calculation from FASTSUM Collaboration [25] shows potential parity doubling of strange resonant states, which could be a signature of chiral symmetry restoration in heavy-ion collisions. The positive-parity masses are largely temperature-independent while negative-parity masses drop and become near-degenerate with the corresponding positive-parity mass close to critical temperature. Because the Ξ(1820) has negative-parity, the comparison of mass shift, width broadening or change in yield to the positive-parity Ξ(1530) can be one of the key measurements to study chiral symmetry restoration.

Journal of Physics: Conference Series **1602** (2020) 012008 doi:10.1088/1742-6596/1602/1/012008

4. Summary

A comprehensive set of resonance and strangeness production measurments in all available collision systems is presented. The p_T-integrated yields normalized by the average charged particle multiplicity are observed to be constant as a function of charged particle multiplicity. The ratios of p_T-integrated particle yields for various resonances are measured and ρ^0/π, K^{*0}/K and $\Lambda(1520)/\Lambda$ exhibit a decrease from pp and peripheral Pb–Pb to central Pb–Pb collisions. This behavior can be explained by the dominance of re-scattering of decay daughters over regeneration in the hadronic phase. The increase of $\langle p_T \rangle$ is observed from low to high multiplicity classes and a mass ordering of $\langle p_T \rangle$ is found in central Pb–Pb collisions. The ϕ meson with hidden strangeness seems to have an effective strangeness between 1 and 2 units. The nuclear modification factors for resonances are suppressed at high p_T for Pb–Pb collisions and the amount of suppression is similar to charged pions, charged kaons, (anti)protons. The first measurement of $\Xi(1820)$ is presented and will be compared to the corresponding positive-parity particle to study chiral symmetry restoration.

References

[1] J. Stachel, A. Andronic, P. Braun-Munzinger, and K. Redlich *J. Phys. Conf. Ser.* **509** (2014) 012019
[2] S. Acharya *et al.* (ALICE Collaboration), arXiv:1610.03001
[3] S. Acharya *et al.* (ALICE Collaboration), *Phys. Rev.* C **99** (2019) 024906
[4] S. Acharya *et al.* (ALICE Collaboration), *Nucl. Phys. A* **971** (2018) 1-20
[5] K. Aamodt *et al.* (ALICE Collaboration), *Int. J. Mod. Phys* **29** (2014) 1430044
[6] M. Tanabashi *et al.* (Particle Data Group Collaboration), *Phys. Rev.* D **98** no. 3,(2018) 030001
[7] S. Acharya *et al.* (ALICE Collaboration), arXiv:1910.14397
[8] S. Acharya *et al.* (ALICE Collaboration), *Eur. Phys. J.* C **76** (2016) 245
[9] S. Acharya *et al.* (ALICE Collaboration), *Phys. Rev.* **C99** 064901
[10] K. Werner, B. Guiot, I. Karpenko, and T. Pierog *Phys. Rev.* **C89** (2016) 064903
[11] M. Bleicher *et al. J. Phys.* **G25** (1999) 1859-1869
[12] B. I. Abelev *et al.* (STAR Collaboration), *Phys. Rev. Lett.* **97** (2006) 132301
[13] S. Acharya *et al.* (ALICE Collaboration), *Phys. Rev.* **C99** 024905
[14] S. Acharya *et al.* (ALICE Collaboration), arXiv:1910.14419
[15] P. Z. Skands, *Phys. Rev.* D **82** (2010) 074018
[16] P. Z. Skands, S. Carrazza, and J. Rojo, *Eur. Phys. J.* C **74** (2014) 3024
[17] T. Pierog *et al. Phys. Rev.* C **92** (2016) 034906
[18] C. Flensburg, G. Gustafson, and L. Lönnblad, *J. High Energy Phys.***8** (2011) 103
[19] J. Adam *et al.* (ALICE Collaboration) *Nat. Phys.* **13** 535-539 (2017)
[20] J. Adam *et al.* (ALICE Collaboration) *Phys. Lett. B* **728** (2014) 25-38
[21] J. Adam *et al.* (ALICE Collaboration) *Phys. Lett. B* **758** (2016) 389-401
[22] C. Shen, U. Heinz, P. Houvinen, and H. Song *Phys. Rev.***C84** (2011) 044903
[23] S. Acharya *et al.* (ALICE Collaboration), *Phys. Rev.* C **99** (2019) 024906
[24] S. Acharya *et al.* (ALICE Collaboration), arXiv:1908.01861
[25] G. Aarts, C. Allton, D. Boni, S. Hands, B. Jäger, C. Praki, and J.I. Skullerud. *EPJ Web Conf.* **171** (2018) 14005

Journal of Physics: Conference Series **1602** (2020) 012009 doi:10.1088/1742-6596/1602/1/012009

Exploring the QCD phase diagram via the collision energy dependence of multi-particle femtoscopy with PHENIX

Máté Csanád for the PHENIX Collaboration
Eötvös Loránd University, H-1117 Budapest, Pázmány P. s. 1/A, Hungary
E-mail: csanad@elte.hu

Abstract. Exploration of the rich structure of the QCD phase diagram is an important topic in the RHIC heavy ion program. One of the ultimate goals of this program is to search for the critical endpoint. Investigation of the space-time structure of hadron emissions at various phase transition points using Bose-Einstein correlations of identical bosons may provide insight on the location of the critical endpoint. PHENIX has performed comprehensive measurements of the Bose-Einstein correlation in Au+Au collisions at $\sqrt{s_{NN}}$ = 15, 19, 27, 39, 62.4, and 200 GeV, where we incorporated Lévy-type source functions to describe the measured correlation functions. We put particular focus on one of the parameters of the Lévy-type source functions, the index of stability α, which is related to one of the critical exponents (the so-called correlation exponent η). We have measured its collision energy and centrality dependence. We have also extended our analysis from two-particle to three-particle correlations to characterize the nature of the hadron emission source. The three particle correlations confirmed the findings of the two-particle correlations, and also provide insight on the pion production mechanism beyond the core-halo model.

1. Introduction

Femtoscopy (coined by Ledniczky [1]) is an important subfield of high energy nuclear and particle physics, as it allows us to investigate the space-time structure of femtometer scale processes. This subfield originates from the work of Hanbury Brown and Twiss, who investigated the angular diameter of stars using radio and optical telescopes, based on intensity correlations; this work was later on understood by Glauber, and the technique independently discovered by Goldhaber and collaborators (see more details and references in Ref. [2]). Nowadays we understand femtoscopic correlations to be caused by Bose-Einstein statistics, and hence use the name Bose-Einstein or quantum statistical correlations as well.

The most important equation utilized in two-particle femtoscopic correlations is one that relates the pair source $D_2(r, K)$ (describing the probability density of creating a pair with spatial separation of r and average momentum K) and the correlation function $C_2(q, K)$ (indicating the amount of correlation of pairs with momentum difference q and average momentum K):

$$C_2(q,K) = 1 + \frac{\widetilde{D_2}(q,K)}{\widetilde{D_2}(0,K)}, \tag{1}$$

where $\widetilde{D_2}$ denotes the Fourier-transform of D_2 in its first variable. The C_2 correlation function can also be written up with the single particle source $S(r,p)$ ($D_2(r,K)$ is the autoconvolution of $S(r,p)$ in the first variable with $p=K$):

$$C_2(q,K) = 1 + \left|\frac{\widetilde{S}(q,K)}{\widetilde{S}(0,K)}\right|^2. \tag{2}$$

There are several approximations behind these equations, see more details and references e.g. in section II.D of Ref. [3]: some of these are related to the kinematics of the pair (e.g. requiring the particle momenta to be close to each other), others to thermal emission. Given the validity of these assumptions, the above equations indicate the importance of femtoscopy: by measuring C_2, one can obtain information on femtometer-scale processes via the above mentioned S or D_2 source functions. If one assumes a Gaussian source for example, then the correlation function will also be Gaussian, and hence the Gaussian source radius can be directly measured. Generally, for expanding sources the obtained radius will be related to the homogeneity length of the source for particles of given momenta [3].

It was observed [3–7] that when investigating two-particle correlation functions, one has to go beyond the Gaussian approximation. Lévy-distributed sources with power-law tails may appear (among other phenomena) due to anomalous diffusion [8–10]. These are characterized by the Lévy index α, and include Gaussian ($\alpha=2$) as well as Cauchy ($\alpha=1$) distributions as well. The three-dimensional Lévy source is given as

$$\mathcal{L}(\alpha,R,\mathbf{r}) = \frac{1}{(2\pi)^3}\int \mathrm{d}^3\mathbf{q}\, e^{i\mathbf{qr}} e^{-\frac{1}{2}|\mathbf{q}\mathbf{R}^2\mathbf{q}|^{\alpha/2}}, \tag{3}$$

where $\mathbf{R^2}$ is the matrix of homogeneity lengths (also known as femtoscopic radii), which (in case of spherical symmetry) can be assumed to be the product of a single Lévy scale squared (R^2) and the 3×3 identity matrix, and then the last exponent will contain $(|\mathbf{q}|R)^\alpha$. In this case $\mathcal{L}(\alpha,R,\mathbf{r})$ becomes spherically symmetric. Such spherically symmetric Lévy sources with various α values are illustrated in Fig. 1. If one aims to describe more general sources, a convenient assumption is to introduce the radii in the Bertsch-Pratt coordinate system $R_{\rm out}$, $R_{\rm side}$ and $R_{\rm long}$, which represent the square roots of the diagonal elements of the $\mathbf{R^2}$ matrix. A 3D femtoscopy analysis usually uses this definition. One may go even beyond this approximation and introduce off-diagonal elements, but we refrain from doing that in the analyses presented in this paper.

Besides anomalous diffusion, other effects can also cause the appearance of Lévy sources. One important possibility is the QCD critical point, where large scale spatial fluctuations can be present, and the spatial correlation function becomes a power-law [11, 12]. In this case the η critical exponent may be identical to the Lévy exponent α. Finite size effects or dynamical criticality may distort this simple picture; nevertheless, this connection between η and α yields a strong motivation to measure the Lévy index precisely. For more details, see Section IV.B of Ref. [3].

As seen in Eqs. (1) and (2), the momentum correlation function takes the value 2 for $q=0$. Experimentally however, one cannot measure correlations down to $q=0$ due to momentum resolution effects. The extrapolated value of $C_2(q\to0)$ differs generally from 2, and takes the value $1+\lambda$. This can be interpreted in the core-halo model easily: it appears that the quantity $\lambda = C_2(q\to0)-1$ is related to the ratio of directly produced hadrons (of the given species used in the measurement) to decay hadrons (of the same species) from long lived resonances. The former particles form the core, while the latter form the halo, and in fact $\lambda = f_c^2$, where $f_c = \text{core}/(\text{core+halo})$. This relation makes it possible to study the in-medium mass modification of the η' meson, as detail in Section IV.A of Ref. [3].

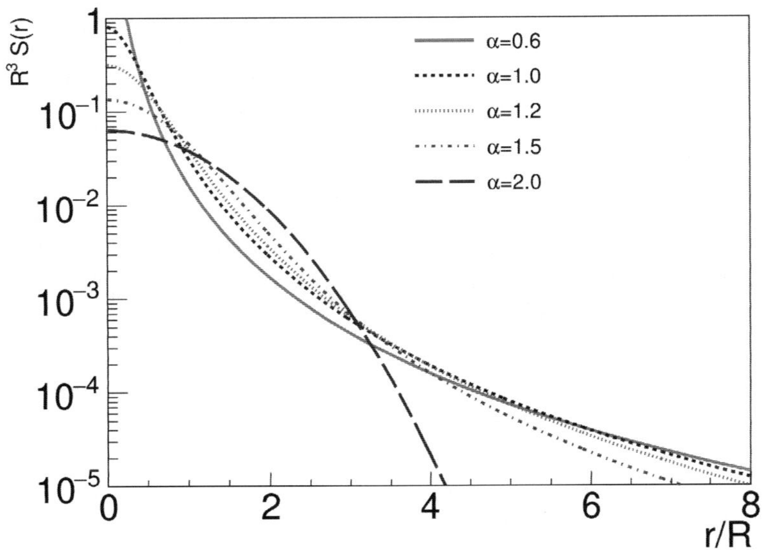

Figure 1. Spherically symmetric Lévy sources $\mathcal{L}(\alpha, R, r)$ for various $\alpha =$ values. The variable on the horizontal axis is r/R, since the R dependence can be scaled out that way.

Furthermore, it is important to note that the above noted λ parameter may differ from unity due to other reasons as well. One such possibility is coherent pion production [13–15]. In this case the above mentioned approximations are not valid, and hence the strength of multi-particle Bose-Einstein correlations (C_n) will be modified; these correlation strengths are given by

$$\lambda_n = \lim_{q_{ij}\to 0} C_n(\{q_{ij}\}) - 1, \tag{4}$$

where $\lambda_2 \equiv \lambda$ for simplicity. In a higher order correlation function, lower order correlations will also give contributions, for example pairs within triplets or quadruplets. Hence a given λ_n has contributions from lower order correlations as well. If p_c is defined as the fraction of coherently produced pions (within the core), then for 2- and 3-particle correlations one obtains [15]

$$\lambda_2 = f_c^2\left[(1-p_c)^2 + 2p_c(1-p_c)\right] = f_c^2(1-p_c^2), \tag{5}$$

$$\lambda_3 = 2f_c^3\left[(1-p_c)^3 + 3p_c(1-p_c)^2\right] + 3f_c^2\left[(1-p_c)^2 + 2p_c(1-p_c)\right]. \tag{6}$$

This means that simultaneous measurement of 2- and 3-particle correlations may shed light on the value of p_c and f_c [2]. Furthermore, one may define a core-halo independent parameter as

$$\kappa_3 = \frac{\lambda_3 - 3\lambda_2}{2\sqrt{\lambda_2}^3} \tag{7}$$

which is independent of the value of f_c, and it takes the value $\kappa_3 = 1$ if $p_c = 0$. Hence by measuring κ_3, one can investigate if the simple core-halo model interpretation of λ_2 is consistent.

2. Results

Two-pion measurements were performed with the Au+Au data taken by the PHENIX detector system during the 2010-2011 running period. Correlation functions were measured using the event mixing technique, and fits based on Lévy-distributed sources were done to obtain the statistically most probable Lévy parameters α, R and λ. Details about the analyses presented in this paper are given in Refs. [2, 3, 16–20].

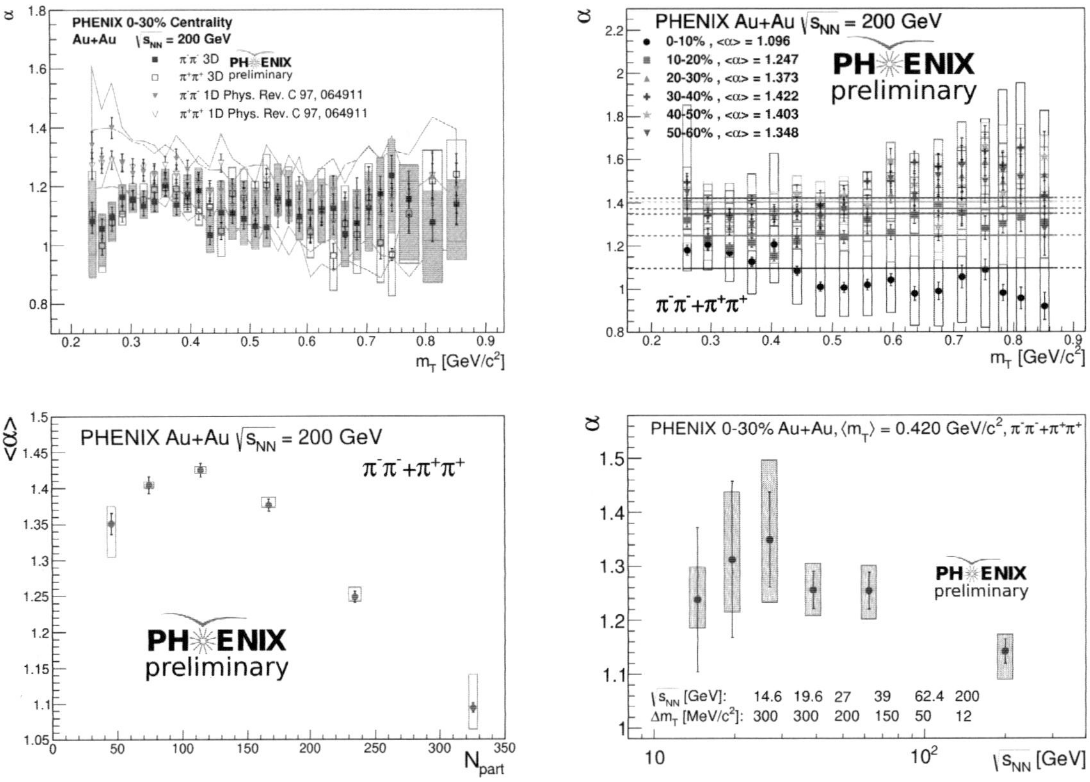

Figure 2. Measurements of the Lévy exponent α in 0-30% centrality $\sqrt{s_{NN}} = 200$ GeV Au+Au collisions (top left panel), various 10% centrality bins also at 200 GeV (versus m_T in the top right panel, versus N_{part} in the bottom left panel), and in 0-30% centrality collisions in for various $\sqrt{s_{NN}}$ values (bottom right panel).

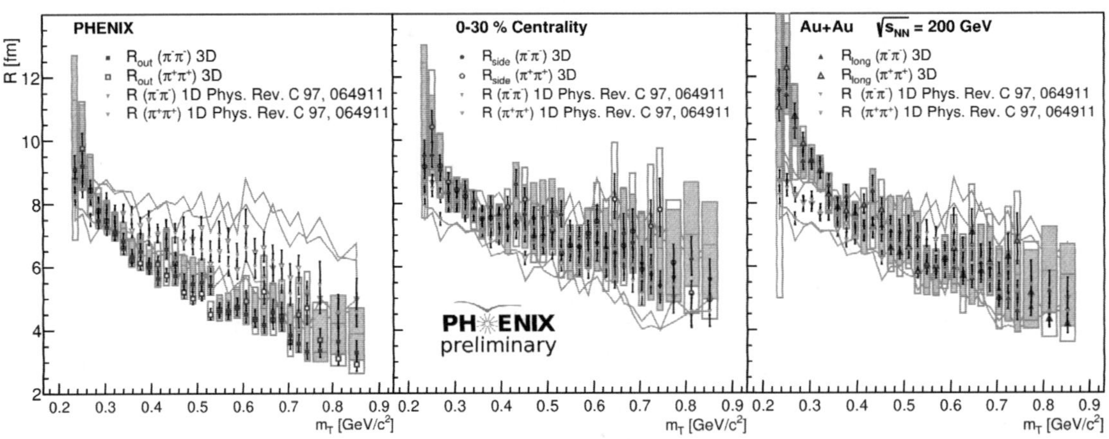

Figure 3. Transverse momentum dependence of the Bose-Einstein correlation radii (Lévy scale parameters) $R_{\text{out,side,long}}$ in $\sqrt{s_{NN}} = 200$ GeV Au+Au collisions. Results of the 1D analysis (with a spherically symmetric source, employing only one Lévy scale R) are also shown in each of the panels for comparison.

Journal of Physics: Conference Series **1602** (2020) 012009 doi:10.1088/1742-6596/1602/1/012009

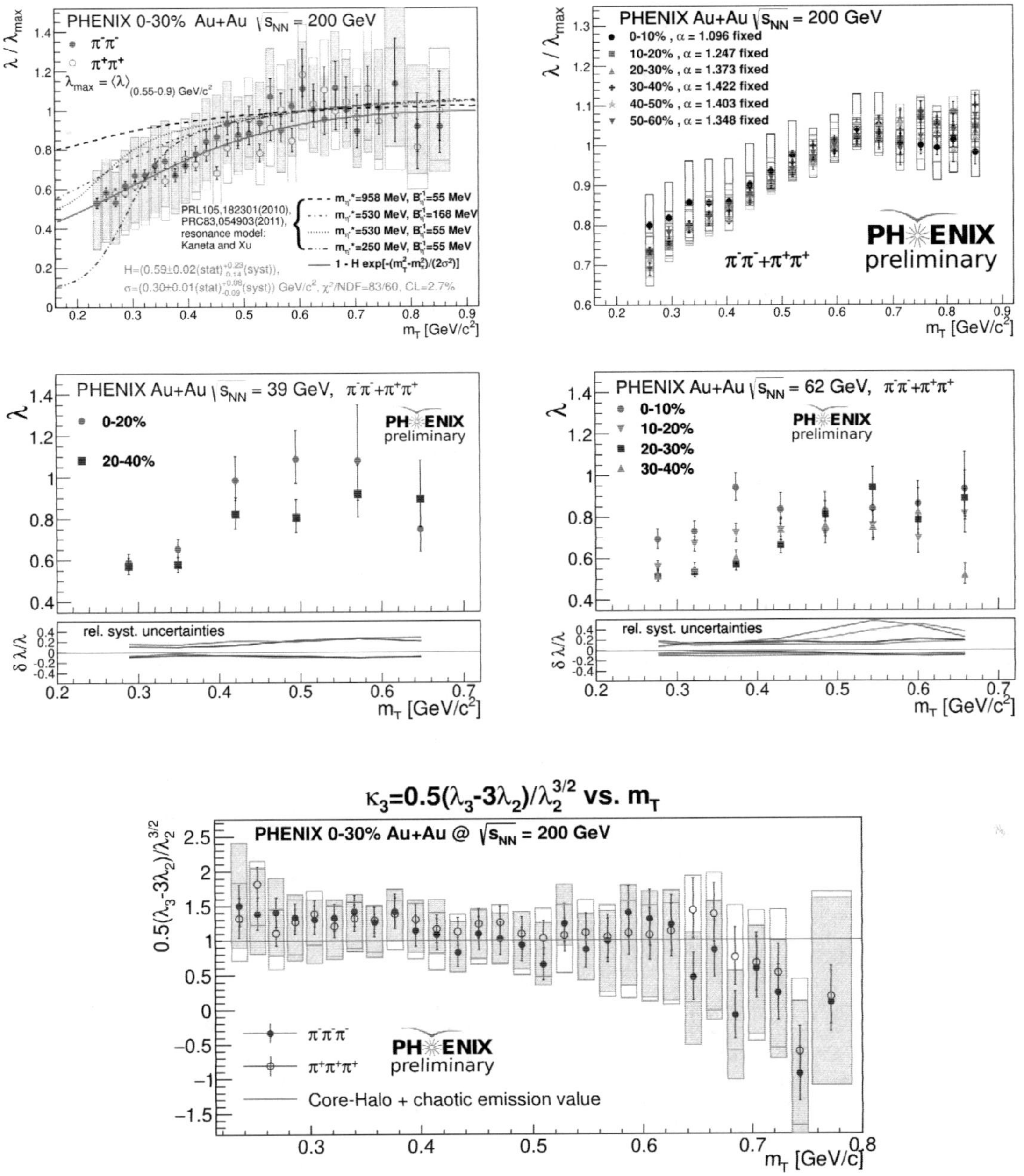

Figure 4. Results on λ normalized by its large m_T average ($\lambda_{\max}$) in $\sqrt{s_{NN}} = 200$ GeV Au+Au collisions are shown in the top two plots. The top left plot also shows calculations based on various η' masses. The second row shows $\lambda(m_T)$ for various centralities in $\sqrt{s_{NN}} = 39$ and 62 GeV Au+Au collisions. The bottom plot shows the core-halo independent κ_3 parameter as a function of m_T (here for clarity, $\lambda_2 \equiv \lambda$ is the strength of two-particle correlations, while λ_3 is the strength of three-particle correlations).

The Lévy exponent α was measured in collisions of various center of mass energies ($\sqrt{s_{NN}}$ ranging from 14.5 GeV to 200 GeV) and with various centrality selections, as shown in Fig. 2. These plots show that:

- Results on α from the 1D and the 3D analyses are compatible with each other, confirming the validity of the applied approach.
- While $\alpha(m_T)$ is approximately constant for each centrality, there is a clear centrality dependence, with α decreasing for more central collisions.
- There does not seem to be a strong $\sqrt{s_{NN}}$ dependence in this range - but this conclusion can be affected by the different m_T windows used for each of the analysis (this had to be done as there was not enough statistics at the lower collision energies).

The Lévy scale R was also measured in both 1D and 3D analyses, and the results are shown in Fig. 3. The 1D (spherically symmetric) analysis is mostly compatible with the 3D results, but there are clear differences in R_{out} in the $m_T \in [0.3 - 0.7]$ GeV/c^2 range, as well as for R_{long} for the $m_T < 0.35$ GeV/c^2 range. This may be due to the non-spherical nature of the source in the longitudinally comoving system (LCMS). This may also be the reason for the small discrepancy between the 1D and the 3D results on $\alpha(m_T)$ seen in Fig. 2.

The correlation strength parameter λ was measured versus m_T from $\sqrt{s_{NN}} = 39$ GeV to 200 GeV, and the results are shown in Fig. 4. The results show a reduced λ and small m_T values, as compared to larger m_T values; this seems to be an universal feature of $\lambda(m_T)$ across collision energies and centralities in the investigated range. As discussed in Ref. [3], this may be attributed to an in-medium mass reduction of the η' meson. This statement relies heavily on the core-halo model assumption. This was further tested in a simultaneous analysis of 2- and 3-particle correlations, in particular the measurement of the κ_3 parameter defined above in Eq. (7). The lowest panel of Fig. 4 shows that this parameter is compatible with $\kappa_3 = 1$, i.e. the value predicted by the core-halo model with fully chaotic emission.

Acknowledgments
The speaker was supported by the ÚNKP-19-4 New National Excellence Program of the Hungarian Ministry for Innovation and Technology, the János Bolyai Research Scholarship of the Hungarian Academy of Sciences, as well as grant FK-123842 of the National Research, Development and Innovation Office of Hungary.

References

[1] Lednicky R 2001 (*Preprint* nucl-th/0112011)
[2] Csanád M (PHENIX Collaboration) 2018 *J. Phys. Conf. Ser.* **1070** 012026 (*Preprint* 1806.05745)
[3] Adare A *et al.* (PHENIX Collaboration) 2018 *Phys. Rev.* **C97** 064911 (*Preprint* 1709.05649)
[4] Adler S S *et al.* (PHENIX Collaboration) 2007 *Phys. Rev. Lett.* **98** 132301 (*Preprint* nucl-ex/0605032)
[5] Afanasiev S *et al.* (PHENIX Collaboration) 2008 *Phys.Rev.Lett.* **100** 232301 (*Preprint* 0712.4372)
[6] Achard P *et al.* (L3 Collaboration) 2011 *Eur.Phys.J.* **C71** 1648 (*Preprint* 1105.4788)
[7] Sirunyan A M *et al.* (CMS Collaboration) 2018 *Phys. Rev. C* **97** 064912 (*Preprint* 1712.07198)
[8] Metzler R, Barkai E and Klafter J 1999 *Phys. Rev. Lett.* **82** 3563–3567
[9] Csörgő T, Hegyi S and Zajc W A 2004 *Eur. Phys. J.* **C36** 67–78 (*Preprint* nucl-th/0310042)
[10] Csanád M, Csörgő T and Nagy M 2007 *Braz. J. Phys.* **37** 1002–1013 (*Preprint* hep-ph/0702032)
[11] Csörgő T, Hegyi S, Novák T and Zajc W A 2006 *AIP Conf. Proc.* **828** 525–532 (*Preprint* nucl-th/0512060)
[12] Csörgő T 2008 *PoS* **HIGH-PTLHC08** 027 (*Preprint* 0903.0669)
[13] Bolz J, Ornik U, Plumer M, Schlei B and Weiner R 1993 *Phys.Rev.* **D47** 3860–3870
[14] Weiner R M 2000 *Phys. Rept.* **327** 249–346 (*Preprint* hep-ph/9904389)
[15] Csörgő T 2002 *Heavy Ion Phys.* **15** 1–80 (*Preprint* hep-ph/0001233)
[16] Kincses D (PHENIX Collaboration) 2018 *Universe* **4** 11 (*Preprint* 1711.06891)
[17] Kincses D (PHENIX Collaboration) 2019 *Acta Phys. Polon. Supp.* **12** 445 (*Preprint* 1811.08311)
[18] Kurgyis B (PHENIX Collaboration) 2019 *Acta Phys. Polon. Supp.* **12** 477 (*Preprint* 1809.09392)
[19] Lökös S (PHENIX Collaboration) 2018 *Universe* **4** 31 (*Preprint* 1801.08827)
[20] Kurgyis B (PHENIX Collaboration) 2020 *Phys. Part. Nucl.* **51** 263–266 (*Preprint* 1910.05019)

Journal of Physics: Conference Series **1602** (2020) 012010 doi:10.1088/1742-6596/1602/1/012010

Characterizing the medium created in Pb–Pb collisions at $\sqrt{s_{\mathrm{NN}}} = 2.76$ TeV by means of the evolution of two-particle transverse momentum correlations

Víctor González on behalf of the ALICE Collaboration

Department of Physics and Astronomy, Wayne State University, Detroit, Michigan 48201, USA

E-mail: `victor.gonzalez@cern.ch`

Abstract. Two-particle correlations are powerful tools for studying the medium produced in heavy-ion collisions. In particular, two-particle transverse momentum correlations enable measurements of the collision dynamics sensitive to momentum currents. Their evolution with collision centrality, which is related to the system lifetime, provides information about the shear viscosity, η/s, and the system relaxation time, τ_π. We report on measurements of two-particle transverse momentum correlations as a function of centrality in Pb–Pb collisions using the ALICE detector at the LHC. The centrality dependence of the near-side peak of the correlation function, particularly in the longitudinal dimension, provides information about the shear viscosity and the system relaxation time of the produced medium. The data are compared to predictions from selected Monte Carlo models. The charge independent momentum correlator exhibits a longitudinal broadening from peripheral to central collisions that is qualitatively consistent with expectations from a model with viscous effects. We will discuss an interpretation of the observed broadening in the context of this model.

1. Introduction

Properties of the medium formed in heavy-ion collisions can be extracted measuring the Fourier harmonic coefficients of the particles produced azimuthal distribution. Comparisons of these coefficients to hydrodynamical model predictions indicate that the medium has a specific shear viscosity, η/s, that nearly vanishes [1], approaching the behavior of a perfect liquid and unveiling the quark–gluon plasma (QGP) as a strongly coupled medium. The specific shear viscosity is the transport coefficient that most directly controls the expansion of the system. It quantifies the resistance the medium presents to its anisotropic deformation. It is responsible for the transfer of momentum between fluid cells as well as the damping of momentum fluctuations. Its effects are expected to grow with the lifetime of the system. Methods to determine η/s rely on quantitative descriptions of heavy-ion collisions using a chain of models which incorporate initial conditions onset, pre-thermalization evolution, hydrodynamical expansion, particlization, and hadronic transport. The precision of model predictions used to be limited by uncertainties in the initial state conditions and by the selection of the concrete model to use. Recent models incorporate the parametrization of the initial conditions which in some way circumvents the high uncertainty initially associated with such model phase. Unfortunately a new source of uncertainties has been also incorporated in the transition from the hydrodynamic phase to particles [2]. On the other hand, it was pointed out [3] that the strength of momentum current correlations may be sensitive to η/s. It was shown, in particular, that the longitudinal broadening with increasing system lifetime of a transverse momentum ($p_{\rm T}$) correlator, hereafter named G_2, is directly sensitive to η/s while it does not have any explicit dependence on the initial conditions of the system.

The G_2 correlator, defined in [3, 4], is experimentally calculated as

$$G_2\left(\eta_1, \varphi_1, \eta_2, \varphi_2\right) = \frac{\left\langle \sum\limits_{\rm i}^{n_{1,1}} \sum\limits_{\rm j\neq i}^{n_{1,2}} p_{\rm T,i}\, p_{\rm T,j} \right\rangle}{\langle n_{1,1}(\eta_1, \varphi_1)\rangle \langle n_{1,2}(\eta_2, \varphi_2)\rangle} - \langle p_{\rm T}\rangle(\eta_1, \varphi_1) \langle p_{\rm T}\rangle(\eta_2, \varphi_2) \tag{1}$$

where $\eta_\alpha, \varphi_\alpha$, $\alpha = 1, 2$, refer to single-track pseudorapidity and azimuthal angle, respectively; $n_{1,1}$ and $n_{1,2}$ are event-wise track counts at (η_1, φ_1) and (η_2, φ_2), respectively; $p_{{\rm T},i}$ and $p_{{\rm T},j}$ the transverse momentum components of particles in a given pair; and $\langle p_{\rm T}\rangle(\eta_i, \varphi_i)$ is the average transverse momentum of particles observed at (η_i, φ_i). Angle brackets, $\langle\cdots\rangle$, refer to event ensemble averages. For inferring the system lifetime the centrality of the collision is used as a proxy, such that the medium created in central collisions lives longer than the medium created in peripheral collisions.

2. Experimental measurements

The presented results (see [5] for more details) correspond to measurements based on 1.1×10^7 minimum bias Pb–Pb collisions at $\sqrt{s_{\rm NN}} = 2.76$ TeV collected during LHC Run 1 by the ALICE experiment [6, 7]. The collision vertex, determined with tracks reconstructed in the Inner Tracking System (ITS) and the Time Projection Chamber (TPC), was required to be within 7 cm, along the beam axis, of the nominal interaction point. The results are reported in nine centrality classes corresponding to 0–5% (most central), 5–10%, 10–20%, . . . , 70–80% (most peripheral) of the total interaction cross section, and whose ranges were estimated based on the hit multiplicity in the V0 detectors. The analysis was restricted to unidentified charged particle tracks which were measured with the TPC detector and required to be within the kinematic ranges $0.2 \leq p_{\rm T} \leq 2.0$ GeV/c and $|\eta| < 0.8$. Individual tracks were corrected for non uniform acceptance and for non uniform efficiency.

Differential charge independent (CI) and charge dependent (CD) two-particle transverse momentum correlators, $G_2^{\rm CI}$ and $G_2^{\rm CD}$, respectively, are measured as a function of pair pseudorapidity and azimuthal angle differences, $\Delta\eta$ and $\Delta\varphi$, respectively, for selected ranges of Pb–Pb collision centrality. Measurements of $G_2(\eta_1, \varphi_1, \eta_2, \varphi_2)$ are averaged across the fiducial acceptance to obtain $G_2(\Delta\eta, \Delta\varphi)$, where $\Delta\eta = \eta_1 - \eta_2$ and $\Delta\varphi = \varphi_1 - \varphi_2$ [8]. $G_2^{\rm CI}$ and $G_2^{\rm CD}$ correlators are extracted as the averaged sum and the averaged difference of the opposite charge and same charge correlators, respectively.

Figure 1 presents the $G_2^{\rm CI}(\Delta\eta, \Delta\varphi)$ correlator measured in 0–5% and 30–40% Pb–Pb collisions. The $G_2^{\rm CI}$ correlators feature sizable $\Delta\varphi$ modulations, dominated in mid-central collisions by a strong elliptic flow ($\cos(2\Delta\varphi)$) component. On the near side, $|\Delta\varphi| < \pi/2$, atop the azimuthal modulation, the $G_2^{\rm CI}$ correlators feature a near-side peak whose amplitude monotonically decreases from peripheral to central collisions while its longitudinal width systematically broadens. Two structures develop from semi-central towards central collisions. On the near side, $|\Delta\varphi| < \pi/2$, two longitudinally extended lobes give place to a depletion around $(\Delta\eta, \Delta\varphi) = (0, 0)$. On the away side, $|\Delta\varphi - \pi| < \pi/2$, a long range depletion develops showing the dilution, in the short longitudinal reach, of the strength of the long range correlations driven by initial fluctuations and collective behavior.

Near-side longitudinal projections and azimuthal projections of the $G_2^{\rm CI}$ correlators are shown in Fig. 2 for the previous centrality collision intervals. The near-side longitudinal projections of

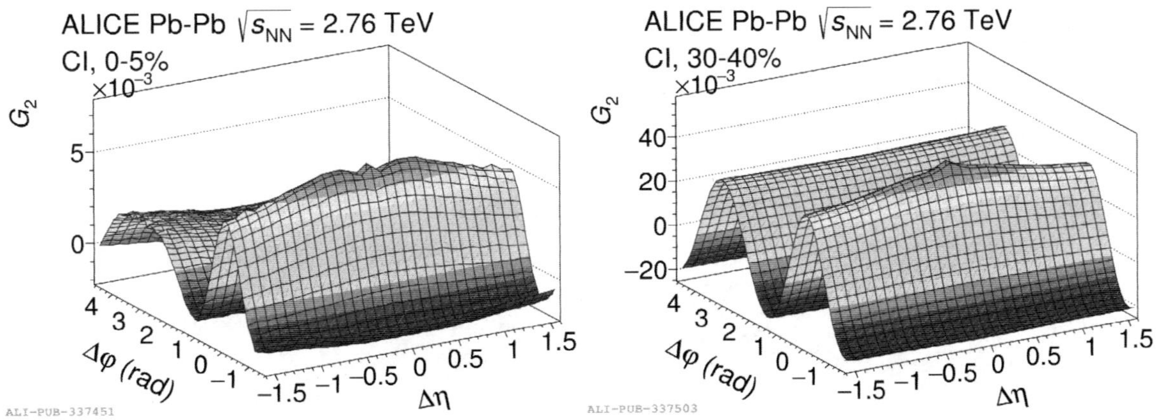

Figure 1. Two-particle transverse momentum correlations $G_2^{\rm CI}$ for the most central (left) and semi-central (right) Pb–Pb collisions at $\sqrt{s_{\rm NN}} = 2.76$ TeV. Under-corrected correlator values at $(\Delta\eta, \Delta\varphi) = (0, 0)$ are not shown [5].

the $G_2^{\rm CI}$ correlator show the monotonic growth of the amplitude of the near-side peak from central to peripheral collisions. The collective behavior, developed along the azimuthal dimension is present just as a uniform baseline on top of which the near-side peak develops. On top of this baseline the near-side peak broadens from peripheral to central collisions. From semi-central to central collisions the near-side peak shows a plateau consistent with the two lobes structure shown in Fig. 1. The azimuthal projections show the evolution with centrality of the long range collective behavior. The dilution on the away side of the long range correlation shown in Fig. 1 is also visible in the progress of the azimuthal projections from semi-central to peripheral collisions.

The focus now is on the evolution with centrality of the width of the the two-particle transverse momentum correlation on the near-side peak. To extract such widths, along the azimuthal and longitudinal dimensions, the two-dimensional correlations are parametrized with a two-component model defined as

$$F(\Delta\eta, \Delta\varphi) = B + \sum_{n=2}^{4} a_n \cos(n\Delta\varphi) + A \frac{\gamma_{\Delta\eta}}{2\,\omega_{\Delta\eta}\,\Gamma\left(\frac{1}{\gamma_{\Delta\eta}}\right)} \mathrm{e}^{-\left|\frac{\Delta\eta}{\omega_{\Delta\eta}}\right|^{\gamma_{\Delta\eta}}} \frac{\gamma_{\Delta\varphi}}{2\,\omega_{\Delta\varphi}\,\Gamma\left(\frac{1}{\gamma_{\Delta\varphi}}\right)} \mathrm{e}^{-\left|\frac{\Delta\varphi}{\omega_{\Delta\varphi}}\right|^{\gamma_{\Delta\varphi}}}, \tag{2}$$

where B and a_n are intended to describe the long-range mean correlation strength and azimuthal anisotropy, while the bidimensional generalized Gaussian, defined by the parameters A, $\omega_{\Delta\eta}$, $\omega_{\Delta\varphi}$, $\gamma_{\Delta\eta}$ and $\gamma_{\Delta\varphi}$, is intended to model the signal of interest. The longitudinal and azimuthal widths of the correlators, denoted $\sigma_{\Delta\eta}$ and $\sigma_{\Delta\varphi}$, respectively, are extracted as the standard deviation of the generalized Gaussian

$$\sigma_{\Delta\eta} = \sqrt{\frac{\omega_{\Delta\eta}^2 \Gamma(3/\gamma_{\Delta\eta})}{\Gamma(1/\gamma_{\Delta\eta})}} \qquad \sigma_{\Delta\varphi} = \sqrt{\frac{\omega_{\Delta\varphi}^2 \Gamma(3/\gamma_{\Delta\varphi})}{\Gamma(1/\gamma_{\Delta\varphi})}} \tag{3}$$

along each dimension. The longitudinal broadening of $G_2^{\rm CI}$ from peripheral to central collisions is interpreted as the increase in the reach of the viscous effects and used to infer the magnitude of the shear viscosity per unit of entropy density of the matter produced in Pb–Pb collisions. The longitudinal and azimuthal widths evolution of $G_2^{\rm CD}$ are used to assess the role of competing effects, including radial flow, diffusion, and the broadening of jets by interactions with the medium. A first measurement of the broadening of the two-particle transverse

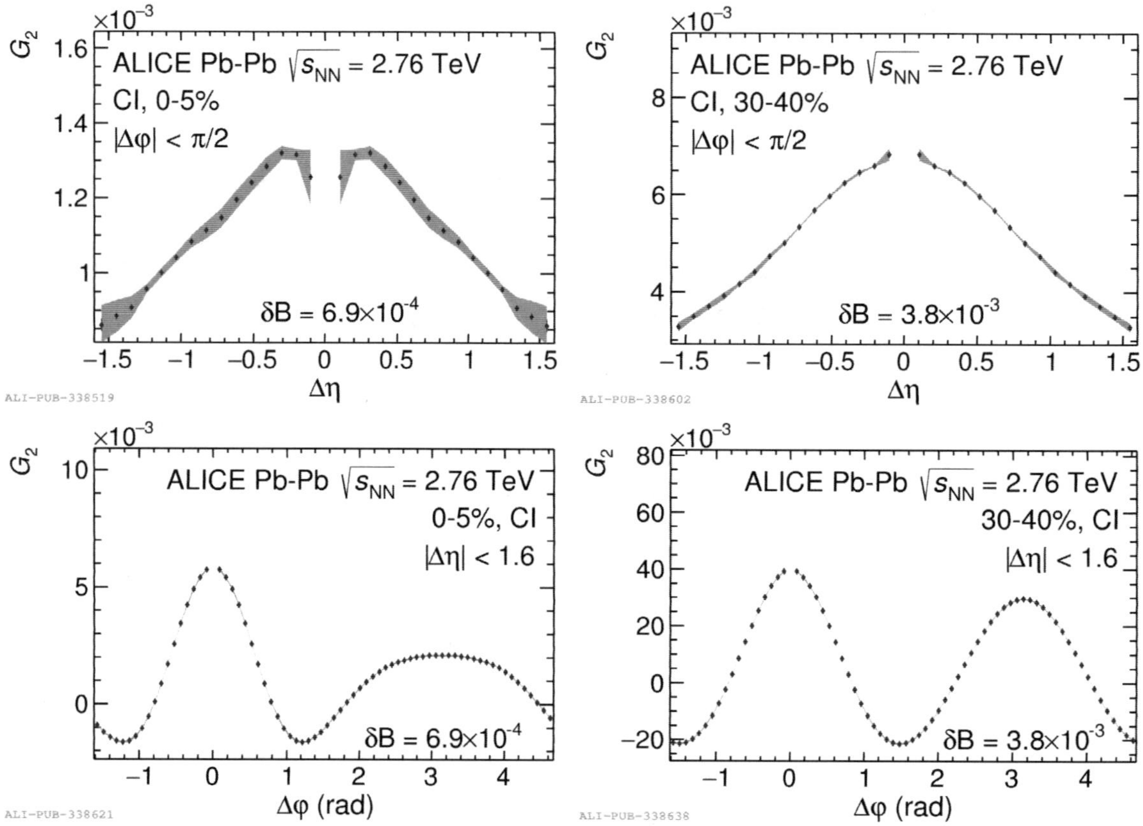

Figure 2. Two-particle transverse momentum correlations $G_2^{\rm CI}$ longitudinal (top) and azimuthal (bottom) projections for the most central (left) and semi-central (right) Pb–Pb collisions at $\sqrt{s_{\rm NN}} = 2.76$ TeV. Vertical bars (mostly smaller than the marker size) and shaded blue bands represent statistical and systematic uncertainties, respectively. The scale uncertainty inherent to the combinatorial nature of the two-particle correlators is also quoted. Under-corrected correlator values at $(\Delta\eta, \Delta\varphi) = (0, 0)$ are not shown [5].

momentum correlator G_2 in Au–Au collisions at $\sqrt{s_{\rm NN}} = 200$ GeV was reported by the STAR collaboration [9].

3. Results and discussion

The evolution with collision centrality of the longitudinal and azimuthal widths of $G_2^{\rm CI}$ and $G_2^{\rm CD}$ correlators is shown in the top panels of Fig. 3. Longitudinally, $G_2^{\rm CI}$ broadens by about 25% from peripheral to semi-central collisions, where its width saturates, while $G_2^{\rm CD}$ narrows by about 10% from peripheral to central collisions. Azimuthally, both $G_2^{\rm CI}$ and $G_2^{\rm CD}$ consistently narrow by about 20%, although $G_2^{\rm CD}$ stays wider. By comparison with the balance functions behavior [12], it is reasonable to infer that radial flow and larger $\langle p_{\rm T}\rangle$, in more central collisions, together with delayed hadronization produce the observed narrowing of $G_2^{\rm CD}$. That might also be the reason for the azimuthal $G_2^{\rm CI}$ narrowing, drawing a scenario where broadening due to viscous effects competes with narrowing due to radial flow and delayed hadronization. The longitudinal broadening observed by STAR, which also saturates, reaches 70% [9] pointing to a strong dependence on the beam energy.

The dependence with collision centrality of the measured longitudinal width of the two-

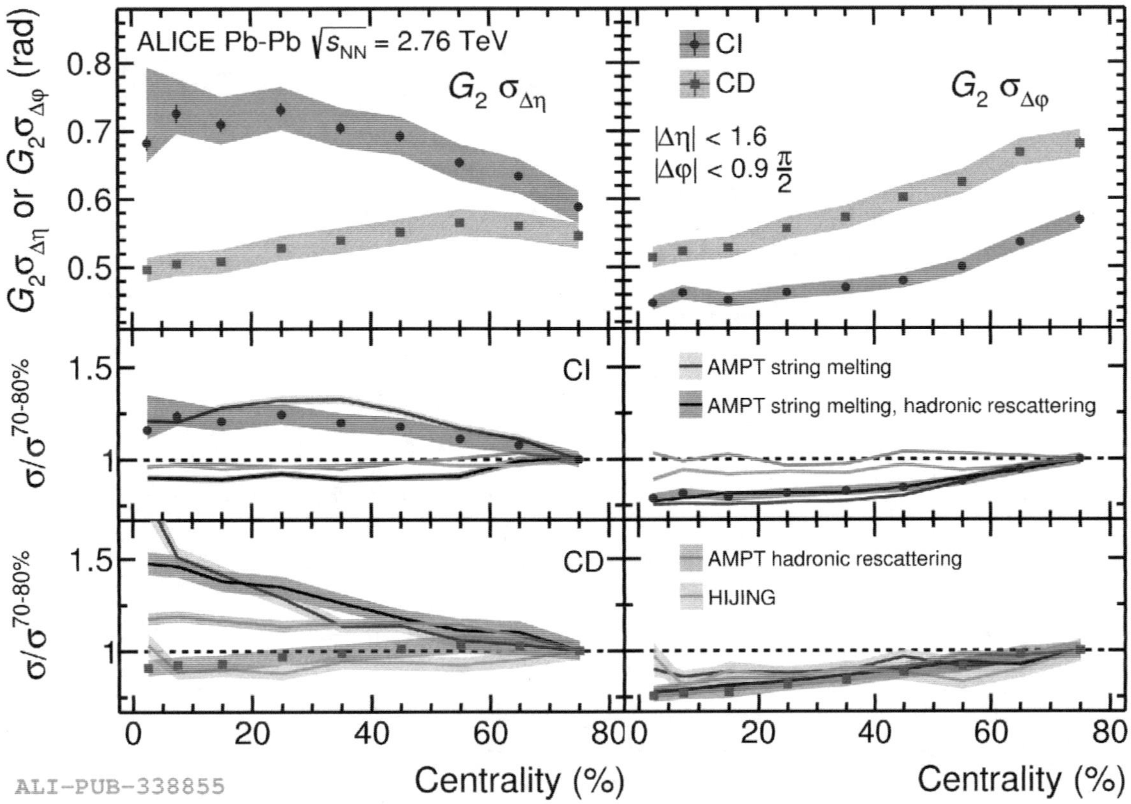

Figure 3. Top panels: collision centrality evolution of the longitudinal (left) and azimuthal (right) widths of the two-particle transverse momentum correlators $G_2^{\rm CI}$ and $G_2^{\rm CD}$ measured in Pb–Pb collisions at $\sqrt{s_{\rm NN}} = 2.76$ TeV. Central and bottom panels: widths evolution relative to the value in the most peripheral collisions of the two-particle transverse momentum correlators $G_2^{\rm CI}$ (central) and $G_2^{\rm CD}$ (bottom) along the longitudinal (left) and azimuthal (right) dimensions. Data are compared to HIJING [10] and AMPT [11] model expectations. In data vertical bars and shaded bands represent statistical and systematic uncertainties, respectively. For models, shaded bands represent statistical uncertainties [5].

particle transverse momentum correlations $G_2^{\rm CI}$ and $G_2^{\rm CD}$ are compared to expectations from HIJING [10] and from three flavors of the AMPT [11] event generator in the central and bottom panels of Fig. 3 for Pb–Pb collisions at $\sqrt{s_{\rm NN}} = 2.76$ TeV. Although grossly failing to reproduce the longitudinal narrowing of $G_2^{\rm CD}$ AMPT with string melting but without hadronic rescattering qualitatively describes the longitudinal broadening of $G_2^{\rm CI}$ pointing to a potential early origin of such a broadening.

Interpreting the longitudinal broadening of $G_2^{\rm CI}$ as originating exclusively from viscous effects, the authors of [3] extract the expression

$$\sigma_{\rm c}^2 - \sigma_0^2 = \frac{4}{T_{\rm c}}\frac{\eta}{s}\left(\frac{1}{\tau_0} - \frac{1}{\tau_{\rm c,f}}\right) \qquad (4)$$

which relates η/s with the difference of the squared longitudinal correlator widths for most central collisions and at formation time, σ_c and σ_0, respectively, with the critical temperature, $T_{\rm c}$, and with the difference of the inverses of the formation time and the freezeout time for most central collisions, τ_0 and $\tau_{\rm c,f}$, respectively. Equation 4, extracted from first order in gradients expansion for the propagation of transverse momentum fluctuations, was used to infer the expected longitudinal widths of the correlator for the most central collisions for different values of η/s as shown in Fig. 4. The $G_2^{\rm CI}$ correlator width measured in the most central

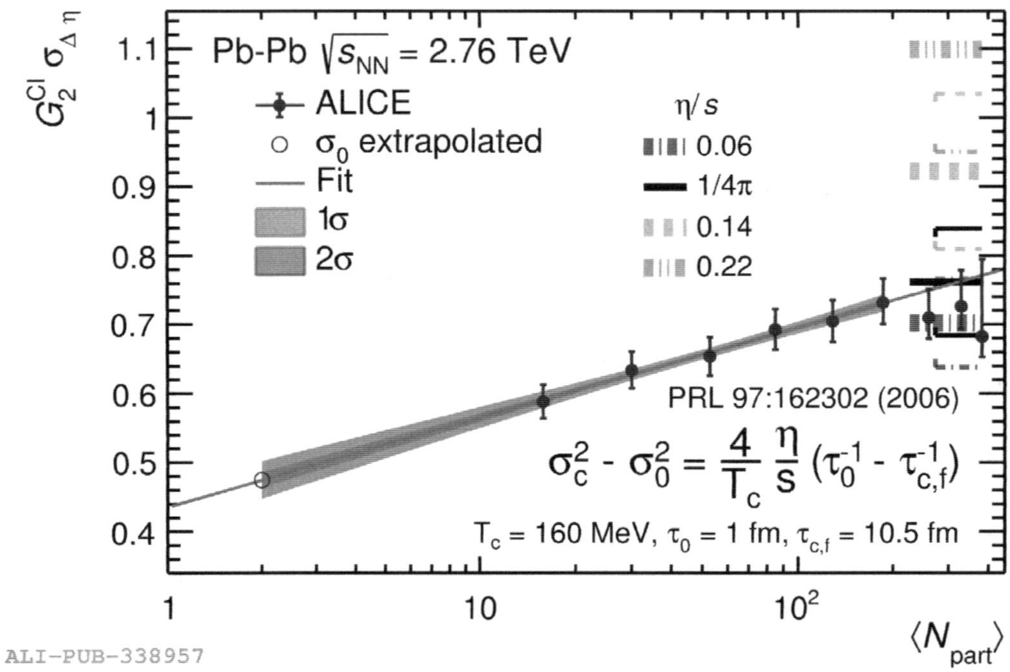

Figure 4. Expected longitudinal widths for the most central collisions of the two-particle transverse momentum correlation $G_2^{\rm CI}$ for different values of η/s by using the expression suggested in [3]. Data point error bars represent total uncertainties obtained by adding in quadrature statistical and systematic uncertainties. In the formula $\sigma_{\rm c}$ is the longitudinal width for the most central collisions inferred by using this expression and represented for each of the η/s values by the color discontinuous bands (continuous for $\eta/s = 1/4\pi$) at the highest number of participants, σ_0 is the longitudinal width for the most peripheral collisions (only two participants) which is obtained by extrapolating the fit, $T_{\rm c}$ is the critical temperature, τ_0 is the formation time and $\tau_{\rm c,f}$ the freeze-out time. Error caps in the same color as the discontinuous bands, represent uncertainties of the inferred longitudinal widths for the most central collisions [5].

collisions favors rather small values of η/s, close to the KSS [13] limit of $1/4\pi$.

4. Conclusion

The near-side peak of the $G_2^{\rm CD}$ correlator is observed to significantly narrow with collision centrality along the longitudinal and azimuthal directions. By contrast, the $G_2^{\rm CI}$ correlator is found to narrow with collision centrality only in the azimuthal direction while broadens toward central collisions in the longitudinal direction. Using the model proposed in [3], which assigns the whole broadening to viscous effects, measured data favor a value of η/s of order $1/4\pi$. String melting AMPT without the hadronic rescattering phase qualitatively reproduces the longitudinal broadening of $G_2^{\rm CI}$, but misses the narrowing of $G_2^{\rm CD}$ along that dimension.

Journal of Physics: Conference Series **1602** (2020) 012010 doi:10.1088/1742-6596/1602/1/012010

References

[1] Adcox K *et al.* (PHENIX) 2005 *Nucl. Phys.* **A757** 184–283 (*Preprint* nucl-ex/0410003)
[2] McNelis M, Everett D and Heinz U 2019 (*Preprint* 1912.08271)
[3] Gavin S and Abdel-Aziz M 2006 *Phys. Rev. Lett.* **97** 162302 (*Preprint* nucl-th/0606061)
[4] Sharma M and Pruneau C A 2009 *Phys. Rev.* **C79** 024905 (*Preprint* 0810.0716)
[5] Acharya S *et al.* (ALICE) 2020 *Phys. Lett. B* **804** 135375 (*Preprint* 1910.14393)
[6] Aamodt K *et al.* (ALICE) 2008 *JINST* **3** S08002
[7] Abelev B *et al.* (ALICE) 2014 *Int.J.Mod.Phys.* **A29** 1430044 (*Preprint* 1402.4476)
[8] Acharya S *et al.* (ALICE) 2019 *Phys. Rev. C* **100**(4) 044903 (*Preprint* 1805.04422)
[9] Agakishiev G *et al.* (STAR) 2011 *Phys. Lett.* **B704** 467–473 (*Preprint* 1106.4334)
[10] Wang X N and Gyulassy M 1991 *Phys. Rev.* **D44** 3501–3516
[11] Lin Z W, Ko C M, Li B A, Zhang B and Pal S 2005 *Phys. Rev.* **C72** 064901 (*Preprint* nucl-th/0411110)
[12] Abelev B *et al.* (ALICE) 2013 *Phys. Lett.* **B723** 267–279 (*Preprint* 1301.3756)
[13] Kovtun P, Son D T and Starinets A O 2005 *Phys. Rev. Lett.* **94** 111601 (*Preprint* hep-th/0405231)

Journal of Physics: Conference Series **1602** (2020) 012011 doi:10.1088/1742-6596/1602/1/012011

The QCD transition line from lattice simulations

Claudia Ratti[a], **Rene Bellwied**[a], **Szabolcs Borsanyi**[b], **Zoltan Fodor**[b,c,d,e], **Jana N. Guenther** [b,f], **Ruben Kara**[b], **Sandor Katz**[c], **Paolo Parotto**[b], **Attila Pasztor**[c], **Kalman Szabo**[b,d]

[a] *Department of Physics, University of Houston, Houston, TX, USA 77204*
[b] *Department of Physics, Wuppertal University, Gaussstr. 20, D-42119 Wuppertal, Germany*
[c] *Inst. for Theoretical Physics, ELTE Eötvös Loránd University, Pázmány P. sétány 1/A, H-1117 Budapest, Hungary*
[d] *Jülich Supercomputing Centre, Forschungszentrum Jülich, D-52425 Jülich, Germany*
[e] *Physics Department, UCSD, San Diego, CA 92093, USA*
[f] *University of Regensburg, Department of Physics, Regensburg D-93053, Germany*

E-mail: `cratti@uh.edu`

Abstract. We present our new results for the QCD transition line at finite chemical potential μ_B from first principle, lattice QCD simulations. We extrapolate our results from imaginary chemical potentials, up to $\mu_B \approx 300$ MeV. We obtain the most precise value for the transition temperature, the curvature of the QCD phase diagram and its fourth order correction. The results are continuum extrapolated, based on $N_t = 10, 12, 16$ lattices. We also study the height and width of the peak of the chiral susceptibility and how they change with increasing chemical potential. We find that both of them are consistent with a constant, which does not indicate any criticality in our explored

1. Introduction

Mapping out the whole phase diagram of strongly interacting matter is one of the main goals of high energy nuclear physics. From the experimental point of view, the deconfined phase of Quantum Chromodynamics (QCD) can be created in relativistic heavy-ion collisions. In particular, the LHC program aims at creating the hottest form of deconfined matter, while the RHIC program can vary the collision energy, to systematically scan the finite-density region. The second Beam Energy Scan (BESII) at RHIC is running between 2019 and 2021, its main goal being the search for the QCD critical point separating crossover from first order phase transition.

From the theoretical point of view, lattice simulations are the best tool to solve QCD in the regime of temperatures and densities where the transition occurs. However, at the moment finite-density simulations are hindered by the sign problem and cannot be directly performed. Two main methods have been proposed to circumvent this problem: Taylor expansion of thermodynamic observables in powers of μ_B [1, 2, 3, 4, 5, 6, 7, 8, 9, 10, 11] or simulations at imaginary μ_B (where the sign problem is absent) and analytical continuation of the results to real μ_B [12, 13, 14, 15, 16, 17, 18, 19, 20, 21, 22, 23, 24, 25, 26, 27, 28].

In our recent manuscript [29], we considered the following expansion for the transition temperature of QCD:

$$\frac{T_c(\mu_B)}{T_c(\mu_B=0)} = 1 - \kappa_2 \left(\frac{\mu_B}{T_c(\mu_B)}\right)^2 - \kappa_4 \left(\frac{\mu_B}{T_c(\mu_B)}\right)^4 \dots \tag{1}$$

and we extracted the most precise values in the literature for $T_c(\mu_B = 0)$, κ_2 and κ_4 along the strangeness neutral crossover line. The latter is defined as the line in the T, μ_B plane at which the phase transition occurs, but with μ_S depending on T and μ_B so that the net-strangeness in the system is identically zero. This is imposed to match the experimental situation in which no net-strangeness is present in the system.

One can evaluate the coefficients κ_2 and κ_4 by either one of the standard extrapolation methods listed above. A direct evaluation of the μ_B-derivatives over $\mu_B = 0$ ensembles was used in Refs. [30, 31, 10]. We choose the analytical continuation from imaginary-μ_B method, which results in a larger signal-to-noise ratio [21, 22]. Results for κ_2 obtained with the two different methods were thoroughly compared in Ref. [27].

In Ref. [29], we also considered the height and width of the peak of the chiral susceptibility as functions of the chemical potential, as an indication of the strength of the crossover transition.

2. Methodology and results

We use the 4-stout staggered fermion action with temporal lattice sizes $N_t = 10,\ 12,\ 16$. We consider the following observables:

$$\begin{aligned} \langle\bar\psi\psi\rangle &= -\left[\langle\bar\psi\psi\rangle_T - \langle\bar\psi\psi\rangle_0\right]\frac{m_{\rm ud}}{f_\pi^4}, \\ \chi &= \left[\chi_T - \chi_0\right]\frac{m_{\rm ud}^2}{f_\pi^4}, \quad \text{with} \\ \langle\bar\psi\psi\rangle_{T,0} &= \frac{T}{V}\frac{\partial \log Z}{\partial m_{\rm ud}} \qquad \chi_{T,0} = \frac{T}{V}\frac{\partial^2 \log Z}{\partial m_{\rm ud}^2}, \end{aligned} \tag{2}$$

where we assumed isospin symmetry, i.e. $m_u = m_d = m_{ud}$. In the above equations, the subscripts $T, 0$ indicate values at finite- and zero-temperature, respectively. In the following, $\langle\bar\psi\psi\rangle$ (the chiral condensate) and χ (the chiral susceptibility) are always shown after applying the correction to satisfy $n_s = 0$ with zero statistical error.

The left and middle panel of Fig. 1 show the chiral condensate and susceptibility as functions of the temperature, for different values of the imaginary chemical potential, respectively. The right panel shows that, when the chiral susceptibility is plotted as a function of the chiral condensate, all curves collapse on a common, smooth line. We exploit this feature in our analysis, which allows us a more precise determination of T_c and, as a consequence, a more precise determination of κ_2 and κ_4.

We proceed as follows:

i) We first determine $\langle\bar\psi\psi\rangle$ and χ as function of T and Imμ_B through lattice simulations.
ii) We extract the curve of chiral susceptibility as a function of the chiral condensate.
iii) We search for the peak of $\chi(\langle\bar\psi\psi\rangle)$ through a low-order polynomial fit for each N_t and Imμ_B obtaining $\langle\bar\psi\psi\rangle_c(N_t, \text{Im}\mu_B)$.
iv) We interpolate $\langle\bar\psi\psi\rangle(T)$ to convert $\langle\bar\psi\psi\rangle_c$ to T_c for each Imμ_B/T.
v) We perform a global fit of $T_c(N_t, \text{Im}\mu_B/T_c)$ to determine its μ_B dependence. This yields κ_2 and κ_4 for $1/N_t^2 = 0$. For this step we use various functions – all containing an independent κ_6 – with coefficients depending linearly on $1/N_t^2$.

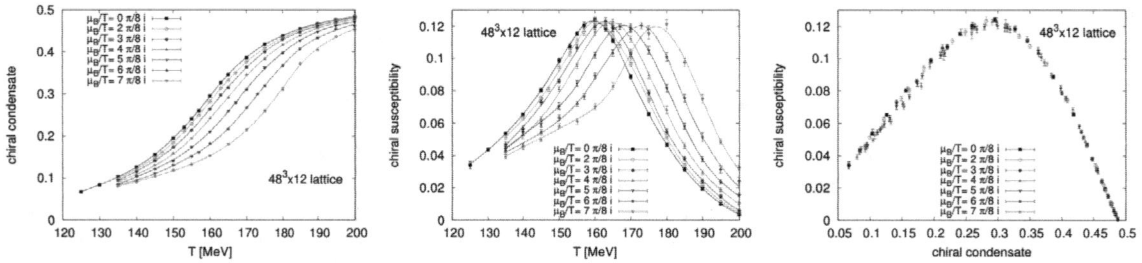

Figure 1. Renormalized chiral condensate $\langle\bar{\psi}\psi\rangle$ (left) and chiral susceptibility χ (middle) as functions of the temperature for the intermediate lattice spacing in this study. The black curves correspond to vanishing baryon density, while results for various imaginary values of the chemical potential are shown in other colors. Finally, in the right panel we show the susceptibility as a function of the condensate. In this representation the chemical potential dependence is very weak.

Each step listed above leads to a few possible choices for e.g. the fit functions, ranges etc. Combining all these choices we get $2^8 = 256$ different analyses. We take this into account in the systematic error. We obtain

$$\begin{aligned} T_c &= 158.0 \pm 0.6 \text{ MeV} \\ \kappa_2 &= 0.0153 \pm 0.0018 \\ \kappa_4 &= 0.00032 \pm 0.00067. \end{aligned} \tag{3}$$

We show the corresponding transition line in Fig. 2.

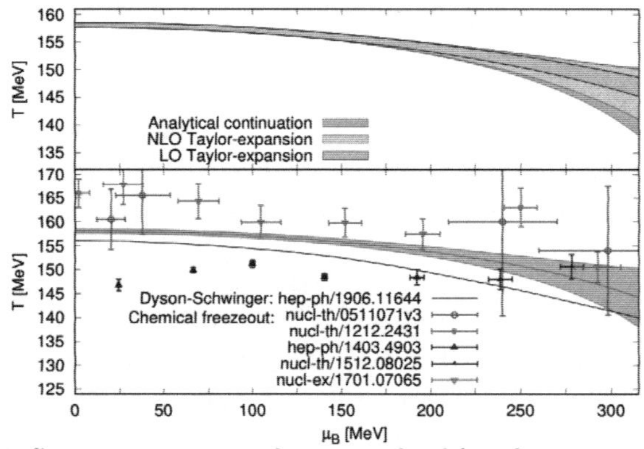

Figure 2. Top: Transition line extrapolated from lattice simulations at imaginary chemical potential using an analytical continuation with the ansätze used in step iv) of our analysis (green band) compared with an extrapolation using the formula in Eq. (1) up to the order of κ_4 (red band) or up to κ_2 (blue band). Bottom: Crossover line from the lattice compared with a prediction from truncated Dyson-Schwinger equations [32] and some estimates of the chemical freezeout parameters in heavy ion collisions [33, 34, 35, 36, 37].

We introduce a width parameter σ as a proxy for the natural definition of the width of the susceptibility peak, given by its second derivative:

$$\langle\bar{\psi}\psi\rangle(T_c \pm \sigma/2) = \langle\bar{\psi}\psi\rangle_c \pm \Delta\langle\bar{\psi}\psi\rangle/2 \ , \tag{4}$$

with $\langle\bar{\psi}\psi\rangle_c = 0.285$ and $\Delta\langle\bar{\psi}\psi\rangle = 0.14$. We show the half width of the transition in the left panel of Fig. 3. As evident from the figure, the half width is consistent with a constant for $\mu_B \approx 300$ MeV.

Finally, as a proxy for the strength of the crossover, we study the value of the chiral susceptibility at the crossover temperature, which corresponds to the height of the peak of the chiral susceptibility. We get this for each Imμ_B and N_t as a byproduct of steps i-ii) of the analysis for κ_2 and κ_4. If one then performs a continuum extrapolation of the resulting values for fixed values of Imμ_B, one gets the right panel of Fig. 3. Again, we see a very mild $\hat{\mu}_B^2$ dependence, consistent with a constant.

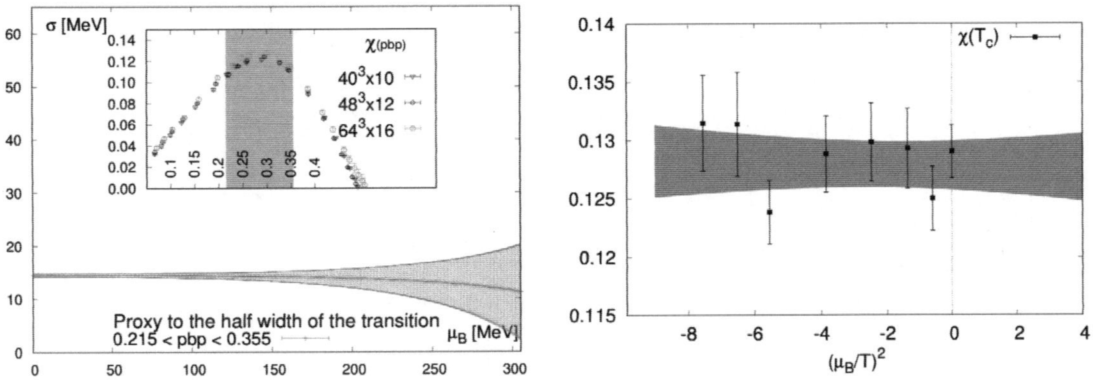

Figure 3. Left: Half width σ of the transition defined in Eq. (4) using the temperature difference of the contours $\langle\bar{\psi}\psi\rangle = 0.31$ and $\langle\bar{\psi}\psi\rangle = 0.19$. In the insert we show a plot of the $\chi(\langle\bar{\psi}\psi\rangle)$ peak, where the shaded region corresponds to $\langle\bar{\psi}\psi\rangle_c \pm \Delta\langle\bar{\psi}\psi\rangle/2$. Both are extrapolated to real μ_B. Right: Result of a μ_B-by-μ_B analysis for the value of the chiral susceptibility at the crossover temperature after continuum extrapolation and including the systematic errors for $LT_c = 4$. The green band shows a linear extrapolation in $\hat{\mu}_B^2$.

3. Conclusions

We have obtained the most precise values for the transition temperature at $\mu_B = 0$, the curvature κ_2 and the fourth-order coefficient κ_4 by studying the chiral condensate and susceptibility as functions of the temperature, for different values of imaginary chemical potential. We have also studied the height and width of the peak of the chiral susceptibility as a function of the chemical potential, and observed that they are both consistent with a constant. For this reason, we conclude that we see no sign of criticality in our explored μ_B range.

Acknowledgements

This project was funded by the DFG grant SFB/TR55. The project also received support from the BMBF Grant No. 05P18PXFCA. This work was also supported by the Hungarian National Research, Development and Innovation Office, NKFIH grants KKP126769 and K113034. A.P. is supported by the J. Bolyai Research Scholarship of the Hungarian Academy of Sciences and by the ÚNKP-19-4 New National Excellence Program of the Ministry for Innovation and Technology. This material is based upon work supported by the National Science Foundation under grants no. PHY-1654219 and by the U.S. DoE, Office of Science, Office of Nuclear Physics, within the framework of the Beam Energy Scan Topical (BEST) Collaboration. This research used resources of the Oak Ridge Leadership Computing Facility, which is a DOE Office of

Science User Facility supported under Contract DE-AC05-00OR22725. The authors gratefully acknowledge the Gauss Centre for Supercomputing e.V. (www.gauss-centre.eu) for funding this project by providing computing time on the GCS Supercomputer JURECA/Booster at Jülich Supercomputing Centre (JSC), on HAZELHEN at HLRS, Stuttgart as well as on SUPERMUC-NG at LRZ, Munich. We acknowledge PRACE for awarding us access to Piz Daint hosted at CSCS, Switzerland. C.R. also acknowledges the support from the Center of Advanced Computing and Data Systems at the University of Houston.

References

[1] C. Allton, S. Ejiri, S. Hands, O. Kaczmarek, F. Karsch, E. Laermann, C. Schmidt and L. Scorzato, Phys. Rev. D **66**, 074507 (2002) doi:10.1103/PhysRevD.66.074507

[2] C. Allton, M. Doring, S. Ejiri, S. Hands, O. Kaczmarek, F. Karsch, E. Laermann and K. Redlich, Phys. Rev. D **71**, 054508 (2005) doi:10.1103/PhysRevD.71.054508

[3] R. Gavai and S. Gupta, Phys. Rev. D **78**, 114503 (2008) doi:10.1103/PhysRevD.78.114503

[4] S. Basak *et al.* [MILC], PoS **LATTICE2008**, 171 (2008) doi:10.22323/1.066.0171

[5] S. Borsanyi, Z. Fodor, S. D. Katz, S. Krieg, C. Ratti and K. Szabo, JHEP **01**, 138 (2012) doi:10.1007/JHEP01(2012)138

[6] S. Borsanyi, G. Endrodi, Z. Fodor, S. Katz, S. Krieg, C. Ratti and K. Szabo, JHEP **08**, 053 (2012) doi:10.1007/JHEP08(2012)053

[7] R. Bellwied, S. Borsanyi, Z. Fodor, S. Katz, A. Pasztor, C. Ratti and K. Szabo, Phys. Rev. D **92**, no.11, 114505 (2015) doi:10.1103/PhysRevD.92.114505

[8] H. T. Ding, S. Mukherjee, H. Ohno, P. Petreczky and H. P. Schadler, Phys. Rev. D **92**, no.7, 074043 (2015) doi:10.1103/PhysRevD.92.074043

[9] A. Bazavov, H. T. Ding, P. Hegde, O. Kaczmarek, F. Karsch, E. Laermann, Y. Maezawa, S. Mukherjee, H. Ohno, P. Petreczky, H. Sandmeyer, P. Steinbrecher, C. Schmidt, S. Sharma, W. Soeldner and M. Wagner, Phys. Rev. D **95**, no.5, 054504 (2017) doi:10.1103/PhysRevD.95.054504

[10] A. Bazavov *et al.* [HotQCD], Phys. Lett. B **795**, 15-21 (2019) doi:10.1016/j.physletb.2019.05.013

[11] A. Bazavov, D. Bollweg, H. T. Ding, P. Enns, J. Goswami, P. Hegde, O. Kaczmarek, F. Karsch, R. Larsen, S. Mukherjee, H. Ohno, P. Petreczky, C. Schmidt, S. Sharma and P. Steinbrecher, Phys. Rev. D **101**, no.7, 074502 (2020) doi:10.1103/PhysRevD.101.074502

[12] Z. Fodor and S. Katz, Phys. Lett. B **534**, 87-92 (2002) doi:10.1016/S0370-2693(02)01583-6

[13] Z. Fodor and S. Katz, JHEP **03**, 014 (2002) doi:10.1088/1126-6708/2002/03/014

[14] P. de Forcrand and O. Philipsen, Nucl. Phys. B **642**, 290-306 (2002) doi:10.1016/S0550-3213(02)00626-0

[15] M. D'Elia and M. P. Lombardo, Phys. Rev. D **67**, 014505 (2003) doi:10.1103/PhysRevD.67.014505

[16] Z. Fodor and S. Katz, JHEP **04**, 050 (2004) doi:10.1088/1126-6708/2004/04/050

[17] M. D'Elia and F. Sanfilippo, Phys. Rev. D **80**, 014502 (2009) doi:10.1103/PhysRevD.80.014502

[18] P. Cea, L. Cosmai and A. Papa, Phys. Rev. D **89**, no.7, 074512 (2014) doi:10.1103/PhysRevD.89.074512

[19] C. Bonati, P. de Forcrand, M. D'Elia, O. Philipsen and F. Sanfilippo, Phys. Rev. D **90**, no.7, 074030 (2014) doi:10.1103/PhysRevD.90.074030

[20] P. Cea, L. Cosmai and A. Papa, Phys. Rev. D **93**, no.1, 014507 (2016) doi:10.1103/PhysRevD.93.014507

[21] C. Bonati, M. D'Elia, M. Mariti, M. Mesiti, F. Negro and F. Sanfilippo, Phys. Rev. D **92**, no.5, 054503 (2015) doi:10.1103/PhysRevD.92.054503

[22] R. Bellwied, S. Borsanyi, Z. Fodor, J. Guenther, S. Katz, C. Ratti and K. Szabo, Phys. Lett. B **751**, 559-564 (2015) doi:10.1016/j.physletb.2015.11.011

[23] M. D'Elia, G. Gagliardi and F. Sanfilippo, Phys. Rev. D **95**, no.9, 094503 (2017) doi:10.1103/PhysRevD.95.094503

[24] J. Guenther, R. Bellwied, S. Borsanyi, Z. Fodor, S. Katz, A. Pasztor, C. Ratti and K. Szabo, Nucl. Phys. A **967**, 720-723 (2017) doi:10.1016/j.nuclphysa.2017.05.044

[25] P. Alba, R. Bellwied, S. Borsanyi, Z. Fodor, J. Guenther, S. D. Katz, V. Mantovani Sarti, J. Noronha-Hostler, P. Parotto, A. Pasztor, I. Portillo Vazquez and C. Ratti, Phys. Rev. D **96**, no.3, 034517 (2017) doi:10.1103/PhysRevD.96.034517

[26] V. Vovchenko, A. Pasztor, Z. Fodor, S. D. Katz and H. Stoecker, Phys. Lett. B **775**, 71-78 (2017) doi:10.1016/j.physletb.2017.10.042

[27] C. Bonati, M. D'Elia, F. Negro, F. Sanfilippo and K. Zambello, Phys. Rev. D **98**, no.5, 054510 (2018) doi:10.1103/PhysRevD.98.054510

[28] S. Borsanyi, Z. Fodor, J. N. Guenther, S. K. Katz, K. K. Szabo, A. Pasztor, I. Portillo and C. Ratti, JHEP **10**, 205 (2018) doi:10.1007/JHEP10(2018)205

[29] S. Borsanyi, Z. Fodor, J. N. Guenther, R. Kara, S. D. Katz, P. Parotto, A. Pasztor, C. Ratti and K. K. Szabo, [arXiv:2002.02821 [hep-lat]].
[30] O. Kaczmarek, F. Karsch, E. Laermann, C. Miao, S. Mukherjee, P. Petreczky, C. Schmidt, W. Soeldner and W. Unger, Phys. Rev. D **83**, 014504 (2011) doi:10.1103/PhysRevD.83.014504
[31] G. Endrodi, Z. Fodor, S. Katz and K. Szabo, JHEP **04**, 001 (2011) doi:10.1007/JHEP04(2011)001
[32] P. Isserstedt, M. Buballa, C. S. Fischer and P. J. Gunkel, Phys. Rev. D **100**, no.7, 074011 (2019) doi:10.1103/PhysRevD.100.074011
[33] A. Andronic, P. Braun-Munzinger and J. Stachel, Nucl. Phys. A **772**, 167-199 (2006) doi:10.1016/j.nuclphysa.2006.03.012
[34] F. Becattini, M. Bleicher, T. Kollegger, T. Schuster, J. Steinheimer and R. Stock, Phys. Rev. Lett. **111**, 082302 (2013) doi:10.1103/PhysRevLett.111.082302
[35] P. Alba, W. Alberico, R. Bellwied, M. Bluhm, V. Mantovani Sarti, M. Nahrgang and C. Ratti, Phys. Lett. B **738**, 305-310 (2014) doi:10.1016/j.physletb.2014.09.052
[36] V. Vovchenko, V. Begun and M. Gorenstein, Phys. Rev. C **93**, no.6, 064906 (2016) doi:10.1103/PhysRevC.93.064906
[37] L. Adamczyk *et al.* [STAR], Phys. Rev. C **96**, no.4, 044904 (2017) doi:10.1103/PhysRevC.96.044904

Exploring the partonic phase at finite chemical potential in and out-of equilibrium

O. Soloveva [1], P. Moreau [2], L. Oliva [1], T. Song [3], Cassing [4] and E. Bratkovskaya [1,3],

[1] Institute for Theoretical Physics, University of Frankfurt, Frankfurt, Germany
[2] Department of Physics, Duke University, Durham, NC 27708, USA
[3] Institute for Theoretical Physics, University of Giessen, Giessen, Germany
[3] GSI Helmholtzzentrum für Schwerionenforschung GmbH, Darmstadt, Germany

E-mail: E.Bratkovskaya@gsi.de

Abstract. We study the influence of the baryon chemical potential μ_B on the properties of the Quark–Gluon–Plasma (QGP) in and out-of equilibrium. The description of the QGP in equilibrium is based on the effective propagators and couplings from the Dynamical QuasiParticle Model (DQPM) that is matched to reproduce the equation-of-state of the partonic system above the deconfinement temperature T_c from lattice Quantum Chromodynamics (QCD). We calculate the transport coefficients such as the ratio of shear viscosity η and bulk viscosity ζ over entropy density s, i.e., η/s and ζ/s in the (T, μ_B) plane and compare to other model results available at $\mu_B = 0$. The out-of equilibrium study of the QGP is performed within the Parton–Hadron–String Dynamics (PHSD) transport approach extended in the partonic sector by explicitly calculating the total and differential partonic scattering cross sections (based on the DQPM propagators and couplings) evaluated at the actual temperature T and baryon chemical potential μ_B in each individual space-time cell of the partonic scattering. The traces of their μ_B dependences are investigated in different observables for relativistic heavy-ion collisions with a focus on the directed and elliptic flow coefficients v_1, v_2 in the energy range 7.7 GeV $\leq \sqrt{s_{NN}} \leq$ 200 GeV.

1. Introduction

The phase diagram of matter is one of the most fascinating subjects in physics, which also has important implications on chemistry and biology. Its phase boundaries and (possibly) critical points have been the focus of physics research for centuries. Apart from the traditional phase diagram in the plane of temperature T and pressure P, its transport properties like the shear and bulk viscosities, the electric conductivity, etc., are also of fundamental interest. These transport coefficients emerge from the stationary limit of correlators and provide additional information on the systems in thermal and chemical equilibrium apart from the equation of state (EoS). In particular the phase diagram (PD) of strongly interacting matter has been the topic of interest for the last decades and substantial experimental and theoretical efforts have been invested to shed light on this issue. The PD contains the information about the properties of our universe from the early beginning—directly after the Big Bang—when the matter was in the quark-gluon plasma (QGP) phase at very high temperature T and practically zero baryon chemical potential μ_B, to the later stages of the universe, where in the expansion phase stars and Galaxies have been formed. Here, the matter is at low temperature, however, large baryon

Journal of Physics: Conference Series **1602** (2020) 012012 doi:10.1088/1742-6596/1602/1/012012

chemical potential. Relativistic and ultra-relativistic heavy-ion collisions (HICs) nowadays offer the unique possibility to study some of these phases, in particular a QGP phase and its phase boundary to the hadronic one. We point out that the phase diagram of strongly interacting matter in the (T, μ_B) plane can also be explored in the astrophysical context at moderate temperatures and high μ_B [1], i.e. in the dynamics of supernovae or—more recently—in the dynamics of neutron-star merges.

In case of relativistic heavy-ion collisions one generates hot and dense matter in the laboratory although within small space-time regions. Whereas in low energy collisions one produces dense nuclear matter with moderate temperature T and large baryon chemical potential μ_B, ultra-relativistic collisions at Relativistic Heavy Ion Collider (RHIC) or Large Hadron Collider (LHC) energies produce extremely hot matter at small baryon chemical potential. In order to explore the phase diagram of strongly interacting matter as a function of T and μ_B both type of collisions are mandatory. According to lattice calculations of quantum chromodynamics (lQCD) [2, 3, 4, 5], the phase transition from hadronic to partonic degrees of freedom (at vanishing baryon chemical potential μ_B=0) is a crossover. This transition is expected to turn into a first order transition at some critical point (T_r, μ_{cr}) in the phase diagram with increasing baryon chemical potential μ_B. Since this critical point cannot be determined theoretically in a reliable way the beam energy scan (BES) program at RHIC aims to find the critical point and the phase boundary by gradually decreasing the collision energy [6, 7]. Furthermore, new facilities such as FAIR (Facility for Antiproton and Ion Research) and NICA (Nuclotron-based Ion Collider fAcility) are under construction to explore in particular the intermediate energy regime where one might study also the competition between chiral symmetry restoration and deconfinement as suggested in Refs. [8, 9].

Current methods to explore QCD in Minkowski space for non-vanishing quark (or baryon) densities (or chemical potential) are effective approaches. Using effective models, one can study the properties of QCD in equilibrium, i.e., thermodynamic quantities as well as transport coefficients. To this aim, the dynamical quasiparticle model (DQPM) has been introduced [10, 11, 12, 13, 14], which is based on partonic propagators with sizeable imaginary parts of the self-energies incorporated. Whereas the real part of the self-energies can be attributed to a dynamically generated mass (squared), the imaginary parts contain the information about the interaction rates of the degrees-of-freedom. Furthermore, the imaginary parts of the propagators define the spectral functions of the 'particles' which might show narrow (or broad) quasiparticle peaks. A further advantage of a propagator based approach is that one can formulate a consistent thermodynamics [15] as well as a causal theory for non-equilibrium configurations on the basis of Kadanoff–Baym equations [16].

Since relativistic heavy-ion collisions start with impinging nuclei in their groundstates, a proper non-equilibrium description of the entire dynamics through possibly different phases up to the final asymptotic hadronic states—eventually showing some degree of equilibration—is mandatory. To this aim, the Parton–Hadron–String Dynamics (PHSD) transport approach [13, 17, 18, 19, 20] has been formulated more then a decade ago (on the basis of the Hadron-String-Dynamics (HSD) approach [21]), and it was found to well describe observables from p+A and A+A collisions from SPS to LHC energies including electromagnetic probes such as photons and dileptons as well as open cham hadrons [13, 22, 23]. In order to explore the partonic systems at higher μ_B, the PHSD approach has been recently extended to incorporate partonic quasiparticles and their differential cross sections that depend not only on temperature T as in the previous PHSD studies, but also on chemical potential μ_B explicitly [24]. Within this extended approach we have previously studied the 'bulk' observables in HICs from AGS to RHIC energies for symmetric and asymmetric systems. However, we have found only a small influence of μ_B dependences of parton properties (masses and widths) and their interaction cross sections in the bulk observables [24].

Journal of Physics: Conference Series **1602** (2020) 012012 doi:10.1088/1742-6596/1602/1/012012

In this contribution we extend our study with respect to the collective flow (v_1, v_2) coefficients for different identified hadrons and their sensitivity to the μ_B dependences of partonic cross sections.

2. The PHSD Approach

We start with recalling the basic ideas of the PHSD transport approach and the dynamical quasiparticle model (DQPM). The Parton–Hadron–String Dynamics (PHSD) transport approach [13, 17, 18, 19, 20] is a microscopic off-shell transport approach for the description of strongly interacting hadronic and partonic matter in and out-of equilibrium. It is based on the solution of Kadanoff–Baym equations in first-order gradient expansion [18] employing 'resummed' propagators from the DQPM [10, 11, 12] for the partonic phase. The DQPM, furthermore, has been introduced in Refs. [10, 11, 12] for the effective description of the QGP in terms of strongly interacting quarks and gluons with properties and interactions which are adjusted to reproduce lQCD results for the equilibrated QGP at finite temperature T and baryon (or quark) chemical potential μ_B. In the DQPM, the quasiparticles are characterized by single-particle Green's functions (in propagator representation) with complex self-energies. The real part of the self-energies is related to the dynamically generated parton masses (squared), whereas the imaginary part provides information about the lifetime and/or reaction rates of the degrees-of-freedom.

In PHSD, the partons (quarks and gluons) are characterized by broad spectral functions ρ_j ($j = q, \bar{q}, g$) and thus are off-shell contrary to the conventional cascade or transport models dealing with on-shell particles, i.e., the δ-functions in the invariant mass squared. The quasiparticle spectral functions are assumed to have a Lorentzian form [13], which are specified by the parton masses and width parameters:

$$\rho_j(\omega, \mathbf{p}) = \frac{\gamma_j}{E_j}\left(\frac{1}{(\omega - E_j)^2 + \gamma_j^2} - \frac{1}{(\omega + E_j)^2 + \gamma_j^2}\right) \tag{1}$$

separately for quarks/antiquarks and gluons ($j = q, \bar{q}, g$). With the convention $E^2(\mathbf{p}^2) = \mathbf{p}^2 + M_j^2 - \gamma_j^2$, the parameters M_j^2 and γ_j are directly related to the real and imaginary parts of the retarded self-energy, e.g., $\Pi_j = M_j^2 - 2i\gamma_j\omega$.

The actual parameters in Eq. (1), i.e. the gluon mass M_g and width γ_g—employed as input in the present PHSD calculations—as well as the quark mass M_q and width γ_q, are depicted in Fig. 1 as a function of the temperature T and baryon chemical potential μ_B (from Ref. [25]). These values for the masses and widths have been fixed by fitting the lattice QCD results from Ref. [26, 27] in thermodynamic equilibrium. One can see that the masses of quarks and gluons decrease with increasing μ_B, and a similar trend holds for the widths of these partons.

A scalar mean-field $U_s(\rho_s)$ for quarks and antiquarks can be defined by the derivative of the potential energy density with respect to the scalar density $\rho_s(T, \mu_B)$,

$$U_s(\rho_s) = \frac{dV_p(\rho_s)}{d\rho_s}, \tag{2}$$

which is evaluated numerically within the DQPM. Here, the potential energy density is defined by

$$V_p(T, \mu_B) = T^{00}_{g-}(T, \mu_B) + T^{00}_{q-}(T, \mu_B) + T^{00}_{\bar{q}-}(T, \mu_B), \tag{3}$$

where the different contributions T^{00}_{j-} correspond to the space-like part of the energy-momentum tensor component T^{00}_j of parton $j = g, q, \bar{q}$ (cf. Section 3 in Ref. [11]). The scalar mean-field $U_s(\rho_s)$ for quarks and antiquarks is repulsive as a function of the parton scalar density ρ_s and shows that the scalar mean-field potential is in the order of a few GeV for $\rho_s > 10$ fm^{-3}. The

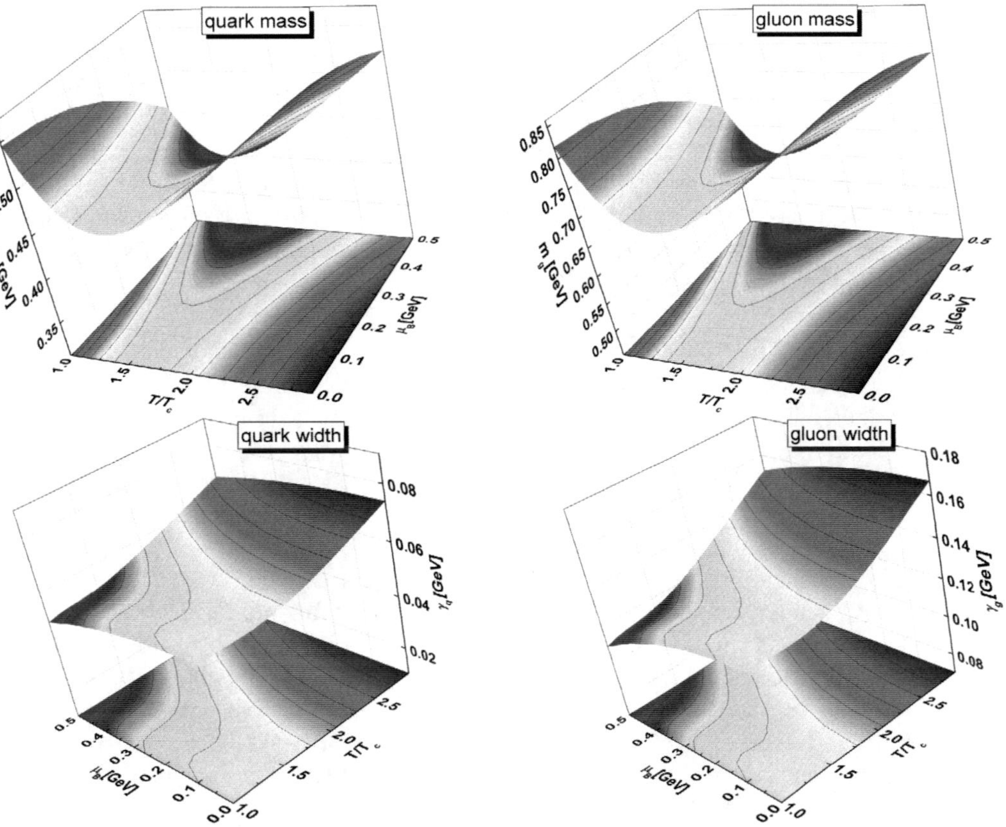

Figure 1. The effective quark (**left**) and gluon (**right**) masses M (**upper row**) and widths γ (**lower row**) from the actual DQPM as a function of the temperature T and baryon chemical potential μ_B [25].

mean-field potential (2) is employed in the PHSD transport calculations and determines the force on a partonic quasiparticle j, i.e., $\sim M_j/E_j \nabla U_s(x) = M_j/E_j\ dU_s/d\rho_s\ \nabla\rho_s(x)$, where the scalar density $\rho_s(x)$ is determined numerically on a space-time grid in PHSD.

Furthermore, a two-body interaction strength can be extracted from the DQPM as well from the quasiparticle widths in line with Ref. [10]. On the partonic side, the following elastic and inelastic interactions are included in the latest version of PHSD (v. 5.0) $qq \leftrightarrow qq$, $\bar{q}\bar{q} \leftrightarrow \bar{q}\bar{q}$, $gg \leftrightarrow gg$, $gg \leftrightarrow g$, $q\bar{q} \leftrightarrow g$, $qg \leftrightarrow qg$, $g\bar{q} \leftrightarrow g\bar{q}$ exploiting 'detailed-balance' with cross sections calculated from the leading order Feynman diagrams employing the effective propagators and couplings $g^2(T/T_c)$ from the DQPM [24]. In Ref. [24], the differential and total off-shell cross sections have been evaluated as a function of the invariant energy of colliding off-shell partons $\sqrt{s}$ for each T, μ_B. We recall that in the previous PHSD studies (using version 4.0 and below) the cross sections depend only on T (cf. the detailed evaluation in Ref. [28]). When implementing the differential cross sections and parton masses into the PHSD5.0 approach, one has to specify the 'Lagrange parameters' T and μ_B in each computational cell in space-time. This has been done by employing the lattice equation of state and a diagonalization of the energy-momentum tensor from PHSD as described in Ref. [24].

The transition from partonic to hadronic degrees-of-freedom (and vice versa) is described by covariant transition rates for the fusion of quark–antiquark pairs or three quarks (antiquarks), respectively, obeying flavor current–conservation, color neutrality as well as energy–momentum conservation [19]. Since the dynamical quarks and antiquarks become very massive close to the phase transition, the formed resonant 'prehadronic' color-dipole states ($q\bar{q}$ or qqq) are of high invariant mass, too, and sequentially decay to the groundstate meson and baryon octets, thus increasing the total entropy during hadronization.

On the hadronic side, PHSD includes explicitly the baryon octet and decouplet, the 0^-- and 1^--meson nonets as well as selected higher resonances as in the Hadron–String–Dynamics (HSD) approach [21]. Note that PHSD and HSD (without explicit partonic degrees-of-freedom) merge at low energy density, in particular below the local critical energy density $\varepsilon_c \approx 0.5$ GeV/fm^3 as extracted from the lQCD results in Ref. [26, 27].

3. Transport Coefficients

The transport properties of the QGP close to equilibrium can be characterized by various transport coefficients. The shear viscosity η and bulk viscosity ζ describe the fluid's dissipative corrections at leading order. Both coefficients are generally expected to depend on the temperature T and baryon chemical potential μ_B. In the hydrodynamic equations, the viscosities appear as dimensionless ratios, η/s and ζ/s, where s is the fluid entropy density. Such specific viscosities are more meaningful than the unscaled η and ζ values because they describe the magnitude of stresses inside the medium relative to its natural scale.

In our recent studies [24, 29, 30], we have investigated the transport properties of the QGP in the (T, μ_B) plane based on the DQPM. One way to evaluate the viscosity coefficients of partonic matter is the Kubo formalism [31, 32, 33, 34], which was used to calculate the viscosities for a previous version of the DQPM within the PHSD transport approach in a box with periodic boundary conditions in Ref. [35] as well as in the more recent study with the DQPM model in Refs. [24, 29]. Another way to calculate transport coefficients (explored also in [24, 29]) is to use the relaxation–time approximation (RTA) as incorporated in Refs. [36, 37, 38, 39].

The shear viscosity based on the RTA (cf. [40]) reads as:

$$\eta^{RTA}(T, \mu_q) = \frac{1}{15T} \sum_{i=q,\bar{q},g} \int \frac{d^3p}{(2\pi)^3} \frac{\mathbf{p}^4}{E_i^2} \tau_i(\mathbf{p}, T, \mu)\ d_i(1 \pm f_i) f_i, \tag{4}$$

where $d_q = 2N_c = 6$ and $d_g = 2(N_c^2 - 1) = 16$ are degeneracy factors for spin and color in case of quarks and gluons , whereas τ_i are relaxation times for particles i. Equation (4) includes the Bose enhancement and Pauli-blocking factors, respectively, which are taken into account in the actual calculations. The pole energy is $E_i^2 = p^2 + M_i^2$, where M_i is the pole mass from the DQPM. The notation $\sum_{j=q,\bar{q},g}$ includes the contribution from all possible partons which in our case are the gluons and the (anti-)quarks of three different flavors (u, d, s).

We consider two cases for the relaxation time for quarks and gluons: (1) $\tau_i(\mathbf{p}, T, \mu) = 1/\Gamma_i(\mathbf{p}, T, \mu)$ and (2) $\tau_i(T, \mu) = 1/2\gamma_i(T, \mu)$, where $\Gamma_i(\mathbf{p}, T, \mu)$ is the parton interaction rate, calculated microscopically from the collision integral using the differential cross sections for parton scattering from Ref. [24], while $\gamma_i(T, \mu)$ are the width parameters in the parton propagators (1).

In the left part of Fig. 2 we show the ratio of the shear viscosity to entropy density as a function of the scaled temperature T/T_c for $\mu_B = 0$ calculated within the Kubo formalism (green solid line) and RTA approach with the interaction rate $\Gamma^{\rm on}$ (red solid line) and the DQPM width 2γ (dashed green line). The RTA approximation (4) of the shear viscosity with the DQPM width 2γ and with the interaction rate $\Gamma^{\rm on}$ are quite close to each other at $\mu_B = 0$ and also very close to the result from the Kubo formalism [24] indicating that the quasiparticle limit ($\gamma \ll M$) holds in the DQPM.

The ratio η/s increases with the scaled temperature. The actual values for the ratio η/s are in a good agreement with the gluodynamic lattice QCD calculations at $\mu_B = 0$ from Ref. [42].

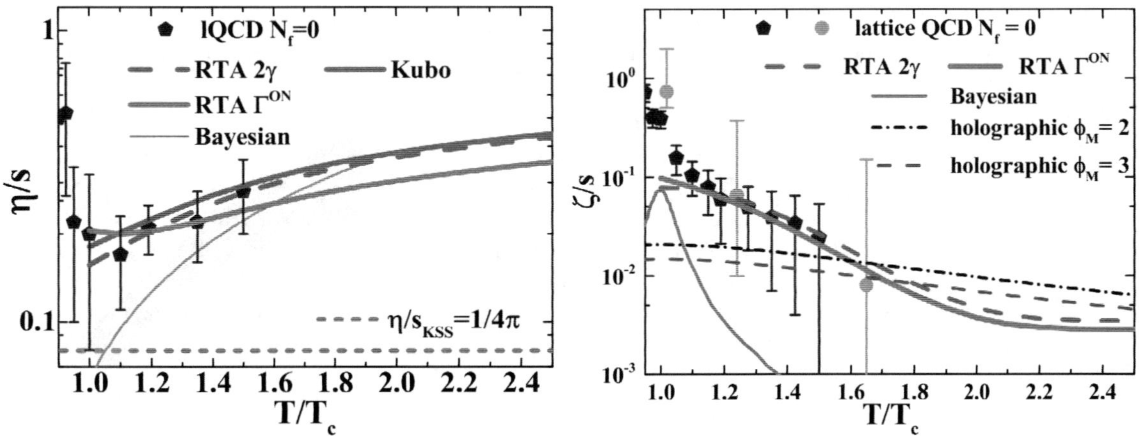

Figure 2. (**lhs**): the ratio of shear viscosity to entropy density η/s as a function of the scaled temperature T/T_c for $\mu_B = 0$ calculated within the Kubo formalism (green solid line) and the RTA approach with the interaction rate $\Gamma^{\rm on}$ (red solid line) and the DQPM width 2γ (dashed green line). The dashed gray line demonstrates the Kovtun–Son–Starinets bound [41] $(\eta/s)_{\rm KSS} = 1/(4\pi)$, and the symbols show lQCD data for pure SU(3) gauge theory taken from Ref. [42] (pentagons). The solid blue line shows the results from a Bayesian analysis of experimental heavy-ion data from Ref. [44]. (**rhs**): the ratio of the bulk viscosity to entropy density ζ/s as a function of the scaled temperature T/T_c for $\mu_B = 0$ calculated using the RTA approach with the on-shell interaction rate $\Gamma^{\rm on}$ (red solid line) and the DQPM width 2γ (dashed green line). The symbols correspond to the lQCD data for pure SU(3) gauge theory taken from Refs. [43] (pentagons) and [45] (circles). The solid blue line shows the results from a Bayesian analysis of experimental heavy-ion data from Ref. [44]. The dot-dashed and dashed lines correspond to the results from the non-conformal holographic model for $\phi_M = 2$ and 3, correspondingly, from Ref. [47].

Moreover, our DQPM results are in qualitative agreement with the results from a Bayesian analysis of experimental heavy-ion data from Ref. [44]. We mention that the DQPM result differs from the recent calculations for the shear viscosity at $\mu_B = 0$ in the quasiparticle model in Ref. [46] where the width of quasiparticles is not considered, which leads to a higher value for the η/s ratio. This shows the sensitivity of this ratio to the modelling of partonic interactions and the properties of partons in the hot QGP medium. We recall also that in Refs. [24, 29, 30] we have found that the ratio η/s shows a very weak dependence on μ_B and has a similar behavior as a function of temperature for all $\mu_B \leq 400$ MeV.

The expression for the bulk viscosity in the partonic phase - derived within the RTA - reads (following Ref. [38])

$$\zeta^{RTA}(T,\mu) = \frac{1}{9T} \sum_{i=q,\bar{q},g} \int \frac{d^3p}{(2\pi)^3} \tau_i(\mathbf{p},T,\mu)\ \frac{d_i(1\pm f_i)f_i}{E_i^2} \left(\mathbf{p}^2 - 3c_s^2 \left(E_i^2 - T^2 \frac{dm_i^2}{dT^2}\right)\right)^2, \quad (5)$$

where c_s^2 is the speed of sound squared, and $\frac{dm_i^2}{dT^2}$ is the DQPM parton mass derivative which becomes large close to the critical temperature T_c.

On the right side of Fig. 2 we show the ratio of the bulk viscosity to entropy density ζ/s as a function of the scaled temperature T/T_c for $\mu_B = 0$ calculated by the RTA approach with the interaction rate $\Gamma^{\rm on}$ (red solid line) and the DQPM width 2γ (dashed green line). The symbols correspond to the lQCD data for pure SU(3) gauge theory taken from Refs. [43] (pentagons)

Journal of Physics: Conference Series **1602** (2020) 012012 doi:10.1088/1742-6596/1602/1/012012

and [45] (circles). The solid blue line shows the results from a Bayesian analysis of experimental heavy-ion data from Ref. [44]. The dot-dashed and dashed lines correspond to the results from the non-conformal holographic model [47] for $\phi_M = 2$ and 3, correspondingly, where ϕ_M is the model parameter which characterizes the non-conformal features of the model. We find that the DQPM result for ζ/s is in good agreement with the lattice QCD results and shows a rise closer to T_c contrary to the holographic results, which show practically a constant behavior independent of model parameters. This rise is attributed to the increase of the partonic mass closer to T_C as shown in Figure 1, thus the mass derivative term in Eq. (5) also grows. The Bayesian result also shows a peak near T_C; however, the ratio drops to zero while lQCD data indicate a positive ζ/s as found also in the DQPM. The explicit μ_B dependence of ζ/s has been investigated within the DQPM in Refs. [24, 29, 30], where it has been shown that it is rather weak for $\mu_B \leq 400$ MeV.

As follows from hydrodynamical calculations, the results for the flow harmonics v_n are sensitive to the transport coefficients [44, 48, 49, 50]. Thus, there are hopes to observe a μ_B sensitivity of v_1, v_2. We will study this sensitivity in the following Section.

4. Heavy-Ion Collisions

In our recent study [24] we have investigated the sensitivity of 'bulk' observables such as rapidity and transverse momentum distributions of different hadrons produced in heavy-ion collisions from AGS to top RHIC energies on the details of the QGP interactions and the properties of the partonic degrees-of-freedom. For that, we have considered the following three cases:

(1) **'PHSD4.0'**: the masses and widths of quarks and gluons depend only on T. The cross sections for partonic interactions depend only on T as evaluated by the 'box' calculations in Ref. [28] in order to merge the QGP interaction rates from all possible partonic channels to the total temperature dependent widths $2\gamma_i$ of the DQPM propagator. This has been used in the PHSD code (version 4.0 or below) for extended studies of many hadronic observables in p+A and A+A collisions at different energies [13, 17, 18, 19, 20, 51] with good success.

(2) **'PHSD5.0 - $\mu_B = 0$'**: the masses and widths of quarks and gluons depend only on T; however, the differential and total partonic cross sections are obtained by calculations of the leading order Feynman diagrams employing the effective propagators and couplings $g^2(T/T_c)$ from the DQPM at $\mu_B = 0$ [24]. Thus, the cross sections depend explicitly on the invariant energy of the colliding partons $\sqrt{s}$ and on T. This is realized in the PHSD5.0 by setting $\mu_B = 0$ (cf. [24]).

(3) **'PHSD5.0 - μ_B'**: the masses and widths of quarks and gluons depend on T and μ_B explicitly; the differential and total partonic cross sections are obtained by calculations of the leading order Feynman diagrams from the DQPM and explicitly depend on invariant energy $\sqrt{s}$, temperature T and baryon chemical potential μ_B. This is realized in the full version of PHSD5.0 (cf. [24]).

The comparison of the 'bulk' observables for A+A collisions within the three cases of PHSD in Ref. [24] has illuminated that they show a very low sensitivity to the μ_B dependences of parton properties (masses and widths) and their interaction cross sections such that the results from PHSD5.0 with and without μ_B were very close to each other. Only in the case of kaons, antiprotons $\bar{p}$ and antihyperons $\bar{\Lambda} + \bar{\Sigma}^0$, a small difference between PHSD4.0 and PHSD5.0 could be seen at top SPS and top RHIC energies. A similar trend has been found for very asymmetric collisions of C+Au: a small sensitivity to the partonic scatterings was found in the kaon and antibaryon rapidity distributions, too. This can be understood as following: at high energies such as top RHIC where the QGP volume is very large in central collisions, the μ_B is very low, while, when decreasing the bombarding energy (or increasing μ_B) the fraction of the QGP is decreasing such that the final observables are dominated by the hadronic phase, i.e., the probability for the hadrons created at the QGP hadronization to rescatter, decay, or be

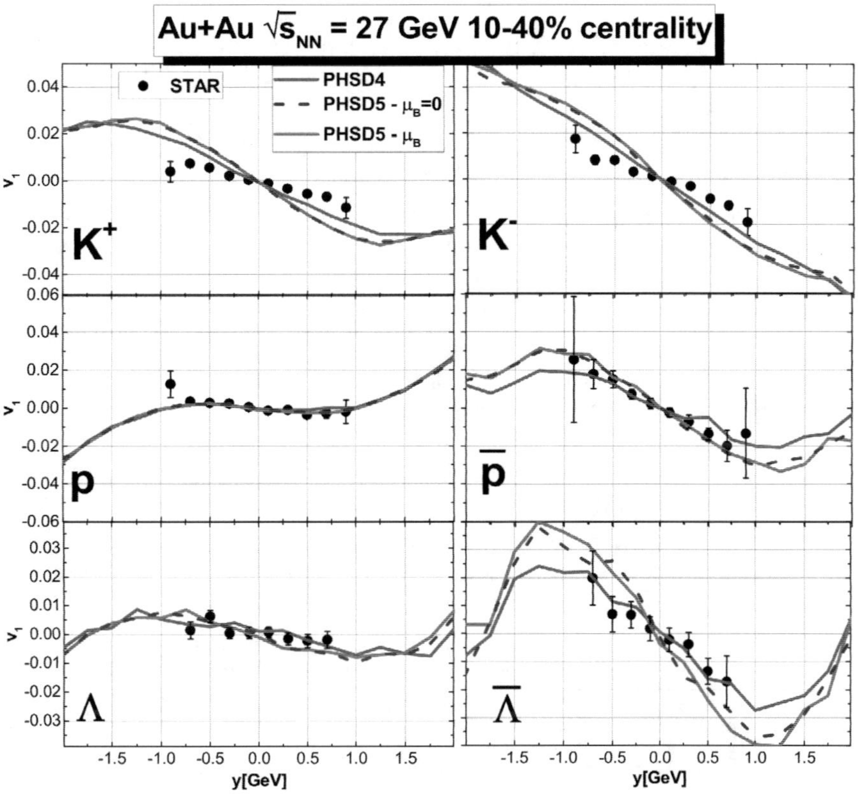

Figure 3. Directed flow of identified hadrons as a function of rapidity for Au+Au collisions at $\sqrt{s_{NN}} = 27$ GeV for PHSD4.0 (green lines), PHSD5.0 with partonic cross sections and parton masses calculated for $\mu_B = 0$ (blue dashed lines) and with cross sections and parton masses evaluated at the actual chemical potential μ_B in each individual space-time cell (red lines) in comparison to the experimental data of the STAR Collaboration [52].

absorbed in hadronic matter increases strongly; as a result the sensitivity to the properties of the QGP is washed out.

4.1. Directed Flow

We here test the traces of μ_B dependences of the QGP interaction cross sections in collective observables such as the directed flow v_1 considering again the three cases for the PHSD as discussed above. We directly report about the actual PHSD results. Fig. 3 depicts the directed flow v_1 of identified hadrons ($K^{\pm}, p, \bar{p}, \Lambda + \Sigma^0, \bar{\Lambda} + \bar{\Sigma}^0$) versus rapidity for Au+Au collisions at $\sqrt{s_{NN}} = 27$ GeV. One can see a good agreement between PHSD results and experimental data from the STAR collaboration [52]. However, the different versions of PHSD for the v_1 coefficients show a quite similar behavior; only antihyperons indicate a slightly different flow. This supports again the finding that strangeness, and in particular anti-strange hyperons, are the most sensitive probes for the QGP properties. Surprizingly, PHSD4.0 performs better for the directed flows of hadrons than PHSD5.0 with the microscopic differential partonic cross sections.

4.2. Elliptic Flow

Hydrodynamic simulations [49, 50] and the Bayesian analysis [44] indicate that the elliptic flow v_2 is sensitive to the transport properties of the QGP as characterized by transport coefficients such as shear η and bulk ζ viscosities. In this subsection we present the results for the elliptic flow of charged hadrons from HICs within the PHSD5.0 with and without μ_B dependence and compare the results with those from PHSD4.0 as before.

In Fig. 4 we display the elliptic flow v_2 of identified hadrons ($K^{\pm}, p, \bar{p}, \Lambda + \Sigma^0, \bar{\Lambda} + \bar{\Sigma}^0$) as a function of p_T at $\sqrt{s_{NN}} = 27$ GeV for PHSD4.0 (green lines), PHSD5.0 with partonic cross sections and parton masses calculated for $\mu_B = 0$ (blue dashed lines) and with cross sections and parton masses evaluated at the actual chemical potential μ_B in each individual space-time cell (red lines) in comparison to the experimental data of the STAR Collaboration [53]. Similar to the directed flow shown in Figure 3, the elliptic flow from all three cases for PHSD shows a rather similar behavior; the differences are very small (within the statistics achieved here). Only antiprotons and antihyperons show a small decrease of v_2 at larger p_T for PHSD5.0 compared to PHSD4.0, which can be attributed to the explicit $\sqrt{s}$-dependence and different angular distribution of partonic cross sections in the PHSD5.0. We note that the underestimation of v_2 for protons and Λ's might be attributed to the details of the hadronic vector potentials involved in this calculations which seem to underestimate the baryon repulsion.

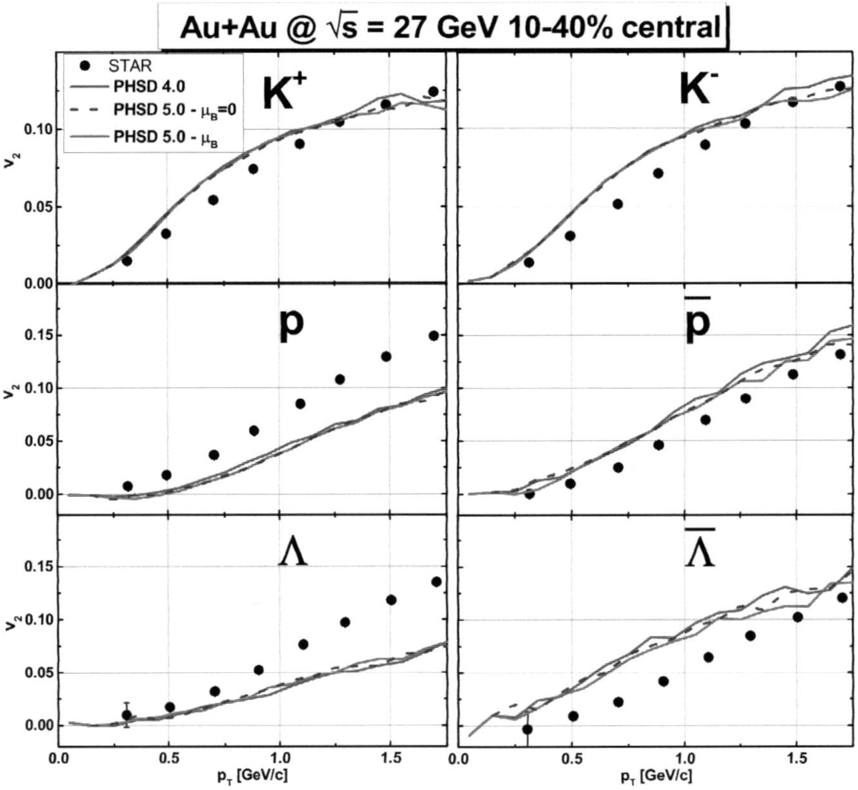

Figure 4. Elliptic flow of identified hadrons ($K^{\pm}, p, \bar{p}, \Lambda + \Sigma^0, \bar{\Lambda} + \bar{\Sigma}^0$) as a function of p_T for Au+Au collisions at $\sqrt{s_{NN}} = 27$ GeV for PHSD4.0 (green lines), PHSD5.0 with partonic cross sections and parton masses calculated for $\mu_B = 0$ (blue dashed lines) and with cross sections and parton masses evaluated at the actual chemical potential μ_B in each individual space-time cell (red lines) in comparison to the experimental data of the STAR Collaboration [53].

5. Conclusions

In this contribution we have reported on the influence of the baryon chemical potential μ_B on the properties of the QGP in equilibrium as well as the QGP created in heavy-ion collisions initially also far from equilibrium. For the description of the QGP we have employed the extended Dynamical QuasiParticle Model (DQPM) that is matched to reproduce the lQCD equation-of-state (versus temperature T) at zero and finite baryon chemical potential μ_B. The DQPM results for transport coefficients, such as shear viscosity η and bulk viscosity ζ, have been compared to available lQCD data, the non-conformal holographic model at $\mu_B = 0$ and with results from a Bayesian analysis of experimental heavy-ion data. We find that the ratios η/s and ζ/s from the DQPM agree very well with the lQCD results from Ref. [43] and show a similar behavior as the ratio obtained from a Bayesian fit [44]. As found previously in Refs. [24, 29] the transport coefficients show only a mild dependence on μ_B.

Our study of the non-equilibrium QGP—as created in relativistic heavy-ion collisions—has been performed within the extended Parton–Hadron–String Dynamics (PHSD) transport approach [24] in which i) the masses and widths of quarks and gluons depend on T and μ_B explicitly; ii) the partonic interaction cross sections are obtained by the leading order Feynman diagrams from the DQPM effective propagators and couplings and explicitly depend on the invariant energy $\sqrt{s}$, temperature T and baryon chemical potential μ_B. This extension is realized in the version PHSD5.0 [24]. In order to investigate the traces of the μ_B dependence of the QGP in observables, the results of PHSD5.0 (with μ_B dependences) have been compared to the results of PHSD5.0 for $\mu_B = 0$ as well as with PHSD4.0, where the masses/widths of quarks and gluons as well as their interaction cross sections depend only on T (as specified in Ref. [28]).

Since the sensitivity (w.r.t. the μ_B-dependence) of hadronic rapidity and p_T distributions of identified hadrons for symmetric and asymmetric heavy-ion collisions from AGS to RHIC energies was found to be very low [24, 29] we have focused on the related sensitivity of the collective flow of hadrons. As a characteristic example we have shown (i) the directed flow v_1 of identified hadrons and (ii) the elliptic flow v_2 of identified hadrons for $Au + Au$ collisions at the invariant energy $\sqrt{s_{NN}} = 27$ GeV. We find only very small differences between the results from PHSD4.0 and from PHSD5.0 on the hadronic flow observables at high as well as at intermediate energies. This is related to the fact that at high energies, where the matter is dominated by the QGP, one probes only a small baryon chemical potential in central collisions at midrapidity, while with decreasing energy (and larger μ_B) the fraction of the QGP drops rapidly, such that in total the final observables are dominated by the hadronic interactions and thus the information about the partonic properties and scatterings is washed out. We have shown that the mild μ_B dependence of QGP interactions is more pronounced in observables for strange hadrons (kaons and especially anti-strange hyperons) which provides an experimental hint for the search of μ_B traces of the QGP for experiments at the future FAIR/NICA facilities and the BESII program at RHIC.

Acknowledgements

The authors acknowledge inspiring discussions with Jörg Aichelin, Vadim Kolesnikov, Ilya Selyuzhenkov and Arkadiy Taranenko. Furthermore, they thank Maximillian Attems for providing the results from Ref. [46] in data form. This work was supported by the LOEWE center "HIC for FAIR", the HGS-HIRe for FAIR and the COST Action THOR, CA15213. LO is supported by the Alexander von Humboldt-Stiftung. Furthermore, PM and EB acknowledge support by DFG through the grant CRC-TR 211 'Strong-interaction matter under extreme conditions'. The computational resources have been provided by the LOEWE-Center for Scientific Computing.

Journal of Physics: Conference Series **1602** (2020) 012012 doi:10.1088/1742-6596/1602/1/012012

References

[1] Klahn T, Blaschke D, Typel S, Van Dalen E, Faessler A, Fuchs C, Gaitanos T, Grigorian H, Ho A, Kolomeitsev E *et al.* 2006 *Phys. Rev.* C **74** 035802
[2] Bernard C *et al.* [MILC Collaboration] 2005 *Phys. Rev.* D **71** 034504
[3] Aoki Y, Endrodi G, Fodor Z, Katz S and D Szabo K K 2006 *Nature* **443** 675
[4] Aoki Y *et al.* 2009 *JHEP* **0906** 088
[5] Bazavov A *et al.* 2012 *Phys. Rev.* D **85** 054503
[6] Mohanty B [STAR Collaboration] 2011 *J. Phys.* G **38** 124023
[7] Kumar L [STAR Collaboration] 2011 *J. Phys.* G **38** 124145
[8] Cassing W, Palmese A, Moreau P and Bratkovskaya E L 2016 *Phys. Rev.* C **93** 014902
[9] Palmese A Cassing W Seifert E Steinert T, Moreau P and Bratkovskaya E L 2016 *Phys. Rev.* C **94** 044912
[10] Peshier A, Cassing W 2005 *Phys. Rev. Lett.* **94** 172301
[11] Cassing W 2007 *Nucl. Phys.* A **795** 70
[12] Cassing W 2007 *Nucl. Phys.* A **791** 365
[13] Linnyk O, Bratkovskaya E, Cassing W 2016 *Prog. Part. Nucl. Phys.* **87** 50
[14] Berrehrah H, Bratkovskaya E, Steinert,T, Cassing W 2016 *Int. J. Mod. Phys.* E **25** 1642003
[15] Vanderheyden B, Baym G 1998 *J. Stat. Phys.* **93** 843
[16] Kadanoff L P, Baym G 1962 *Quantum Statistical Mechanics*; W. A. Benjamin, Inc.: New York, NY, USA
[17] Cassing W, Bratkovskaya E L 2008 *Phys. Rev.* C **78** 034919
[18] Cassing W 2009 *Eur. Phys. J. Spec. Top.***168** 3
[19] Cassing W, Bratkovskaya E L 2009 *Nucl. Phys.* A **831** 215
[20] Bratkovskaya E L, Cassing W, Konchakovski V P, Linnyk O 2011 *Nucl. Phys.* A **856** 162
[21] Cassing W, Bratkovskaya E L 1999 *Phys. Rep.* **308** 65
[22] Linnyk O, Konchakovski V P, Cassing W, Bratkovskaya E L 2013 em Phys. Rev. C **88** 034904
[23] Song T, Cassing W, Moreau P, Bratkovskaya E L 2018 *Phys. Rev.* C **97** 064907
[24] Moreau P, Soloveva O, Oliva L, Song T, Cassing W, Bratkovskaya E 2019 *Phys. Rev.* C **100** 014911
[25] Soloveva O, Moreau P, Oliva L, Voronyuk, V Kireyeu, V *et al.* 2020 *Particles* **3** 178
[26] Borsanyi S, Endrodi G, Fodor Z, Katz S D, Krieg S, Ratti C, Szabo K K 2012 *J. High Energy Phys.* **1208** 053
[27] Borsanyi S, Fodor Z, Hoelbling C, Katz S D, Krieg S, Szabo K K 2014 *Phys. Lett.* B **730** 99
[28] Ozvenchuk V, Linnyk O, Gorenstein M I, Bratkovskaya E L, Cassing W 2013 *Phys. Rev.* C **87** 024901
[29] Soloveva O, Moreau P, Bratkovskaya E 2020 *Phys. Rev.* C **101** 045203
[30] Soloveva O, Moreau P, Oliva L, Song T, Cassing W, Bratkovskaya E 2019 *arXiv*:1911.03131
[31] Kubo R 1957 *J. Phys. Soc. Jpn.* **12** 570
[32] Aarts G, Martinez Resco J M 2002 *J. High Energy Phys.* **204** 053
[33] Fernandez-Fraile D, Gomez Nicola A 2006 *Phys. Rev.* D **73** 045025
[34] Lang R, Kaiser N, Weise W 2015 *Eur. Phys. J.* A **51** 127
[35] Ozvenchuk V, Linnyk O, Gorenstein M I, Bratkovskaya E L, Cassing W 2013 *Phys. Rev.* C **87** 064903
[36] Hosoya A, Kajantie K 1985 *Nucl. Phys.* B **250** 666
[37] Chakraborty P, Kapusta J I 2011 *Phys. Rev.* C **83** 014906
[38] Albright M, Kapusta J I 2016 *Phys. Rev.* C **93** 014903
[39] Gavin S 1985 *Nucl. Phys.* A **435** 826
[40] Sasaki C, Redlich K 2009 *Phys. Rev.* C **79** 055207
[41] Kovtun P K, Son D T, Starinets A O 2005 *Phys. Rev. Lett.* **94** 111601
[42] Astrakhantsev N, Braguta V, Kotov A 2017 *Phys. J. High Energy Phys.* **4** 101
[43] Astrakhantsev N Y, Braguta V V, Kotov A Y 2018 *Phys. Rev.* D **98** 054515
[44] Bernhard J E, Moreland J S, Bass S A, Liu J, Heinz U 2016 *Phys. Rev.* C **94** 024907
[45] Meyer H B 2008 *Phys. Rev. Lett.* **100** 162001
[46] Mykhaylova V, Bluhm M, Redlich K, Sasaki C 2019 *Phys. Rev.* D **100** 034002
[47] Attems M, Casalderrey-Solana J, Mateos D, Papadimitriou I, Santos-Oliván D, Sopuerta C F, Triana M, Zilhão Z 2016 *J. High Energy Phys.* **1610** 155
[48] Marty R, Bratkovskaya E, Cassing W, Aichelin J, Berrehrah H 2013 *Phys. Rev.* C **88** 045204
[49] Romatschke P, Romatschke U 2007 *Phys. Rev. Lett.* **99** 172301
[50] Song H, Heinz U W 2008 *Phys. Rev.* C **78** 024902
[51] Konchakovski V P, Bratkovskaya E L, Cassing W, Toneev V D, Voronyuk V 2012 *Phys. Rev.* C **85** 011902
[52] Adamczyk L, Adams J R, Adkins J K, Agakishiev G, Aggarwal M M, Ahammed Z, Ajitanand N N, Alekseev I, Anderson D M, Aoyama R *et al.* 2018 *Phys. Rev. Lett.* **120** 062301
[53] Adamczyk L, Adkins J K, Agakishiev G, Aggarwal M M, Ahammed Z, Alekseev I, Aparin A, Arkhipkin D, Aschenauer E C, Averichev G S *et al.* 2016 *Phys. Rev.* C **93** 014907

3-Dimensional QCD Phase Diagrams for Strange Matter

V. Dexheimer, K. Aryal, C. Constantinou, J. Peterson
Department of Physics, Kent State University, Kent, OH 44243 USA

R. L. S. Farias
Departamento de Física, Universidade Federal de Santa Maria, 97105-900 Santa Maria, RS, Brazil

E-mail: vdexheim@kent.edu

Abstract.
In this work, we examine in detail the difference between constraining the electric charge fraction and isospin fraction when calculating the deconfinement phase transition in the presence of net strangeness. We present relations among charge and isospin fractions and the corresponding chemical potentials and draw 3-dimensional QCD phase diagrams for matter out of weak equilibrium. Finally, we briefly discuss how our results can be applied to comparisons of matter created in heavy ion collisions and binary neutron star mergers.

1. Introduction and Formalism

When presenting phase diagrams for quantum chromodynamics (QCD), most studies only look in two dimensions at once; usually, temperature and either baryon chemical potential or isospin chemical potential. In the former case, the charged chemical potential is set to minus the electron chemical potential $\mu_Q = -\mu_e$ (to describe cold neutron stars [1]) or the isospin chemical potential is set to zero $\mu_I = 0$ (to describe very energetic relativistic heavy-ion collisions [2]). In the latter case, the baryon chemical potential is set to zero $\mu_B = 0$ [3]. See Refs. [4, 5] for details on changes that appear in 2-dimensional phase diagrams in which either the charge fraction is fixed at $Y_Q = 0.3$ or the lepton fraction at $Y_L = 0.4$.

Several studies have compared extreme astrophysical environments (e.g. binary neutron star mergers, core collapse supernovae, and proto-neutron stars) to heavy ion collisions (HIC) [6, 7], motivated by the high (compared to Fermi) temperatures generated in all cases. However, recent simulations [8] suggest that neutron star mergers are incapable of reaching the high charge fractions typically associated with HICs and young/hot neutron stars. Acknowledging this variety of conditions, we explore the consequences of varying the charge and isospin fractions and use either of these quantities as a third axis in the QCD phase diagram, focusing on changes to the first order phase transition coexistence line between the hadronic and the quark phases.

Complimentary to what was done recently in Ref. [9], where 3-dimensional phase diagrams were shown for non-strange matter, here we introduce 3-dimensional phase diagrams with net strangeness, a case consistent with astrophysical processes occurring over timescales much longer than those of weak interactions. We do not impose weak equilibrium, but vary freely the charged or isospin chemical potential of the system (depending on the respective chosen constrain, charge or isospin fraction) within physically meaningful values.

For this purpose, we make use of the Chiral Mean Field (CMF) formalism. It is a non-linear realization of the SU(3) linear sigma model for hadronic matter, which in its present version also contains contributions from quarks [1] [1]. As a result, this formalism reproduces chirally symmetric matter with quarks at large densities and/or temperatures, the phase transition being of first order, except when it becomes a crossover at large temperatures. The hadronic sector of the model was fitted to various physical properties including particle vacuum masses, decay constants, nuclear saturation properties, symmetry energy, and reasonable values for hyperon potentials. The quark sector of the model reproduces expectations of the phase diagram from heavy-ion collisions and lattice QCD. In addition, at high densities this formalism was shown to be in good agreement with perturbative QCD for the conditions expected to be found in different astrophysical scenarios [5].

2. Results and Discussion

Fig. 1 is comprised of different 3-dimensional phase diagrams built using the CMF model. The top two panels show the temperature vs. baryon chemical potential vs. charge fraction on either the hadronic side of the deconfinement phase transition (left panel) or on the quark side (right panel). The differences between the two are due to the extra constraint of conserved charge fraction, which is varied between 0 and 0.5. For this work, we choose to stop all our phase diagrams at a temperature $T = 160$ MeV (around the critical temperature for the first order phase transition), as a more thorough examination of this feature is beyond the scope of the current discussion.

The charge fraction is defined as the total electric charge of a system normalized by the total particle number of the system. For a baryon/quark mixture, the former quantity can be expressed as a sum (over all baryon and quark species) of the individual electric charge - number density products. Similarly, the latter is a sum of all baryon number - number density products

$$Y_Q = \frac{Q}{B} = \frac{\sum_i Q_i \, n_i}{\sum_i Q_{B,i} \, n_i}. \tag{1}$$

In particular, having $Y_Q = 0$ means that there is no net charge in the system, even though the presence of charged particles is not prohibited insofar as the sum of their charges is zero. Having $Y_Q = 0.5$ corresponds to the situation in which the total number of baryons of the system is twice as large as its net charge (in units of the electron charge). When charge neutrality is imposed, for example to model neutron stars, the (hadronic and quark) charge fraction must be equal to the lepton fraction. In this work, we are not showing contributions from leptons, as they are not in weak equilibrium with hadrons and quarks for the large temperatures in which we are interested.

In the bottom panel of Fig. 1, the temperature vs. free energy $\widetilde{\mu}$ vs. charge fraction phase diagram is shown. The Gibbs free energy per baryon (shortened here to free energy) is by the requirement of chemical stability the same on the hadronic and quark sides of the deconfinement phase transition. In our case when, besides baryon number, electric charge or isospin are conserved, it is

$$\widetilde{\mu} = \mu_B + Y_Q \mu_Q, \tag{2}$$

[1] Note that an alternative version of the CMF model includes in addition the chiral partners of the baryons and gives the baryons a finite size [10, 11].

Journal of Physics: Conference Series **1602** (2020) 012013 doi:10.1088/1742-6596/1602/1/012013

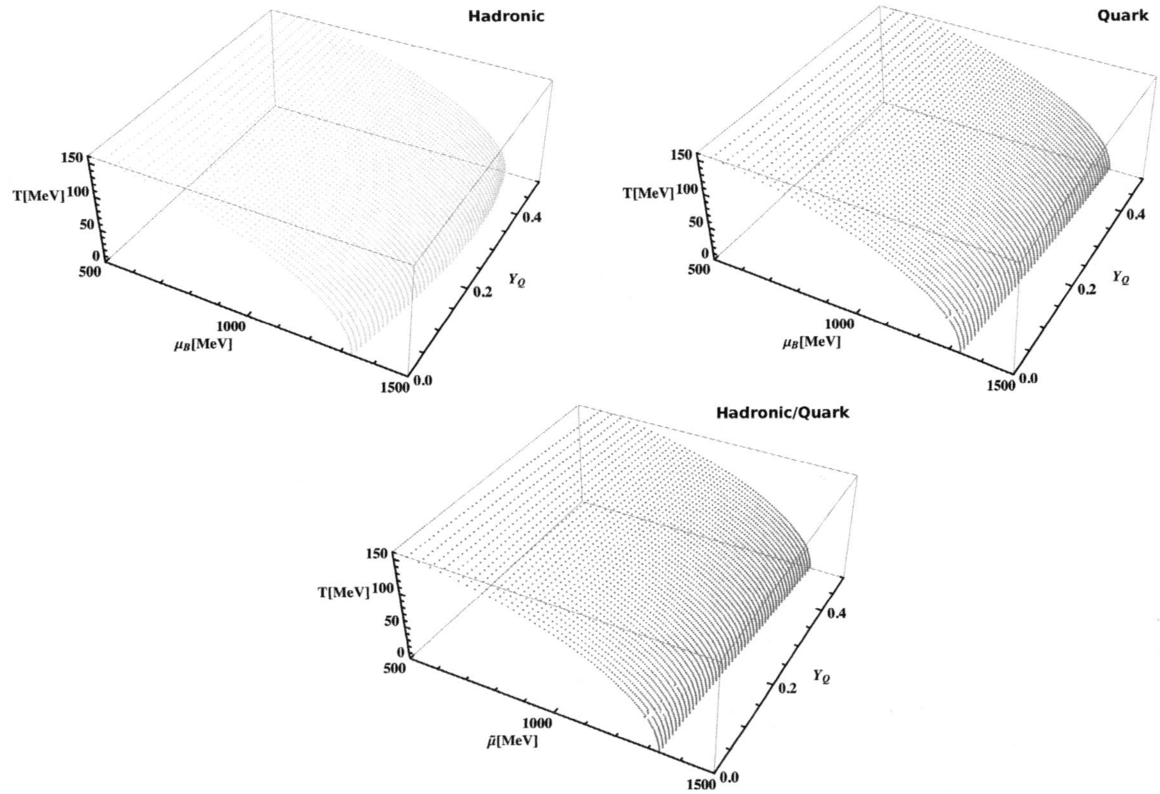

Figure 1. Top panels: the temperature T vs. baryon chemical potential μ_B vs. charge fraction Y_Q phase diagram on the hadronic side of the deconfinement phase transition (left panel) and on the quark side (right panel). Bottom panel: the temperature T vs. free energy $\widetilde{\mu}$ vs. charge fraction Y_Q phase diagram either on the hadronic or quark side of the deconfinement phase transition. All curves were calculated varying the charge fraction between $Y_Q = 0$ and 0.5.

or

$$\widetilde{\mu} \;=\; \mu_B + (Y_I + 1/2)\mu_I, \tag{3}$$

respectively, where μ_B, μ_Q and μ_I are the baryon, charged, and isospin chemical potentials. Y_I is the isospin fraction defined as

$$Y_I = \frac{I}{B} = \frac{\sum_i Q_{I,i}\; n_i}{\sum_i Q_{B,i}\; n_i}. \tag{4}$$

Note that the free energy is equal to the baryon chemical potential only in the particular cases of zero electric charge or zero charged/isospin chemical potential. This is the case for deleptonized cold neutron stars (charge neutral in weak equilibrium $Y_Q = 0$) and relativistic HICs ($Y_I = 0$ with $\mu_I = 0$). Eq. 2 was derived and discussed in detail in the Appendix D of Ref. [4], Eqs. 2 and 3 were derived with an extra term for the general case in which net strangeness is constrained in Ref. [9], and Eq. 2 was derived with an extra term for the case with trapped neutrinos and fixed lepton fraction in Ref. [5].

The bottom panel of Fig. 1 shows how the free energy at the deconfinement coexistence line increases as a function of Y_Q. This behavior is related to the softening of nuclear matter

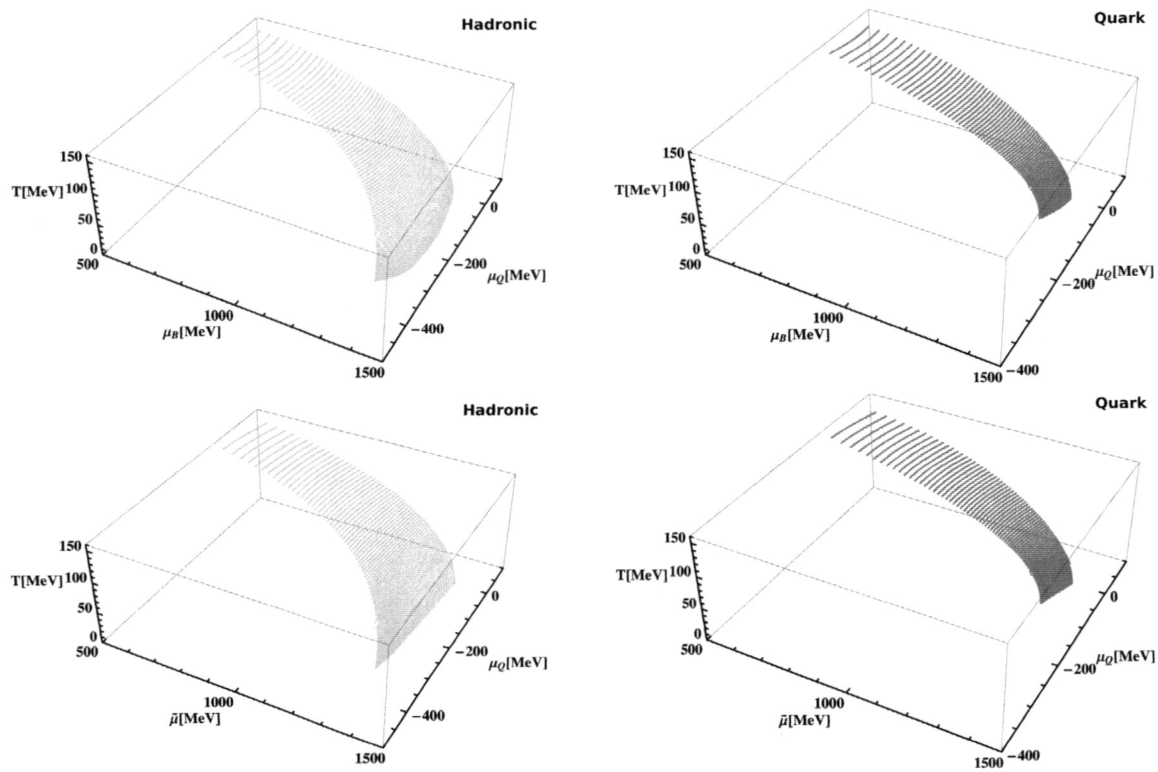

Figure 2. Same as Fig. 1 but showing the charged chemical potential μ_Q. The separate bottom panels show the hadronic (left panel) and quark (right panel) sides of the deconfinement phase transition

with increased net charge (larger proton-to-neutron or up-to-down quark ratio), the effect being stronger for hadronic matter. A softening of the equation of state (pressure vs. energy density) of hadronic matter corresponds to an increase in pressure at a given free energy (with respect to the quark phase) and, therefore, the hadronic phase persists to larger free energies.

The difference between the top panels of Fig. 1 comes from the fact that the baryon chemical potential is calculated from the free energy using Eq. 2 with a different charged chemical potential on either side of the phase transition. When comparing the top left panel with the bottom one, there is a significant difference for all cases corresponding to $\mu_B \neq \tilde{\mu}$ (for all Y_Q other than 0 and 0.5, the latter implying $\mu_Q = 0$). Note that the difference is much smaller between the top right panel of Fig. 1 and the bottom one, as the charged chemical potential μ_Q is always small (in absolute value) in the quark phase (see Fig. 3 of Ref. [5] for the particular case of chemically equilibrated matter).

The charged chemical potentials are shown in Fig. 2 on both sides of the coexistence line, reaching positive values at large temperatures and more negative values in the hadronic phase (left panels) in comparison to the quark phase (right panels). A larger (in absolute value) charged chemical potential means a larger difference between, for example, the number density of protons and neutrons or up and down quarks. In Fig. 2, the bottom panels are always different from each other, since the charged chemical potential itself is discontinuous across the first-order phase transition.

In order to draw 3-dimensional phase diagrams using isospin axes, one either rewrites the formalism using isospin as a conserved quantity (and corresponding computer codes), or uses

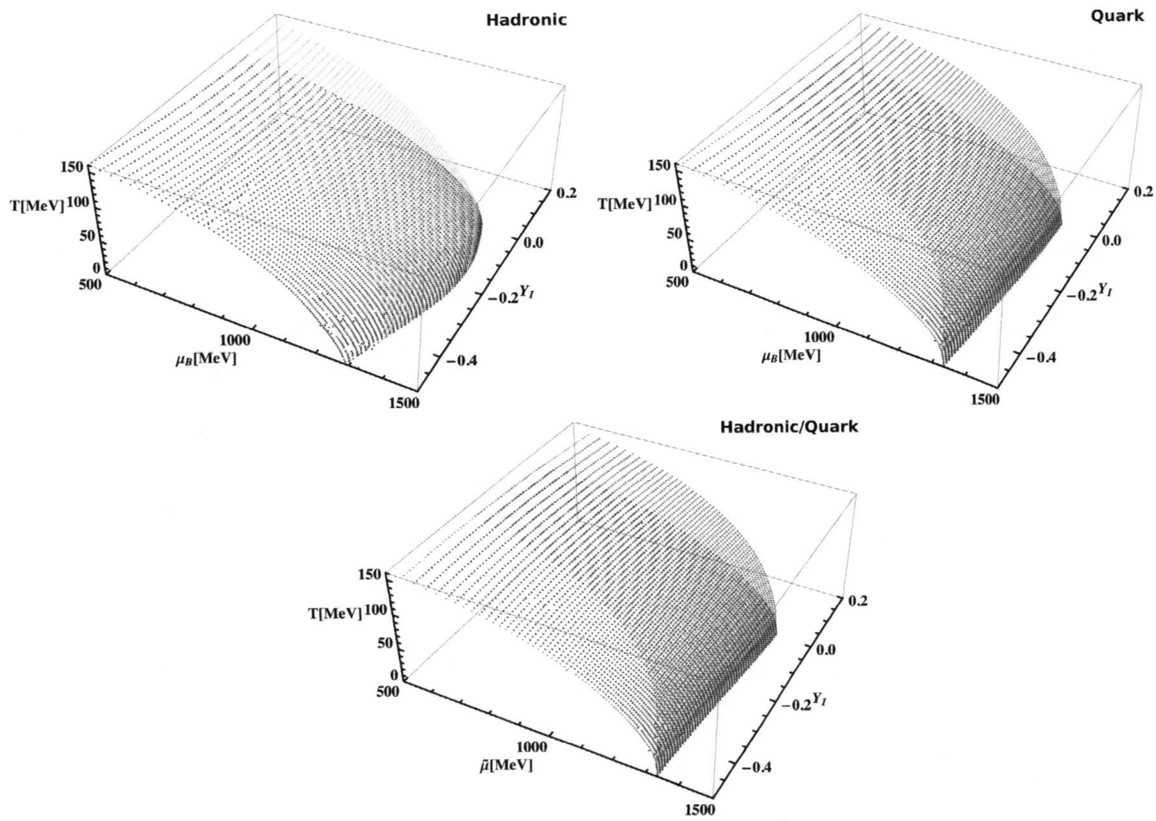

Figure 3. Same as Fig. 1 but showing the isospin charge fraction Y_I. Additional blue regions show results obtained by varying the isospin fraction in the range $Y_I = -0.5 \to 0$ (instead of varying the charge fraction in the range $Y_Q = 0 \to 0.5$).

the following transformation

$$Y_I = Y_Q - \frac{1}{2} + \frac{1}{2}Y_S, \tag{5}$$

which was derived in detail in Ref. [9]. The strangeness fraction Y_S is defined as the other fractions but now accounting for the strangeness of each hadron. In our notation, we consider the strangeness of particles to be positive, otherwise, all strangeness related quantities would have their signs reversed. Eq. 5 implies that for non-strange matter ($Y_S = 0$), the change of variable from Y_Q to Y_I introduces only a simple shift of the phase diagram relative to that axis, $Y_I = Y_Q - 0.5$. When net strangeness is finite ($Y_S \neq 0$), and to an increasing extent as temperature is raised, phase diagrams drawn against an isospin axis receive nontrivial modifications allowing the deconfinement coexistence line to cross to positive Y_I values for large charge fractions $Y_Q \simeq 0.5$. This can be seen in the green, red, and brown regions of Fig. 3.

Additionally, all panels of Fig. 3 display blue regions. These were calculated by rewriting our computer code in terms of the isospin fraction Y_I, which runs from -0.5 to 0 (instead of varying the charge fraction in the range $Y_Q = 0 \to 0.5$). As a result, these blue regions cover a different part of the QCD phase diagram. Note that, as $Y_S \to 0$ for low temperatures, the blue regions are deformed, such that they become identical with the other ones.

As can be seen in Fig. 4 (when compared to Fig. 2), both charged and isospin chemical potentials are equal. This is the case when using the following (non-unique) definitions based

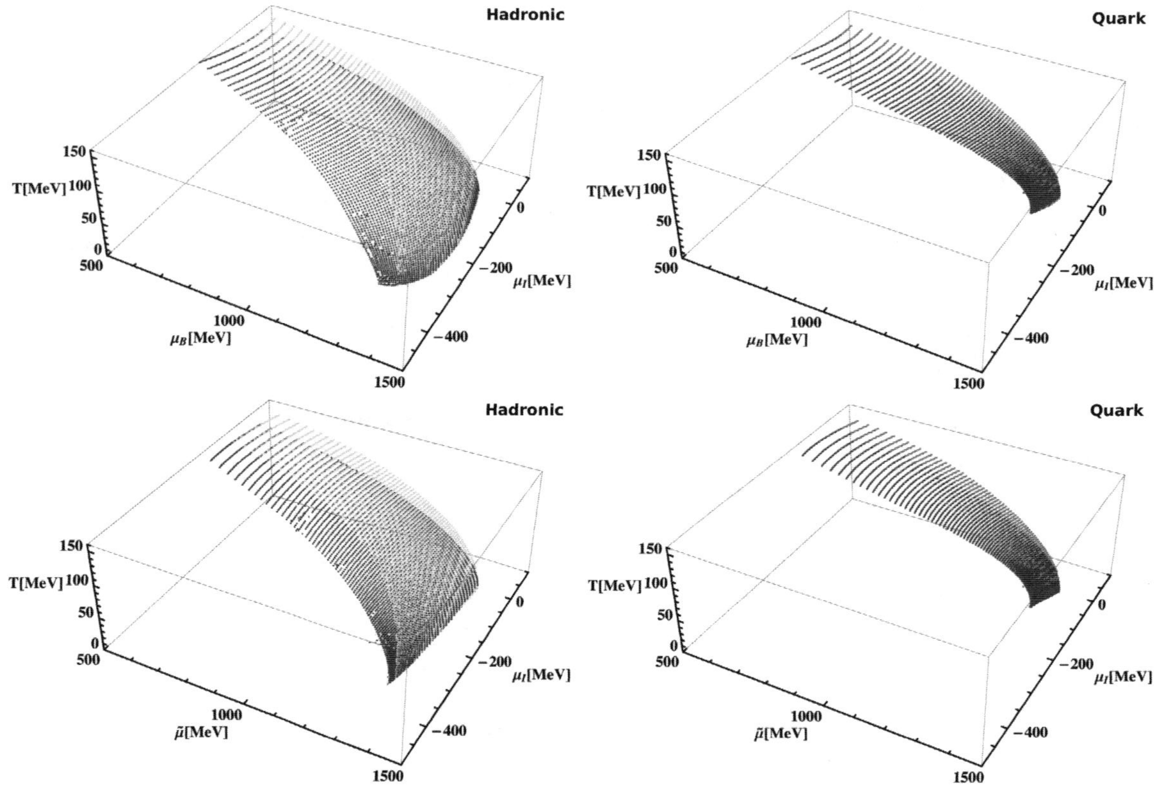

Figure 4. Same as Fig. 2 but showing the isospin chemical potential μ_I. Additional blue regions show again results obtained by varying the isospin fraction in the range $Y_I = -0.5 \to 0$ (instead of varying the charge fraction in the range $Y_Q = 0 \to 0.5$).

on the conserved quantities baryon and charge or isospin to calculate the chemical potential for each particle

$$\mu_i \quad = \quad Q_{B,i}\, \mu_B + Q_i\, \mu_Q, \tag{6}$$

or

$$\mu_i \quad = \quad Q_{B,i}\, \mu_B + (Q_{I,i} + 1/2 Q_{B,i} - 1/2 Q_{S,i})\mu_I, \tag{7}$$

where Q_I and Q_S refer to the isospin and strangeness of each hadron or quark. Note that, even in the case that strangeness is not conserved, Eq. 7 (derived in detail in Ref. [9]) still contains a strangeness related term. The extra terms in the parentheses are required to obtain the same standard chemical equilibrium equations for all hadrons and quarks (shown in Appendix A of Ref. [9]), independently of the chosen formulation. As in Fig. 3, all panels of Fig. 4 display blue regions calculated by rewriting our computer code in terms of the isospin fraction Y_I (instead of Y_Q). Within the blue regions, the isospin chemical potential μ_I remains negative.

3. Final Remarks

In this work, we presented 3-dimensional QCD phase diagrams with the extra axis being the charge fraction/chemical potential or the isospin fraction/chemical potential. We discussed these two different approaches for conserved quantities following different common practices in the astrophysics and heavy-ion collision communities. We found that changing the charge or isospin fraction by 0.5 can change the position of deconfinement phase transition significantly. The baryon chemical potential can change up to 130 MeV (at zero temperature on the hadronic side), the free energy up to 50 MeV (at zero temperature), and the charge/isospin chemical potential by 330 MeV (at zero temperature on the hadronic side). Our results show that comparisons among results from heavy-ion collision and hot astrophysical scenarios concerning the position of the deconfinement phase transition have to be interpreted carefully, as their different characteristics can change considerably the position of the phase-transition coexistence line.

Acknowledgements
Support for this research comes from the National Science Foundation under grant PHY-1748621, PHAROS (COST Action CA16214), the LOEWE-Program in HIC for FAIR, Conselho Nacional de Desenvolvimento Científico e Tecnológico - CNPq under grant 304758/2017-5 (R.L.S.F), and Fundação de Amparo à Pesquisa do Estado do Rio Grande do Sul - FAPERGS under grants 19/2551-0000690-0 and 19/2551-0001948-3 (R.L.S.F.).

References

[1] Dexheimer V and Schramm S 2008 *Astrophys. J.* **683** 943–948 (*Preprint* 0802.1999)
[2] Stephanov M 2006 *PoS* **LAT2006** 024 (*Preprint* hep-lat/0701002)
[3] Brandt B B, Endrődi G and Schmalzbauer S 2018 *Phys. Rev. D* **97**(5) 054514 URL https://link.aps.org/doi/10.1103/PhysRevD.97.054514
[4] Hempel M, Dexheimer V, Schramm S and Iosilevskiy I 2013 *Phys. Rev.* **C88** 014906 (*Preprint* 1302.2835)
[5] Roark J and Dexheimer V 2018 *Phys. Rev.* **C98** 055805 (*Preprint* 1803.02411)
[6] Hanauske M, Steinheimer J, Bovard L, Mukherjee A, Schramm S, Takami K, Papenfort J, Wechselberger N, Rezzolla L and Stcker H 2017 *J. Phys. Conf. Ser.* **878** 012031
[7] Bastian N U F, Blaschke D, Fischer T and Rpke G 2018 *Universe* **4** 67 (*Preprint* 1804.10178)
[8] Most E R, Jens Papenfort L, Dexheimer V, Hanauske M, Stoecker H and Rezzolla L 2020 *Eur. Phys. J. A* **56** 59 (*Preprint* 1910.13893)
[9] Aryal K, Constantinou C, Farias R and Dexheimer V 2020 (*Preprint* 2004.03039)
[10] Steinheimer J, Schramm S and Stocker H 2011 *Phys. Rev.* **C84** 045208 (*Preprint* 1108.2596)
[11] Motornenko A, Steinheimer J, Vovchenko V, Schramm S and Stoecker H 2020 *Phys. Rev.* **C101** 034904 (*Preprint* 1905.00866)

New channels of prompt-photon production by magnetic fields in heavy-ion collisions

Alejandro Ayala[1,2], Jorge David Castaño-Yepes[1], Isabel Dominguez Jimenez[3], Jordi Salinas San Martín[1], María Elena Tejeda-Yeomans[4*].

[1]Instituto de Ciencias Nucleares, Universidad Nacional Autónoma de México, Apartado Postal 70-543, México Distrito Federal 04510, Mexico. [2]Centre for Theoretical and Mathematical Physics, and Department of Physics, University of Cape Town, Rondebosch 7700, South Africa. [3]Facultad de Ciencias Físico-Matemáticas, Universidad Autónoma de Sinaloa, Avenida de las Américas y Boulevard Universitarios, Ciudad Universitaria, C.P. 80000, Culiacán, Sinaloa, México. [4]Facultad de Ciencias - CUICBAS, Universidad de Colima, Bernal Díaz del Castillo No. 340, Col. Villas San Sebastián, 28045 Colima, Mexico.

E-mail: *matejeda@ucol.mx

Abstract. In this work I present recent results on the photon yield and elliptic flow coefficient in relativistic heavy-ion collisions for different centralities. In our calculation, we take into account gluon fusion and splitting channels which become available due to an intense and short-lived magnetic field. We show that with reasonable values for the saturation scale, our calculation helps to better describe the experimental results obtained at RHIC energies for the lowest part of the transverse photon momentum at different centralities.

1. Introduction

Photons have proven to be a versatile tool to study relativistic heavy-ion collisions. They are created in various mechanisms occurring from the initial moments of the fireball and through the space-time evolution of this strongly interacting system. Under these conditions, photons are a special probe of the system since they mostly escape without much interaction, but this means that the information they carry is hard to disentangle. One of the current open questions in the field is precisely the puzzling situation in which direct photon yields are large and show strong azimuthal asymmetries, which were both theoretically unexpected [1]. The anomalous excess of direct photons produced in these reactions [2, 3, 4], referred to as the *direct photon puzzle*, can be studied in terms of the new open channels of photoproduction that become available with the intense magnetic fields generated in peripheral heavy-ion collisions [5]. These magnetic fields become both a source for an excess in photon yield, and they enhance the second harmonic coefficient (v_2) of the Fourier expansion in the azimuthal photon distribution [6]. In the literature, there are some hydrodynamic [7, 8] and transport [9] calculations that show partial agreement with ALICE and PHENIX measurements of low and intermediate transverse momentum photons, but further studies are still needed to have a more robust result for different observables [10]. Also, in Au+Au and Cu+Cu collisions at different centralities and beam energies, PHENIX [11] has found that the yield of low p_T photons ($\lesssim 2$ GeV) scales with a given power of $N_{\rm coll}$, which may mean that the source of these photons is very similar across beam energies and colliding species.

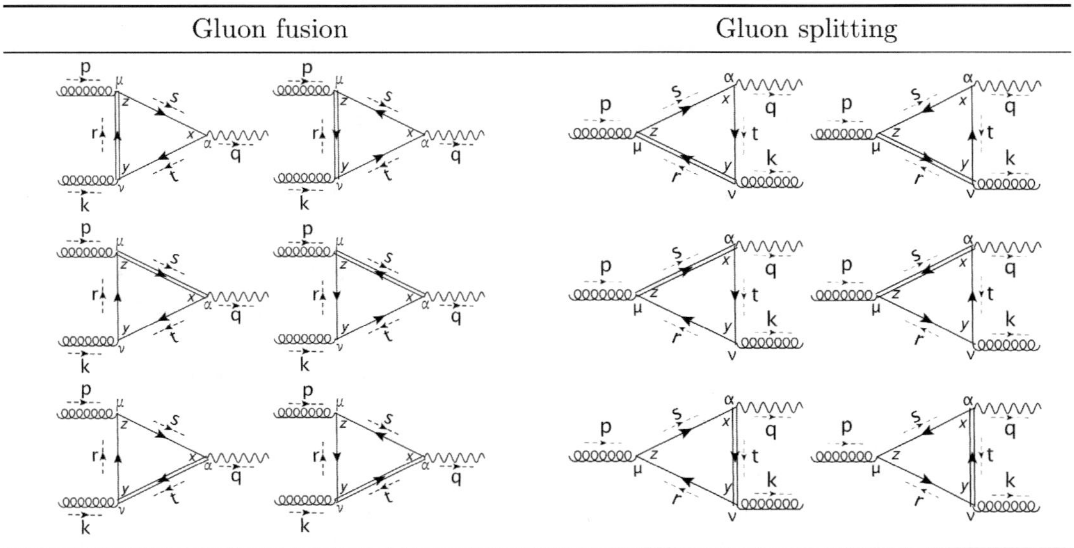

Table 1. Leading order contribution for photon production from gluon fusion (left) and gluon splitting (right) channels in the presence of a magnetic field. The double lines represent a quark propagator in the 1LL and the single lines represent the propagator in the LLL. The arrows in the propagators represent the charge flow and the arrows on the sides represent the momentum direction.

In this work I present a summary of our findings as reported in Refs.[12, 13]. We use UrQMD [14] to obtain the time evolution of the magnetic field and the volume produced by the collision participants and spectators in Au+Au and Cu+Cu collisions at $\sqrt{s_{\rm NN}} = 200$ GeV for different centrality ranges. We use these results for the megnetic field and interaction volume to compute both the photon yield and v_2 through the gluon fusion and gluon splitting channels, and we perform a centrality dependence study for the excess photon yield and v_2 comparing to the recent PHENIX measurements. We find a relatively good agreement for the low p_T region which improves for peripheral collisions and in the case where the magnetic field strength includes contributions from both spectators and participants.

2. Gluon fusion and splitting channels in the presence of a strong magnetic field

In this work we include both the gluon fusion and gluon splitting channels, depicted in Table 1, where each row of diagrams includes the different charge flow inside the fermion loop. These diagrams contribute to the total photon production amplitude at lowest order in the strong, $\alpha_{\rm s} = g^2/4\pi$, and electromagnetic, $\alpha_{\rm em} = e^2/4\pi$, couplings. Due to the magnetic field, translational invariance is broken and we calculate the total amplitude for photon production in coordinate space and we later integrate over space-time. I will summarize the results of this calculation but further details can be found in Ref. [12, 13]. We consider a constant magnetic field in the z-direction, obtained from a vector potential A^μ in the symmetric gauge $A^\mu = \frac{B}{2}(0, -y, x, 0)$. Also, for a four-momentum p^μ, we have defined $p_\perp^\mu = (0, p_1, p_2, 0)$, $p_\parallel = (p_0, 0, 0, p_3)$, $p_\perp^2 = p_1^2 + p_2^2$, $p_\parallel^2 = p_0^2 - p_3^2$, and therefore $p^2 = p_\parallel^2 - p_\perp^2$.

Following the Feynman diagrams of Table 1 we can write the expression for the amplitude

$\widetilde{\mathcal{M}}_{gg\to\gamma}$, and this is explicitly given by

$$\begin{aligned}
\widetilde{\mathcal{M}}_{gg\to\gamma} &= -\int d^4x d^4y d^4z \int \frac{d^4r}{(2\pi)^4}\frac{d^4s}{(2\pi)^4}\frac{d^4t}{(2\pi)^4} \\
&\times\ e^{-it\cdot(y-x)}e^{-is\cdot(x-z)}e^{-ir\cdot(z-y)}e^{-ip\cdot z}e^{-ik\cdot y}e^{iq\cdot x} \\
&\times\ \Big\{\mathrm{Tr}\left[ieq_f\gamma_\alpha iS_{ac}(s)ig\gamma_\mu t^c iS_{cd}(r)ig\gamma_\nu t^d iS_{da}(t)\right] \\
&+\ \mathrm{Tr}\left[ieq_f\gamma_\alpha iS_{ad}(t)ig\gamma_\nu t^d iS_{dc}(r)ig\gamma_\mu t^c iS_{ca}(s)\right]\Big\} \\
&\times\ \Phi(x,y)\Phi(y,z)\Phi(z,x)\epsilon^\mu(\lambda_p)\epsilon^\nu(\lambda_k)\epsilon^\alpha(\lambda_q), \qquad (1)
\end{aligned}$$

where p and k are the gluon and q the photon four-momenta, t^c, t^d are Gell-Mann matrices, and the polarization vectors for the gluons and the photon are $\epsilon^\mu(\lambda_p)$, $\epsilon^\nu(\lambda_k)$, $\epsilon^\alpha(\lambda_q)$, respectively. The product of phase factors is $\Phi(x,y)\Phi(y,z)\Phi(z,x) = \exp\{i\epsilon_{mj}(z-x)_m(x-y)_j\ |eq_fB|/2\}$, where the indices $m, j = 1, 2$ and ϵ_{mj} is the Levi-Civita tensor.

It can be shown that if all the quarks in the triangle are in the lowest Landau Level (LLL), the amplitude vanishes. The leading order contribution is obtained when at least one of the quarks in the loop is in the first excited Landau level (1LL). Thisis shown in Table 1, where the single line in the loop represents a quark in the LLL and the double line represents a quark in the 1LL. While we aim to explore the most general case, here we focus on the study of the low p_T region of the photon spectrum so that $|eq_fB|$ is large with respect to the loop momenta. Also for central rapidity and for gluons in the shattered glasma with momenta much less than the saturation scale (Λ_s), it is possible to further simplify the denominators of Eq. (1) considering that $2|eq_fB| \gg t_\parallel^2,\ s_\parallel^2,\ r_\parallel^2$[1]. After some steps into the calculation, the matrix element can be written as

$$\begin{aligned}
\widetilde{\mathcal{M}}_{gg\to\gamma} &= -i(2\pi)^4\delta^{(4)}(q-k-p)\frac{eq_fg^2\delta^{cd}e^{f(p_\perp,k_\perp)}}{32\pi(2\pi)^8}\left\{\left(g_\parallel^{\mu\alpha}-\frac{p_\parallel^\mu p_\parallel^\alpha}{p_\parallel^2}\right)h^\nu(a)\right. \\
&-\ \left(g_\parallel^{\mu\nu}-\frac{p_\parallel^\mu p_\parallel^\nu}{p_\parallel^2}\right)h^\alpha(a)+\left(g_\parallel^{\mu\nu}-\frac{k_\parallel^\mu k_\parallel^\nu}{k_\parallel^2}\right)h^\alpha(b)-\left(g_\parallel^{\alpha\nu}-\frac{k_\parallel^\alpha k_\parallel^\nu}{k_\parallel^2}\right)h^\mu(b) \\
&+\ \left.\left(g_\parallel^{\alpha\nu}-\frac{q_\parallel^\alpha q_\parallel^\nu}{q_\parallel^2}\right)h^\mu(c)-\left(g_\parallel^{\mu\alpha}-\frac{q_\parallel^\mu q_\parallel^\alpha}{q_\parallel^2}\right)h^\nu(c)\right\}\epsilon_\mu(\lambda_p)\epsilon_\nu(\lambda_k)\epsilon_\alpha(\lambda_q), \qquad (2)
\end{aligned}$$

where the trace over the Gell-Mann matrices has also been performed and $h^\mu(a) = -(i/\pi)\epsilon_{ij}a^ig_\perp^{j\mu}$, $a_i = p_i + 2k_i + i\epsilon_{im}p_m$, $b_i = 2p_i + k_i - i\epsilon_{im}k_m$, $c_i = k_i - p_i + i\epsilon_{im}(p_m + k_m)$, with $f(p_\perp, k_\perp) = \frac{1}{8|eq_fB|}(p_m - k_m + i\epsilon_{mj}(p_j + k_j))^2 - \frac{1}{2|eq_fB|}\left(p_m^2 + k_m^2 + 2i\epsilon_{jm}p_mk_j\right)$, where $g_\perp = \mathrm{diag}(1,1)$ and $g_\parallel = \mathrm{diag}(1,-1)$ are the metric tensors in the transverse and longitudinal spaces.

The matrix element for gluon splitting $\tilde{\mathcal{M}}_{g\to g\gamma}$., can be calculated with the diagrams on the right columns of Table 1. This process is related to the gluon fusion channel, $\tilde{\mathcal{M}}_{gg\to\gamma}$, by the crossing symmetry $\widetilde{\mathcal{M}}_{g\to g\gamma}(p,k,q) = \widetilde{\mathcal{M}}_{gg\to\gamma}(p,-k,q)$. Since the processes do not interfere, we sum incoherently the two matrix elements squared, obtaining

$$\overline{\sum_{\mathrm{c,p,f}}}|\widetilde{\mathcal{M}}|^2 = \mathcal{V}\Delta\tau(2\pi)^4\overline{\sum_{\mathrm{c,p,f}}}\left[\delta^{(4)}(q-k-p)\,|\mathcal{M}_{gg\to\gamma}|^2+\delta^{(4)}(q+k-p)\,|\mathcal{M}_{g\to g\gamma}|^2\right], \qquad (3)$$

[1] The more general case is currently being explored and will be reported elsewhere.

where

$$\begin{aligned}\overline{\sum_{\text{c,p,f}}}|\mathcal{M}_{gg\to\gamma}|^2 &= \overline{\sum_{\text{c,p,f}}}|\mathcal{M}_{g\to g\gamma}|^2 = \frac{2\alpha_{\text{em}}\alpha_s^2 q_\perp^2}{\pi\omega_q^2}\sum_f q_f^2\left(2\omega_p^2+\omega_k^2+\omega_p\omega_k\right)\\ &\times\ \exp\left\{-\frac{q_\perp^2}{|eq_fB|\,\omega_q^2}\left(\omega_p^2+\omega_k^2+\omega_p\omega_k\right)\right\}.\end{aligned} \tag{4}$$

Notice that the factor $\mathcal{V}\Delta\tau$ in Eq. 3 comes from squaring the delta function for energy-momentum conservation in Eq. (2). This is the space-time volume where the reaction takes place. We will discuss more details about this factor in Sec. 4.

3. Photon yield and v_2

The invariant photon momentum distribution is given by

$$\begin{aligned}\omega_q\frac{dN^{\text{mag}}}{d^3q} &= \frac{\mathcal{V}\Delta\tau}{2(2\pi)^3}\int\frac{d^3p}{(2\pi)^3 2\omega_p}\int\frac{d^3k}{(2\pi)^3 2\omega_k}\\ &\times\Bigg\{(2\pi)^4\delta^{(4)}\left(q-k-p\right)n(\omega_p)n(\omega_k)\overline{\sum_{\text{c,p,f}}}|\mathcal{M}_{gg\to\gamma}|^2\\ &+(2\pi)^4\delta^{(4)}\left(q+k-p\right)n(\omega_p)\left[1+n(\omega_k)\right]\overline{\sum_{\text{c,p,f}}}|\mathcal{M}_{g\to g\gamma}|^2\Bigg\},\end{aligned} \tag{5}$$

where $n(\omega)$ represents the distribution of gluons coming from the shattered glasma. We use for this distribution a simple model that accounts for the high occupation gluon number given by [15] $n(\omega)=\eta/(e^{\omega/\Lambda_s}-1)$, where η represents the high gluon occupation factor. So the initial state gluon with energy ω_k comes weighed with an occupation factor $n(\omega_k)$ and the final state one comes weighed with an enhanced occupation factor $1+n(\omega_k)$. With this in mind, we finally arrive to the expression we will use to compare with data

$$\begin{aligned}\frac{1}{2\pi\omega_q}\frac{dN^{\text{mag}}}{d\omega_q} &= \mathcal{V}\Delta\tau\frac{\alpha_{\text{em}}\alpha_s^2\pi}{2(2\pi)^6\omega_q}\sum_f q_f^2\int_0^{\omega_q}d\omega_p\left(2\omega_p^2+\omega_q^2-\omega_p\omega_q\right)e^{-g_f(\omega_p,\omega_q)}\\ &\times\left\{I_0\left[g_f(\omega_p)\right]-I_1\left[g_f(\omega_p)\right]\right\}\left\{n(\omega_p)n(|\omega_q-\omega_p|)+n(\omega_p)\left[1+n(|\omega_q-\omega_p|)\right]\right\},\end{aligned} \tag{6}$$

where $g_f(\omega_p,\omega_q)=(\omega_p^2+\omega_q^2-\omega_p\omega_q)(2\,|eq_fB|)$, and I_0, I_1 are the modified Bessel function of the first kind. Notice that $\Delta\tau$ in Eq. (6) corresponds to one of the small time intervals that we used to divide the whole time interval $\Delta\mathcal{T}$ over which the magnetic pulse is appreciable. The whole yield thus corresponds to the sum of the yields in each of the small time intervals represented by $\Delta\tau$. Furthermore, given that the azimuthal distribution with respect to the reaction plane can be written in terms of a Fourier decomposition as

$$\frac{dN^{\text{mag}}}{d\phi}=\frac{N^{\text{mag}}}{2\pi}\left[1+\sum_{n=1}^{\infty}2v_n(\omega_q)\cos(n\phi)\right]\qquad\text{where}\qquad N^{\text{mag}}=\int d^3q\frac{dN^{\text{mag}}}{d^3q}, \tag{7}$$

Journal of Physics: Conference Series **1602** (2020) 012014 doi:10.1088/1742-6596/1602/1/012014

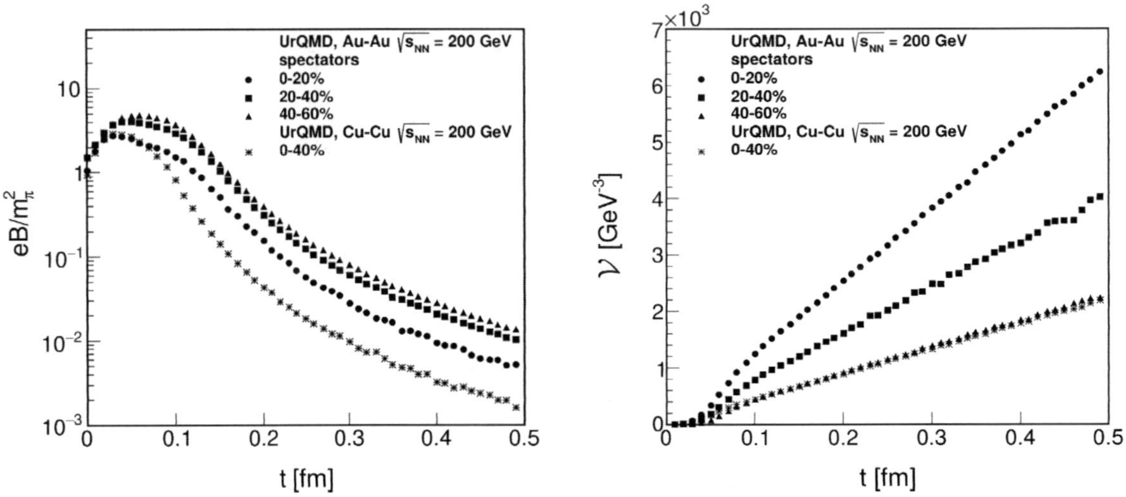

Figure 1. On the left, the mean magnetic field strength produced by spectators at the middle of the interaction region as a function of time and on the right, the volume $\mathcal{V}$ as a function of time. Both results are shown for three centrality classes 0-20%, 20-40% and 40-60% in Au+Au collisions and one centrality class 0-40% in Cu+Cu collisions at $\sqrt{s_{\rm NN}} = 200$ GeV.

we find that the second flow coefficient is given by

$$
\begin{aligned}
v_2^{\rm mag}(\omega_q) &= \frac{\alpha_{\rm em}\alpha_s^2\pi\mathcal{V}\Delta\mathcal{T}}{2(2\pi)^5 N^{\rm mag}} \sum_f q_f^2 \int_0^{\omega_q} d\omega_q' \int_0^{\omega_q'} d\omega_p \\
&\times \left(2\omega_p^2 + {\omega_q'}^2 - \omega_p\omega_q'\right) e^{-g_f(\omega_p,\omega_q')} \left\{ I_0\left[g_f(\omega_p)\right] - \left[1 + \frac{1}{g_f(\omega_p)}\right] I_1\left[g_f(\omega_p)\right] \right\} \\
&\times \left\{ n(\omega_p) n(|\omega_q' - \omega_p|) + n(\omega_p)\left[1 + n(|\omega_q' - \omega_p|)\right] \right\}. \qquad (8)
\end{aligned}
$$

We use the results of Eqs. (6) and (8) to compare with data in Sec. 5, after we discuss the procedure we use to obtain estimates for $\mathcal{V}$ and $\Delta\mathcal{T}$ and the magnetic field strength, from simulations.

4. Magnetic field pulse and space-time volume from simulations

In Eq. 3 the factor $\mathcal{V}\Delta\tau$ appeared and consists of the product of the spatial volume of the nuclear overlap region $\mathcal{V}(t)$ at time t and the time interval $\Delta\tau$. In this time interval, the magnetic field can be taken as having a constant intensity $B(t)$. Furthermore, for a given centrality class, the spatial volume can be estimated from the fraction of the number of participants to the total number of nucleons. Since the reaction stops as soon as the magnetic field becomes negligible, the overall lifetime $\Delta\mathcal{T}$ can be estimated calculating the duration of the magnetic pulse. We perform Monte Carlo simulations using UrQMD [14] to estimate these factors, and to have a blueprint of the magnetic field pulse. First we simulate relativistic Au+Au collisions at $\sqrt{s_{\rm NN}} = 200$ GeV for three centrality classes, 0-20%, 20-40% and 40-60%. Next, for Cu+Cu collisions at the same collision energy, we simulate events in the centrality class 0-40%. Then for each event, the magnetic field at position $\mathbf{x}$ and at time t can be calculated using the Liénard-Wiechert

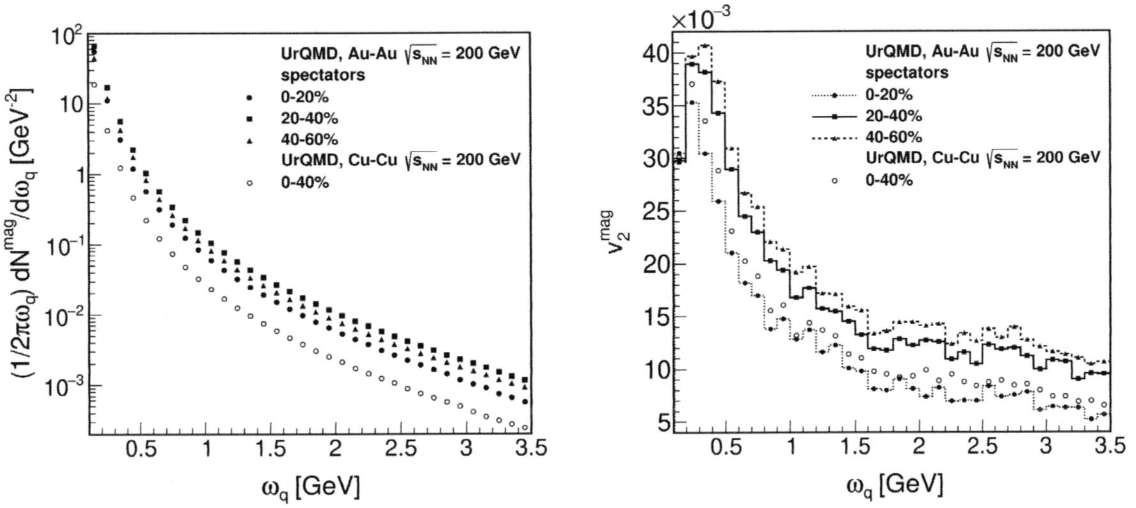

Figure 2. On the left, the invariant momentum distribution, and on the right $v_2^{\rm mag}$ for Au+Au collisions in the 0-20%, 20-40% and 40-60% centrality classes and Cu+Cu collisions in the 0-40% centrality class at $\sqrt{s_{\rm NN}} = 200$ GeV considering only the magnetic field generated by spectators.

potential generated by non-accelerated charges moving along the beam direction [16] as

$$e\mathbf{B}(\mathbf{x},t) = \alpha_{em}\sum_j \frac{(1-v_j^2)\ \mathbf{v}_j \times \mathbf{R}_j}{R_j^3\left[1-\frac{(\mathbf{v}_i\times\mathbf{R}_j)^2}{R_j^2}\right]^{3/2}}, \tag{9}$$

where $\mathbf{R}_j = \mathbf{x} - \mathbf{x}_j(t)$, $\mathbf{x}_j(t)$ is the position of the j-th charge moving with velocity $\mathbf{v}_j$, R_j is the magnitude of $\mathbf{R}_j$ and the sum runs over charged particles per event. The magnetic field computed in the middle of the interaction region $\mathbf{x} = 0$ is shown on the left of Figure 1 produced by spectators, for different centrality ranges in Au+Au and Cu+Cu collisions at $\sqrt{s_{\rm NN}} = 200$ GeV. On the right hand side of the same figure we show the volume $\mathcal{V}(t)$ calculated for each centrality class as $\mathcal{V}(t) = 2t\pi r_{\rm A}^2(N_{\rm part}/2N)^{2/3}$, where $\pi r_{\rm A}^2$ corresponds to the transverse area of a cylinder having a radius $r_{\rm A}$ ($r_{\rm Au} = 6.38$ fm, $r_{\rm Cu} = 3.9$ fm [18]).

5. Results

We now summarize some of our findings where we use throughout $\alpha_s = 0.3$, $\Lambda_s = 2$ GeV, $\eta = 3$. We use the magnetic field strength generated only by spectators[2] and we calculate the net yield by adding the yields for all time intervals within $\Delta\mathcal{T}$ given by $\frac{dN^{\rm mag}}{d\omega_q} = \sum_{i=1}\left[\frac{dN^{\rm mag}}{d\omega_q}\right]_i$, where $\left[\frac{dN^{\rm mag}}{d\omega_q}\right]_i$ is the yield corresponding to the ith-time interval $\Delta\tau_i$, given by Eq. (6). Also, we calculate the corresponding net v_2 as a weighted average which accounts for the time-varying yield, as

$$v_2^{\rm mag}(\omega_q) \;=\; \frac{\sum_{i=1}\left[\frac{dN^{\rm mag}}{d\omega_q}(\omega_q)\right]_i [v_2^{\rm mag}(\omega_q)]_i}{\sum_{i=1}\left[\frac{dN^{\rm mag}}{d\omega_q}\right]_i} \tag{10}$$

[2] The results for an analysis taking into account both participants and spectators, can be found in Ref. [13].

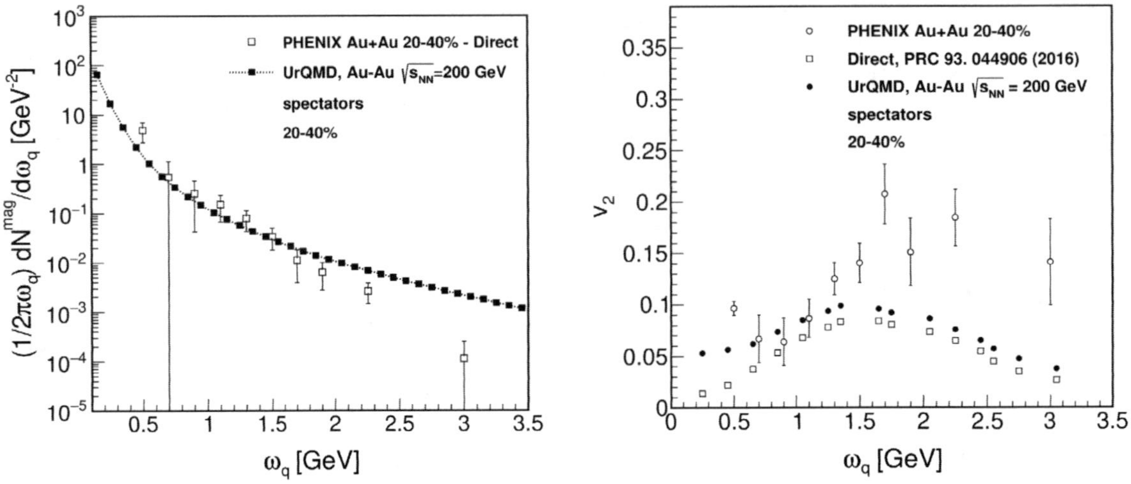

Figure 3. On the left, difference between PHENIX data [2] and the hydrodynamical calculation of Ref. [7] (open symbols) for Au+Au collisions at $\sqrt{s_{NN}} = 200$ GeV for the 20-40% centrality classes, compared to the calculation of the invariant yield (filled symbols) considering only the contribution to the magnetic field strength produced by the spectators. On the right, v_2 as a function of the photon energy ω_q where we show the contributions from the hydrodynamical calculation of Ref. [7] (open squares) and the magnetic field-dependent calculation summed as a weighed average including the field strength coming from the spectators. The experimental data (open symbols) correspond to Au+Au collisions at $\sqrt{s_{NN}} = 200$ GeV in the 20-40% centrality class from Ref. [2].

where $[v_2^{\rm mag}(\omega_q)]_i$ is the harmonic coefficient corresponding to the ith-time interval $\Delta\tau_i$, given by Eq. (8).

Figure 2 shows the invariant momentum distribution and corresponding $v_2^{\rm mag}$, for Au+Au collisions at $\sqrt{s_{\rm NN}} = 200$ GeV in the 0-20%, 20-40% and 40-60% centrality ranges and Cu+Cu collisions in the 0-40% centrality range at the same energy.

In order to compare these results to experimental data, we note that since the magnetic contribution represents an excess over calculations not including these effects, we can compare the magnetic yield we calculated, to the difference between data and a hydrodynamical simulation of the direct photon yield. For v_2, we use a weighed average between v_2 from photons produced by the magnetic field and v_2 contribution from the direct photons as

$$v_2(\omega_q) = \frac{\sum_{i=1}^{m} \left[\frac{dN}{d\omega_q}\right]_i [v_2^{\rm mag}(\omega_q)]_i + \frac{dN^{\rm direct}}{d\omega_q}(\omega_q)\; v_2^{\rm direct}(\omega_q)}{\sum_{i=1}^{m} \left[\frac{dN}{d\omega_q}\right]_i + \frac{dN^{\rm direct}}{d\omega_q}(\omega_q)}, \tag{11}$$

where $dN^{\rm direct}/d\omega_q$ and $v_2^{\rm direct}$ are the (ω_q-dependent) spectrum and second harmonic coefficient of direct photons from Ref. [7], respectively.

Figure 3 shows on the left, the difference between PHENIX data [2] and the hydrodynamical calculation of Ref. [7] –open symbols– compared to our calculation of the invariant yield –filled symbols– considering only the contribution to the magnetic field strength produced by the spectators, in Au+Au collisions at $\sqrt{s_{\rm NN}} = 200$ GeV for the 20-40% centrality class. On the right, this figure shows v_2 as the weighted average accounting for the magnetic and direct photons, compared to the experimental data for Au+Au collisions at $\sqrt{s_{NN}} = 200$ GeV in the 20-40% centrality class from Ref. [2] with magnetic field effects coming only from the spectators. Notice that the magnetic field contribution improves the agreement with experimental data for the low part of the spectrum helping to describe the rise of v_2 as the photon energy decreases.

6. Summary and conclusions

We have computed the contribution to the photon yield and v_2 from gluon fusion and splitting induced by a magnetic field during the early stages of a relativistic heavy-ion collision, where there is a large gluon occupation number below the saturation scale Λ. The magnetic field strength and volume are computed using UrQMD simulations and the results compared with recent data from PHENIX. For the yield, the excess coming from the magnetic field induced processes is compared with the difference between PHENIX data and the hydrodynamical calculation of Ref. [7]. v_2 is computed as a weighed average accounting for magnetic and direct photons. The results show a relatively good agreement for the lower part of the spectra which improves for peripheral collisions. Our results point into the direction of enhancing the yield and v_2 for $p_T \lesssim 1$ GeV/c, a trend not incompatible with data [1]. Several avenues for improvement of our work and extensions of these results to other experimental regimes are currently being pursued and will be reported elsewhere.

Acknowledgements

The authors thank L.A. Hernández and S. Hernández-Ortiz for useful discussions. Support for this work has been received in part by UNAM-DGAPA-PAPIIT grant number IG100219 and by Consejo Nacional de Ciencia y Tecnología grant numbers A1-S-7655 and A1-S-16215.

References

[1] Gabor David, Rept.Prog.Phys. 83 (2020) 4, 046301.
[2] A. Adare *et al.* [PHENIX Collaboration], Phys. Rev. C **91**, 064904 (2015); A. Adare *et al.* [PHENIX Collaboration], Phys. Rev. C **98**, 054902 (2018).
[3] A. Adare *et al.* [PHENIX Collaboration], Phys. Rev. C **94**, 064901 (2016).
[4] J. Adam *et al.* [ALICE Collaboration], Phys. Lett. B **754**, 235 (2016).
[5] D. E. Kharzeev, L. D. McLerran and H. J. Warringa, Nucl. Phys. A **803**, 227 (2008); V. Skokov, A. Y. Illarionov, V. Toneev, Int. J. Mod. Phys. A **24**, 5925 (2009); V. Voronyuk, V. D. Toneev, W. Cassing, E. L. Bratkovskaya, V. P. Konchakovski, S. A. Voloshin, Phys. Rev. C **83**, 054911 (2011); L. McLerran, V. Skokov, Nucl. Phys. A **929**, 184-190 (2014); A. Bzdak, V. Skokov, Phys. Lett. B **710**, 171-174 (2012). G. Basar, D. E. Kharzeev, V. Skokov, Phys. Rev. Lett. **109**, 202303 (2012); G. Basar, D. E. Kharzeev, E. V. Shuryak, Phys. Rev. C **90**, 014905 (2014). B. G. Zakharov, Eur. Phys. J. C **76**, 609 (2016). K. Tuchin, Phys. Rev. C **91**, 014902 (2015).
[6] A. Adare *et al.* [PHENIX Collaboration], e-Print: arXiv:1509.07758 [nucl-ex]; D. Lohner, *et al.* (ALICE Collaboration), e-Print: arXiv:1212.3995 [nucl-ex].
[7] J.-F. Paquet, C. Shen, G. S. Denicol, M. Luzum, B. Schenke, S. Jeon, C. Gale, Phys. Rev. C **93**, 044906 (2016).
[8] H. van Hees, M. He, R. Rapp, Nucl. Phys. A **933**, 256-271 (2015).
[9] O. Linnyk, V. Konchakovski, T. Steinert, W. Cassing, E. L. Bratkovskaya, Phys. Rev. C **92**, 054914 (2015).
[10] For a recent review see C. Shen, Nucl. Phys. A **956**, 184-191 (2016).
[11] A. Adare *et al.* [PHENIX Collaboration], e-Print: arXiv:1805.04066 [hep-ex] (2018); A. Adare *et al.* [PHENIX Collaboration], e-Print: arXiv:1805.04084 [hep-ex] (2018); T. Sakaguchi, J. Phys. Conf. Ser. **1070** (2018) no.1, 012012 (2018).
[12] A. Ayala, J. D. Castaño-Yepes, C. A. Dominguez, L. A. Hernández, S. Hernández-Ortíz, M. E. Tejeda-Yeomans, Phys. Rev. D **96**, 014023 (2017) [Erratum: Phys.Rev. D **96**, 119901 (2017)].
[13] A. Ayala, J. D. Castaño-Yepes, I. Dominguez Jimenez, J. Salinas San Martín, M. E. Tejeda-Yeomans. Eur.Phys.J. **A** 56 (2020) 2, 53.
[14] S. A. Bass, M. Belkacem, M. Bleicher, M. Brandstetter, L. Bravina, C. Ernst, L. Gerland, M. Hofmann, S. Hofmann, J. Konopka, G. Mao, L. Neise, S. Soff, C. Spieles, H. Weber, L. A. Winckelmann, H. Stocker, W. Greiner, C. Hartnack, J. Aichelin and N. Amelin Prog. Part. Nucl. Phys. **41**, 225-370 (1998); M. Bleicher,

Journal of Physics: Conference Series **1602** (2020) 012014 doi:10.1088/1742-6596/1602/1/012014

E. Zabrodin, C. Spieles, S.A. Bass, C. Ernst, S. Soff, H. Weber, H. Stöcker, W. Greiner. J. Phys. **G25**, 1859-1896 (1999).
[15] L. McLerran and B. Schenke, Nucl. Phys. A **929**, 71-82 (2014).
[16] L. D. Landau, E.M. Lifshitz, *The Classical Theory of Fields*, Pergamon Press, Oxford, England, (1984).
[17] C. Shen, U. W. Heinz, J.-F. Paquet, C. Gale, Phys. Rev. C **89**, 044910 (2014).
[18] H. De Vries, C.W. De Jager, and C. De Vries, Atom. Data Nucl. Data Tabl. **36**, 495 (1987).

Journal of Physics: Conference Series **1602** (2020) 012015 doi:10.1088/1742-6596/1602/1/012015

Direct photon production and scaling properties in large and small system collisions

Vladimir Khachatryan for the PHENIX Collaboration
Department of Physics and Astronomy, Stony Brook University, Stony Brook, New York 11794-3800, USA
E-mail: vladimir.khachatryan@stonybrook.edu

Abstract. We present recent measurement results of low momentum direct photons obtained by the PHENIX collaboration from Cu+Cu at 200 GeV and Au+Au at 200 GeV, 62.4 GeV, 39 GeV, as well as from p+p and p+Au at 200 GeV. In these measurements a large excess over the scaled reference p+p yield of direct photons is observed in A+A collisions, and a non-zero excess over the scaled p+p yield is found within systematic uncertainties in central p+Au collisions. Based upon the comparison between this recent data and the PHENIX published Au+Au data at 200 GeV, along with including the ALICE Pb+Pb data at 2760 GeV, PHENIX has found that the direct photon p_T-spectra and their integrated yield, dN_γ/dy, above 1 GeV/c in p_T follow a universal scaling as a function of the charged-hadron multiplicity, $(dN_{ch}/d\eta)^\alpha$, with $\alpha = 1.25$. Within experimental uncertainties, the direct photon scaling properties observed in large collision systems is independent of the colliding species, collision centrality selection and center-of-mass energy. One possible explanation is that the photon radiation sources may have similar properties for all colliding large systems across different energies. There is also another trend coming from integrated direct photon p_T-spectra of small systems, which suggests that the QGP turn off/on transition region may exist between large and small system collisions.

1. Introduction

Direct photons produced in (ultra)relativistic heavy ion collisions are considered to be a unique probe for studying the entire space-time evolution of colliding systems, with which one can extract important information on thermal properties and collective behavior of the hot fireball of Quark Gluon Plasma (QGP). These photons being color blind probes, contrary to partons, directly probe the conditions of their production environment at the time of production, and due to the small cross section with the collision-produced medium like QGP, they leave the collision region virtually unmodified with no final state interactions.

Direct photons describe the excess yield, which one can obtain by subtracting the hadronic decay photon yield (mostly from π^0 and η decays) from the total inclusive photon yield. In general, the direct photons have thermal and non-thermal origins. According to current understanding, the thermal photons originate from the pre-equilibrium phase, the QGP phase and the late hadronic gas (HG) phase. This type of photons dominate over the low transverse momentum region. The non-thermal photons, dominating over the high-p_T region, come from initial hard scattering processes (prompt photons) like quark anti-quark annihilation or QCD Compton scattering among the incoming and outgoing partons, and from the pre-equilibrium phase as well.

PHENIX has measured low momentum direct photon p_T-spectra in Au+Au collisions at $\sqrt{s_{NN}} = 200\,\text{GeV}$, with a large excess yield observed at $p_T < 3\,\text{GeV/c}$ above the hard scattering contribution given by a reference $p+p$ yield scaled by the number of binary collisions, N_{coll}, [1, 2]. PHENIX has also measured large anisotropies of direct photons in Au+Au collisions at $\sqrt{s_{NN}} = 200\,\text{GeV}$ with respect to the reaction plane [3]. After these large yield and large anisotropy measurements, a challenging problem arose, known as "direct (thermal) photon flow puzzle", where different theoretical models basically fall short in describing these two quantities simultaneously. The best models and mechanisms have mixed success in doing so [4, 5, 6, 7, 8, 9, 10, 11]. In order to help resolve this puzzle, and in particular to reconcile the assumed photon sources with the calculated invariant yields, PHENIX has continued to measure low momentum direct photons by varying the system size and geometry, going from large to small collision systems, along with changing the center-of-mass (beam) energy for large systems. In this proceedings we report on these results and other related findings by PHENIX.

2. Direct photon measurement techniques and their p_T-spectra from large systems

So far PHENIX has analyzed wealth of its data sets available for direct photon analysis by exploiting three different analysis methods – the internal conversion method [1], the external conversion method [2] and the calorimeter method [12].

- In the internal conversion method one uses virtual photons that internally convert to low mass dilepton pairs ($\gamma^* \to e^+e^-$).
- In the external conversion method one uses real photons that externally convert to dilepton pairs ($\gamma \to e^+e^-$) in a well selected detector material.
- In the calorimeter method the photons are reconstructed directly from their energy deposition in electromagnetic calorimeters.

Direct photon p_T-spectra in Au+Au collisions at 200 GeV in Ref. [1] have been obtained with the internal conversion method, which recently has also been used to obtain the spectra in Cu+Cu at 200 GeV [13] (see the left panel of Fig. 1). The PHENIX measurements with 2010 Au+Au datasets were accomplished via the external conversion method, where the photons were measured through their conversions to e^+e^- pairs in the readout plane and electronics of the PHENIX detector subsystem called Hadron Blind Detector, or in short, HBD backplane. Here one used conversions happening at fixed radius equal to 60 cm from the event vertex. The HBD backplane had $X/X_0 \sim 3\%$ radiation length. So the direct photon invariant yield results in Au+Au collisions at 200 GeV from Ref. [2] have been obtained by the external conversion method with the HBD backplane, which has also been used to analyze 2010 Au+Au datasets for obtaining the yield at lower beam energies – 62.4 GeV and 39 GeV [14, 15, 16].

STAR collaboration has reported direct photon measurements in Au+Au at 200 GeV [17], with invariant yields significantly lower than those from the available PHENIX results. In order to address this observed discrepancy between the two experiments, an independent measurement was accomplished by PHENIX, utilizing the Au+Au dataset with high statistics taken in 2014 [18]. The external conversion method has been used again, however, in that analysis the photons were measured through their conversions to e^+e^- pairs at the Silicon Vertex Tracker (VTX), which had the radiation length of $X/X_0 \sim 14\%$.

In order to construct the direct photon p_T-spectra in the external conversion method, first by using the *Double Ratio Tagging Method* we should measure the direct photon relative yield, R_γ, which is an excess photon ratio defined as the ratio of the yield of true inclusive photons, γ^{incl}, to the yield of true photons coming from hadronic decays, γ^{hadron}. By dividing γ^{incl} and

γ^{hadron} by the yield of true photons from π^0 decays, we eventually get the following expression:

$$R_\gamma = \frac{\gamma^{incl}}{\gamma^{hadron}} = \frac{\frac{\gamma^{incl}}{\gamma^{\pi^0}}}{\frac{\gamma^{hadron}}{\gamma^{\pi^0}}} = \frac{\langle \epsilon f \rangle \left(\frac{N_\gamma^{incl}}{N_\gamma^{\pi^0}} \right)}{\frac{\gamma^{hadron}}{\gamma^{\pi^0}}}, \tag{1}$$

where the factor $\langle \epsilon f \rangle$ in the numerator in the right hand side is the efficiency and acceptance correction for pion tagging, which is obtained by running a full GEANT3-based simulation to account for the detector efficiency and acceptance effects. This factor quantifies the efficiency with which one can successfully tag a conversion pair as coming from a π^0 decay given that conversion pair is already reconstructed. And for measuring the π^0 yield, we pair the inclusive photons reconstructed from conversions with other photons reconstructed in the PHENIX electromagnetic calorimeter. N_γ^{incl} is the experimentally measured inclusive photon yield, and $N_\gamma^{\pi^0}$ is the measured tagged photo yield. Therefore the ratio $\frac{N_\gamma^{incl}}{N_\gamma^{\pi^0}}$ comes from data analysis. The second ratio in the denominator $\frac{\gamma^{hadron}}{\gamma^{\pi^0}}$, called Cocktail Ratio, is generated in a Monte Carlo simulation for estimating the ratio of hadronic decay photons to those of π^0 decays. The advantage of this method of double ratio tagging reduces the systematic uncertainty in R_γ measurements because the systematic uncertainties from the acceptance/efficiency corrections and conversion probability cancel out explicitly in the double ratio of Eq. (1).

After R_γ is measured, one can obtain the invariant yield using the following formula:

$$\text{Invariant Yield} = (R_\gamma - 1)\,\gamma^{hadron}. \tag{2}$$

The middle and right plots of Fig. 1 show the invariant yield of low momentum direct photons in minimum bias Au+Au at 62.4 GeV and 39 GeV. Fig. 2 shows the invariant yields of low momentum direct photons in Au+Au 0-20% and 20-40% centrality bins at 200 GeV obtained with the external conversion method with VTX. Fig. 3 shows similar yields but in 40-60% and 60-93% centrality bins. In Fig. 2 and Fig. 3 we see how four direct photon measurement results,

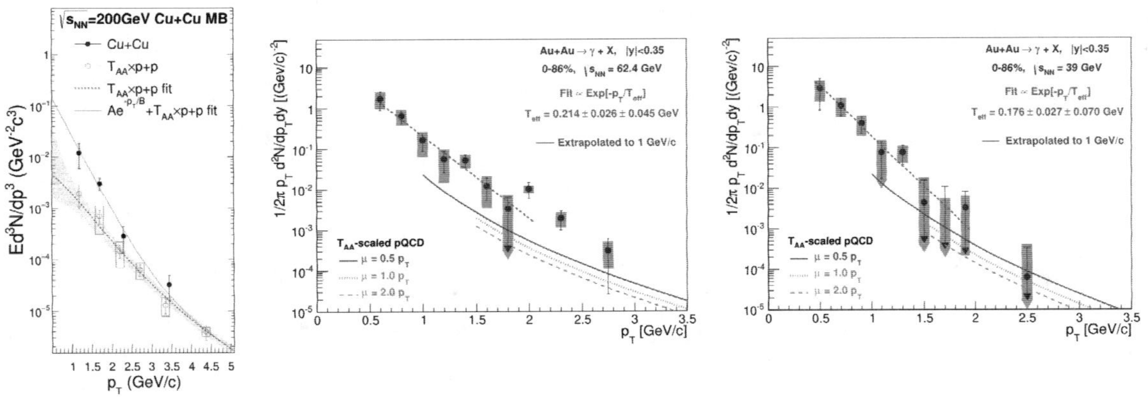

Figure 1. Direct photon p_T-spectra in Cu+Cu collisions at $\sqrt{s_{NN}} = 200$ GeV (left), measured by the internal conversion method [13]; in Au+Au collisions at $\sqrt{s_{NN}} = 62.4$ GeV (middle) and at $\sqrt{s_{NN}} = 39$ GeV (right), both measured by the external conversion method with HBD [14, 15, 16]. All the data are in minimum bias selection. The left plot also shows the N_{coll} scaled p+p fit, the middle and right plots show the N_{coll} scaled curves from pQCD calculations [19].

obtained independently of the analysis methods, are consistent with each other, which can be considered as validation of the PHENIX measurements.

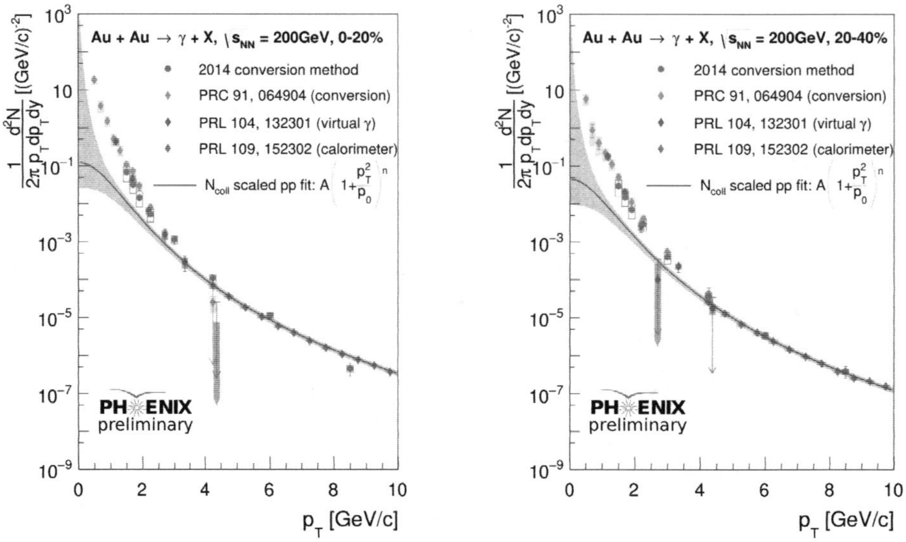

Figure 2. Direct photon p_T-spectra in Au+Au collisions in 0-20% (left) and 20-40% (right) centrality bins at $\sqrt{s_{NN}} = 200\,\mathrm{GeV}$, measured by the external conversion method with VTX. A significant excess over the N_{coll} scaled p+p yield is observed below $p_T \sim 3\,\mathrm{GeV/c}$ in these (semi-)central Au+Au collisions. The data are also consistent with the N_{coll} scaled p+p fit at high-p_T. This figure is from [18].

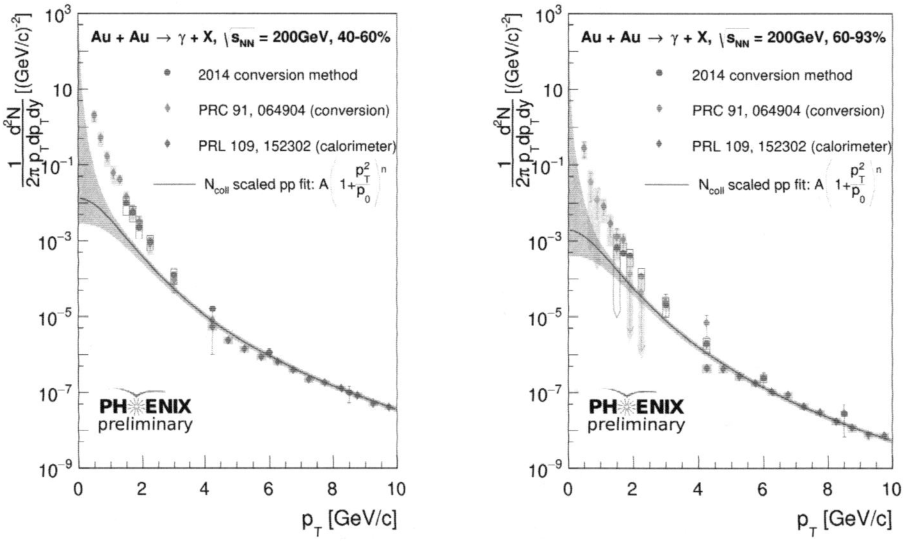

Figure 3. Direct photon p_T-spectra in Au+Au collisions in 40-60% (left) and 60-93% (right) centrality bins at $\sqrt{s_{NN}} = 200\,\mathrm{GeV}$, measured by the external conversion method with VTX. The enhancement observed in Fig. 2 persists below $p_T \sim 3\,\mathrm{GeV/c}$ in these (semi-)peripheral Au+Au collisions as well. This figure is from [18].

3. Direct photon p_T-spectra from small systems

PHENIX has also measured direct photon invariant cross sections in small system collisions, such as in p+p and minimum bias d+Au at 200 GeV [20]. These measurements have been accomplished by the internal conversion method. Nevertheless, there are recent measurements by the external conversion method with VTX, using p+p and p+Au data sets taken in 2015. These results are shown in Fig. 4. The left plot shows that consistent results are obtained in p+p collisions independent of the analysis method. The right plot shows a non-zero excess yield of low momentum direct photons within systematic uncertainties ($\sim 1\ \sigma$) in 0-5% p+Au collisions above the scaled p+p fit. This observation is consistent with a possible formation of QGP droplets in small systems [22]. See the caption of Fig. 4 for more details.

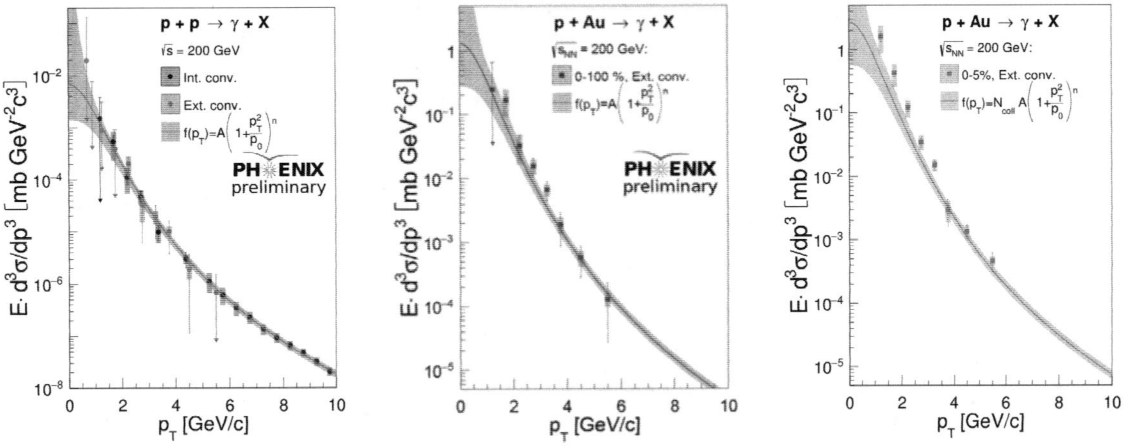

Figure 4. Invariant cross section in p+p and p+Au [15, 21]. The left plot shows the comparison of invariant cross sections measured in p+p collisions at $\sqrt{s_{NN}} = 200\,\mathrm{GeV}$ from the internal conversions (black data points below 5 GeV/c), calorimeter method (black data points above 5 GeV/c), and external conversions (red data points). The shown fit is the most recent PHENIX fit obtained with all these p+p data points. The middle plot shows the invariant cross section from the external conversions with VTX in minimum bias p+Au collisions at $\sqrt{s_{NN}} = 200\,\mathrm{GeV}$; the right plot correspondingly shows the cross section in 0-5% p+Au at $\sqrt{s_{NN}} = 200\,\mathrm{GeV}$. The fit used in these plots are also used in the left panel of Fig. 1 as well as in Figs. 2 and 3.

4. Scaling properties of direct photons

Now we can discuss the scaling properties of direct photons, after having shown the p_T-spectra from large and small system collisions in the previous sections.

If we use the number of nucleons participating in A+A collisions, N_{part}, or the number of binary nucleon-nucleon interactions, N_{coll}, in this case one can compare direct photon data from different centrality classes (or system sizes) for a given center-of-mass energy. Meanwhile, when we compare data at different energies, we can first be based upon measuring the system size through the pseudorapidity density of produced charge particles at midrapidity, $dN_{ch}/d\eta|_{\eta\approx 0}$. This charged-hadron multiplicity can be used in such a case because it has an advantage of being an experimental observable. Moreover, it measures not only the volume but also the energy density. Besides, $dN_{ch}/d\eta|_{\eta\approx 0}$ itself has an interesting scaling relation with N_{coll}, as shown in Fig. 5. By this relation we can better understand the interplay between the system size and hard scattering processes. In Fig. 5 four datasets are simultaneously fitted by a power-law, with vertical and horizontal uncertainties of N_{coll} and $dN_{ch}/d\eta|_{\eta\approx 0}$, respectively [23, 24].

We see that N_{coll} scales like $(dN_{ch}/d\eta|_{\eta\approx 0})^{\alpha}$ with a $\sqrt{s_{NN}}$ -dependent proportionality factor called specific yield, SY. It is a logarithmically slowly increasing function with $\sqrt{s_{NN}}$ shown in the figure inset, and has the following form: $\mathrm{SY} \approx 0.98 \log\left(\sqrt{s_{NN}}\right) - 1.83$. For the power α we find 1.25 ± 0.02. Thereby, one can use $(dN_{ch}/d\eta|_{\eta\approx 0})^{\alpha}$ in order to scale direct photon data for

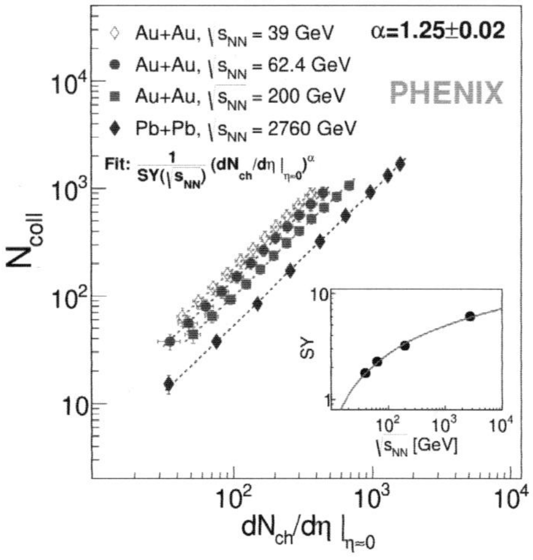

Figure 5. The scaling relation of N_{coll} vs. $dN_{ch}/d\eta|_{\eta=0}$ for four data sets in Au+Au at $\sqrt{s_{NN}} = 200$ GeV, 62.4 GeV, 39 GeV and in Pb+Pb at $\sqrt{s_{NN}} = 2760$ GeV. Across different beam energies N_{coll} is approximately proportional to $(dN_{ch}/d\eta|_{\eta\approx 0})^{1.25}$ with an energy-dependent proportionality factor shown in the inset. This figure is from [14].

different colliding species, centrality bins and beam energies.

Let us take, e.g., the direct photon p_T-spectra in minimum bias Au+Au collisions at 62.4 GeV and 39 GeV with the corresponding pQCD curves from Fig. 1, and normalize them by $(dN_{ch}/d\eta|_{\eta\approx 0})^{\alpha}$. It results in normalized data falling on top of each other at low-p_T, as shown in the panel (a) of Fig. 6. We also see that the normalized ISR p+p data coincide with the pQCD calculations at high-p_T within the quoted uncertainties. If we now take the direct photon data in Au+Au at 200 GeV from [1, 2, 12], and normalize them in the same way (panel (b)), then one can see that the normalized data are on top of each other at low- and high-p_T. At low-p_T they are clearly above the normalized p+p data, p+p fit and pQCD curve at 200 GeV. Finally, in the panel (c) we show the normalized data compared for different beam energies from 62.4 GeV to 2760 GeV. Again the conclusion is that all the data coincide at low-p_T within experimental uncertainties, while the expected difference with $\sqrt{s_{NN}}$ and N_{coll} scaling at high-p_T is seen too.

In Fig. 6 all the normalized data, p+p fit, pQCD curves are from [14, 15, 16, 21]. And they are obtained with Au+Au data from [1, 2, 12], Pb+Pb data from [25], p+p data at 200 GeV from [20], p+p data at 62.4/63 GeV from [26]/[27]. The empirical fit to the p+p data at 200 GeV is from [13, 14, 15, 21], the pQCD calculations at different beam energies are from [6, 19], and the data on $dN_{ch}/d\eta$ is from [23, 24]. Each error bar is the quadratic sum of the corresponding systematic and statistical uncertainties, and uncertainties on $dN_{ch}/d\eta$ are not included.

Another way to bring out more clearly some of the features of the discussed scaled data is to integrate the direct photon p_T-spectra above some p_T threshold value. The result, which is another representation of direct photon scaling is shown in Fig. 7 and Fig. 8, where the integrated yield dN_{γ}/dy grows faster than the multiplicity $dN_{ch}/d\eta|_{\eta\approx 0}$. The figures also show the integrated yield of the fit to p+p data, and of the three different pQCD calculations scaled by N_{coll} (which are extrapolated down to $p_T = 1$ GeV/c). One of the interesting features is that the prompt photons (the purple band) and integrated pQCD lines have nearly the same slopes as the slope describing the large systems. The integrated A+A yield is a factor of ~ 10 larger than the expected yield from p+p.

Journal of Physics: Conference Series **1602** (2020) 012015 doi:10.1088/1742-6596/1602/1/012015

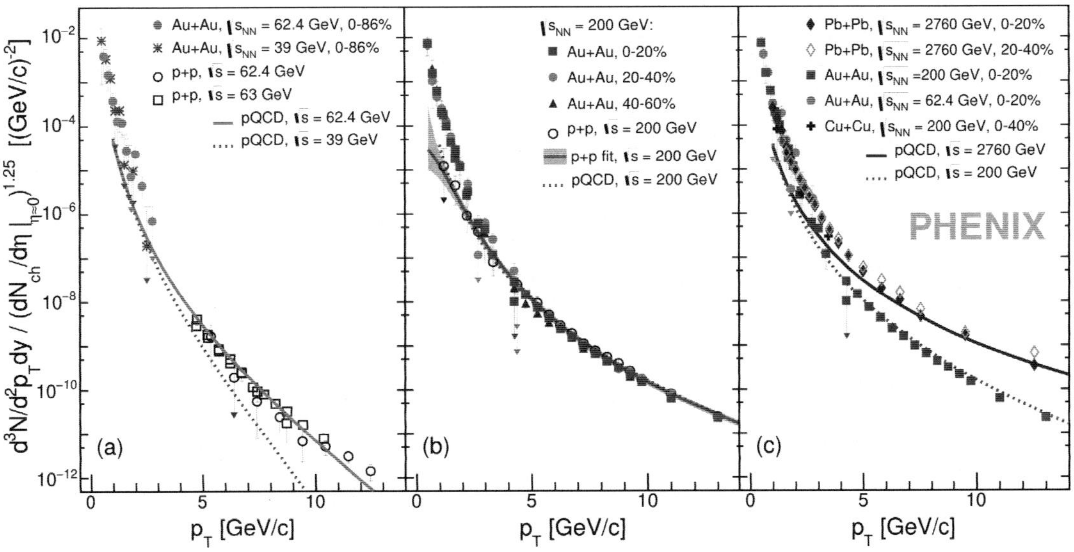

Figure 6. Direct photon p_T-spectra normalized by $(dN_{ch}/d\eta)^{\alpha}$ from [14, 15, 16]. The comparison is shown for minimum bias Au+Au collisions at 62.4 GeV and 39 GeV in the panel (a); for Au+Au data in three centrality bins at 200 GeV in the panel (b); and for different A+A systems at three beam energies in the panel (c). The panels (a) and (b) also show p+p data, and all the panels show pQCD calculations at respective energies.

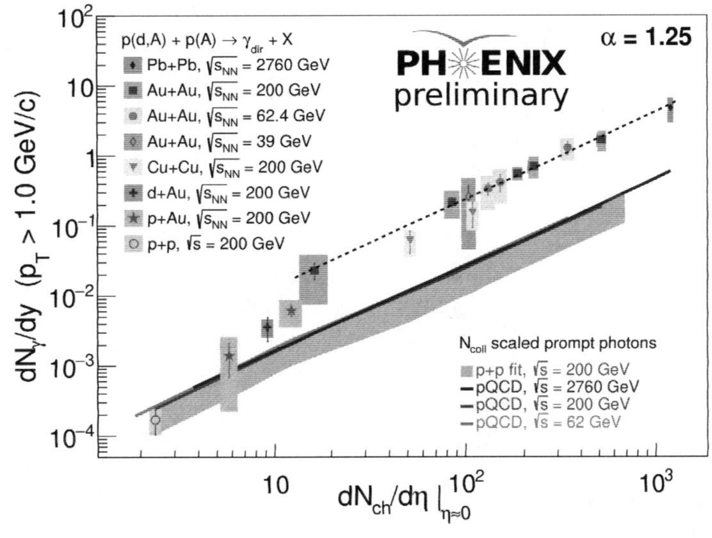

Figure 7. The integrated direct photon yield at $p_T > 1\,\mathrm{GeV/c}$ from various colliding systems as a function of the charged-hadron multiplicity at mid-rapidity. For integration some of the data are taken from Fig. 6. The other data are Cu+Cu from [13], p+Au data from Fig. 4 [15, 21], and d+Au data from [20]. The magenta band represents the N_{coll} scaled p+p data at $\sqrt{s_{NN}} = 200\,\mathrm{GeV}$, and the colored lines represent the N_{coll} scaled pQCD calculations at different energies. This figure is from [15, 16, 21].

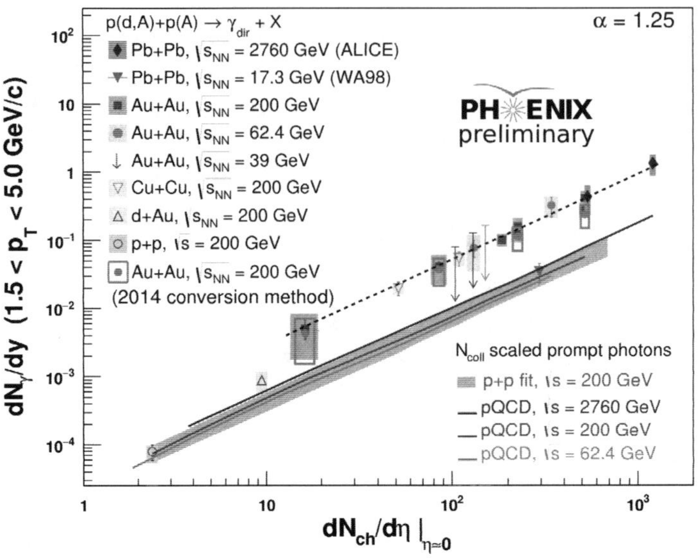

Figure 8. The integrated direct photon yield at $p_T > 1.5\,\mathrm{GeV/c}$ from various colliding systems as a function of the charged-hadron multiplicity at mid-rapidity. Most of the data are already shown in Fig. 7 but integrated above 1 GeV/c. One of the additions here is the p_T-integrated Au+Au data in four centrality bins at $\sqrt{s_{NN}} = 200\,\mathrm{GeV}$, the p_T-spectra of which are shown in Figs. 2 and 3. This figure is from [18].

Another quite interesting feature, which can be seen in Fig. 7 and Fig. 8, is that the PHENIX data from p+A and d+Au indicate a rapid onset of direct photon excess around $dN_{ch}/d\eta|_{\eta\approx 0}$ in the region of $\sim 5 \div 20$. There seems to be another trend stemming from small systems different from that of large systems. Both trends suggest the existence of a QGP turn off or turn on transition region between both types of collision systems.

5. Summary and outlook

The PHENIX collaboration has measured low momentum direct photons in large system collisions, such as Au+Au at 200 GeV, 62.4 GeV, 39 GeV and in Cu+Cu at 200 GeV. Also in small system collisions, such as p+p, p+Au and d+Au at 200 GeV. Recent measurements in Au+Au at 200 GeV with higher statistical precision is consistent with the previous measurements in Au+Au at 200 GeV within experimental uncertainties. Having all the available data, including the data from the ALICE experiment, PHENIX has observed a surprising scaling behavior of low momentum direct photons in large systems: namely, the low- and high-p_T direct photon invariant yields in A+A collisions scale with N_{coll} at a given center-of-mass energy; meanwhile, across different energies N_{coll} is proportional to $(dN_{ch}/d\eta|_{\eta\approx 0})^{\alpha}$; and for all energies the low-p_T yields scale like $(dN_{ch}/d\eta|_{\eta\approx 0})^{\alpha}$. PHENIX has also discovered a non-zero excess of low momentum direct photons (within systematic uncertainties) at low-p_T in the most central p+Au collisions above N_{coll} scaled p+p fit, which is consistent with a formation of QGP droplets in small systems. The data in this low and narrow multiplicity region suggests the existence of a thermal transition region when we go from p+p to A+A systems or vice versa. More direct photon data should be analyzed before we have better understanding on what happens in this region. Besides, having more data from various colliding species and collision centrality selections at different beam energies may help constrain the sources of low momentum direct photons.

Journal of Physics: Conference Series **1602** (2020) 012015 doi:10.1088/1742-6596/1602/1/012015

References

[1] Adare A *et al.* (PHENIX Collaboration) 2010 *Phys. Rev. Lett.* **104** 132301 (*Preprint* 0804.4168)
[2] Adare A *et al.* (PHENIX Collaboration) 2015 *Phys. Rev.* C **91** 064904 (*Preprint* 1405.3940)
[3] Adare A *et al.* (PHENIX Collaboration) 2016 *Phys. Rev.* C **94** 064901 (*Preprint* 1509.07758)
[4] van Hees H, He M and Rapp R 2015 *Nucl. Phys.* A **933** 256 (*Preprint* 1404.2846)
[5] Rapp R, van Hees H and He M 2014 *Nucl. Phys.* A **931** 696 (*Preprint*1408.0612)
[6] Paquet J-F *et al.* 2016 *Phys. Rev.* C **93** no. 4 044906 (*Preprint* 1509.06738)
[7] Shen C, Heinz U, Paquet J-F and Gale C 2014 *Phys. Rev.* C **89** 044910 (*Preprint* 1308.2440)
[8] Linnyk O, Cassing W and Bratkovskaya E 2014 *Phys. Rev. C* **89** 034908 (*Preprint* 1311.0279)
[9] Linnyk O, Konchakovski V, Steinert T, Cassing W and Bratkovskaya E 2015 *Phys. Rev.* C **92** 054914 (*Preprint* 1504.05699)
[10] Kim Y M, Lee C-H, Teaney D and Zahed I 2017 *Phys. Rev.* C **96** no. 1 015201 (*Preprint* 1610.06213)
[11] Lee C-H and Zahed I 2014 *Phys. Rev.* C **90** 025204 (*Preprint* 1403.1632)
[12] Afanasiev S *et al.* (PHENIX Collaboration) 2012 *Phys. Rev. Lett.* **109** 152302 (*Preprint* 1205.5759)
[13] Adare A *et al.* (PHENIX Collaboration) 2018 *Phys. Rev.* C **98** no. 5 054902 (*Preprint* 1805.04066)
[14] Adare A *et al.* (PHENIX Collaboration) 2019 *Phys. Rev. Lett.* **123** 022301 (*Preprint* 1805.04084)
[15] Khachatryan V (PHENIX Collaboration) 2019 *Nucl. Phys.* A **982** 763 (*Preprint* 1812.01841)
[16] Drees A (PHENIX Collaboration) 2019 *PoS* **HardProbes2018** 176
[17] Adamczyk L *et al.* (STAR Collaboration) 2017 *Phys. Lett.* B **770** 451 (*Preprint* 1607.01447)
[18] Fan W (for PHENIX Collaboration) Quark Matter 2019 - the XXVIIIth International Conference on Ultra-relativistic Nucleus-Nucleus Collisions (*Thermal photon production in Au+Au collisions*)
[19] Paquet J.-F 2017 *Private communication*
[20] Adare A *et al.* (PHENIX Collaboration) 2013 *Phys. Rev.* C **87** 054907 (*Preprint* 1208.1234)
[21] Novitzky N (PHENIX Collaboration) 2019 *PoS* **HardProbes2018** 185
[22] Aidala C *et al.* (PHENIX Collaboration) 2019 *Nature Phys.* **15** no. 3 214 (*Preprint* 1805.02973)
[23] Adare A *et al.* (PHENIX Collaboration) 2016 *Phys. Rev.* C **93** 024901 (*Preprint* 1509.06727)
[24] Aamodt K *et al.* (ALICE Collaboration) 2011 *Phys. Rev. Lett.* **106** 032301 (*Preprint* 1012.1657)
[25] Adam J *et al.* (ALICE Collaboration) 2016 *Phys. Lett.* B **754** 235 (*Preprint* 1509.07324)
[26] Angelis A S *et al.* (CCOR Collaboration) 1980 *Phys. Lett.* B **94** 106
[27] Angelis A S *et al.* (CMOR Collaboration) 1989 *Nucl. Phys.* B **327** 541; Akesson T *et al.* (AFS Collaboration) 1990 *Sov. J. Nucl. Phys.* **51** 836

Dilepton measurements with the ALICE experiment at the LHC

Elisa Meninno on behalf of the ALICE Collaboration
Stefan Meyer Institute for Subatomic Physics, Boltzmanngasse 3, 1090, Wien
E-mail: elisa.meninno@cern.ch

Abstract. The production of low-mass dielectrons and dimuons is a powerful tool for the understanding of the properties of the quark–gluon plasma (QGP) created in ultra-relativistic heavy-ion collisions. Since such pairs do not interact strongly and are emitted during all stages of the collisions, they provide information about the full time evolution and dynamics of the medium created. Measurements in pp and p–Pb collisions are the necessary reference for heavy-ion studies. In addition, they can be used to extract charm and beauty cross sections.
In this contribution, we present the latest measurements of e^+e^- and $\mu^+\mu^-$ pair production in pp, p–Pb and Pb–Pb collisions performed by ALICE at different energies.

1. Introduction

Photons and lepton pairs (electrons or muons) are penetrating probes that allow us to investigate the full time evolution and dynamics of the dense and hot medium created in ultra-relativistic heavy-ion collisions, since they don't undergo strong interactions in the final state. Dileptons can be emitted during all the stages of the collision and a differential study as a function of the invariant mass $m_{l^+l^-}$ and transverse momentum p_{T,l^+l^-} (with $l = e, \mu$) allow one to disentangle the various sources. At low invariant masses ($m_{\mathrm{ee}} < 1$ GeV/c^2) Dalitz decays of pseudoscalar mesons (π^0, η, η',..) and direct decays of vector mesons (ρ, ω, ϕ) are the main dilepton sources. Measurements of ϕ mesons, due to their $s\bar{s}$ quark content, are useful to study the strangeness production, which is one of the key observables for the QGP. Studying hadron production in different transverse momentum (p_{T}) regions gives important information on the hadronization processes. Strangeness production, indeed, can be studied in two regions. While at high p_{T} it is dominated by hard QCD processes, such as partonic scattering, which can be described by perturbative calculations, at low p_{T} is dominated by soft non-perturbative processes, described by phenomenological models. The ρ meson is potentially sensitive to chiral symmetry restoration, which is expected to happen close to the phase transition to the QGP in heavy-ion collisions [1]. This meson has a shorter lifetime than that of the medium. Therefore, its spectral function, which can be inferred from the study of its dielectron decay channel, is affected by the dense medium and the predicted restoration of chiral symmetry. Also the ϕ meson may be sensitive to the chiral symmetry restoration, that could be observed by measuring a broadening of its spectral function, although no evidence of such an effect has been observed so far. Dileptons also originate from thermal QGP radiation from the medium in heavy-ion collisions. The thermal radiation, from both the QGP and the hadronic phase, contributes over a broad mass range. At intermediate masse ranges (IMR) ($1 < m_{\mathrm{ee}} < 3$ GeV/c^2) measurements

of thermal radiation are very challenging, because of the large contribution of dileptons from semileptonic decays of correlated heavy-flavour hadrons [2]. Therefore, it is important to improve the knowledge of the contribution from charm- and beauty-hadron decays.
Measurements in small systems, pp and p–Pb collisions, are the necessary reference for studies in Pb–Pb collisions.
In addition to dileptons produced by hadronic interactions, also coherent photo-production of dileptons via electromagnetic interactions plays a role. This process is usually studied in ultra-peripheral collisions, where no strong interactions between colliding nuclei take place and the condition of coherence, i.e. the assumption that the produced electromagnetic field probes the whole nucleus, is easier to be maintained.

2. Experimental apparatus and data samples

ALICE [3, 4] is the heavy-ion dedicated experiment at CERN with an apparatus characterized by unique capabilities for Particle Identification (PID) and tracking at low transverse momenta, aimed to study the properties of the QGP. The detector consists of a central barrel at mid-rapidity, a muon spectrometer at forward rapidity and a set of detectors (forward and backward rapidity) for global collision characterization and triggering purposes.
Electrons are reconstructed and identified at mid-rapidity using the Inner Tracking System (ITS) [5] (for vertexing and tracking), the Time Projection Chamber [6] (for tracking and PID) and the Time-Of-Flight [7] (for PID). Low-mass dimuon production is studied with the muon spectrometer [8] at forward rapidity.
Dilepton measurements have been performed by ALICE in different collision systems (pp , p–Pb and Pb–Pb collisions) at different centre-of-mass energies. In this contribution, the most recent dielectron measurements in pp and p–Pb collisions at $\sqrt{s_{\mathrm{NN}}}$= 5.02 TeV, in pp collisions at $\sqrt{s}$= 13 TeV and in Pb–Pb collisions at $\sqrt{s_{\mathrm{NN}}}$= 5.02 TeV are presented. At $\sqrt{s}$= 13 TeV pp collisions were collected with a particular detector setup, characterized by a reduced solenoidal magnetic field to explore dielectron regions at very low $p_{\mathrm{T,ee}}$ and m_{ee}, which was not accessible with higher magnetic field. Results related to the production of ϕ mesons, studied through its decay in muon pairs, in Pb–Pb collisions at $\sqrt{s_{\mathrm{NN}}}$= 2.76 TeV and in pp collisions at different centre-of-mass energies ($\sqrt{s}$= 2.76, 5.02, 8 and 13 TeV) are also presented.

3. Analysis strategy

First of all, selections on the track quality in the reconstruction procedure are applied. Electron candidates are selected from tracks reconstructed in the ITS and in the TPC. Electrons are identified using information on specific energy loss in the TPC and time-of-flight measurements in the TOF detector. For the dimuon analysis, tracks are reconstructed in the muon spectrometer. Tracks reconstructed in the muon tracking chambers have to match track segments reconstructed in the trigger chambers.
An opposite-sign invariant mass distribution is created combining e^+ (μ^+) and e^- (μ^-) tracks. This distribution contains not only signal, but also background from combinatorial pairs, as well as residual correlations from jets and from electron-positron pairs from photon conversion in the detector material. The background contribution is estimated from the distribution of same-sign pairs from the same event in the dielectron analysis, as explained in [20], or using the event mixing technique, combining muon tracks from different events [9].
The signal is then extracted subtracting the background contribution, corrected for the different acceptance of opposite-sign and same-sign pairs (in the dielectron analysis), from the opposite-sign spectrum. Photon conversion pairs are removed, for example, with selections on the opening angle of the pairs relative to the magnetic field. The dilepton measurement is compared with the expected contributions from light and heavy-flavour hadron contributions, the so-called *hadronic cocktail*.

4. Production of ϕ mesons in pp collisions

The production of ϕ mesons has been studied via the dimuon decay channel in pp collisions at $\sqrt{s} = 5.02$, 8 and 13 TeV at forward rapidity ($2.5 < y < 4$). In Fig. 1 the dimuon invariant mass spectrum in pp collisions at $\sqrt{s} = 13$ TeV after the subtraction of the combinatorial background is shown. The spectrum is fitted with a superposition of the hadronic cocktail (light-flavour resonances) and a continuum contribution due to semi-muonic decays of charm and beauty hadrons, described with an empirical function. For each p_T and rapidity interval, the raw number of ϕ mesons is determined via a fit procedure based on a χ^2 minimisation.

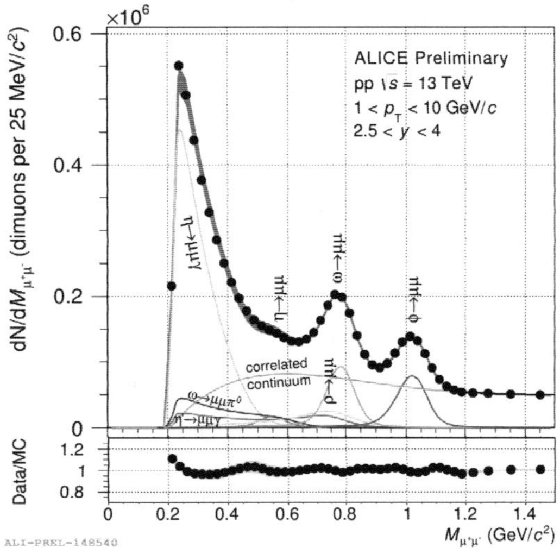

Figure 1. Dimuon invariant mass spectrum in pp collisions at $\sqrt{s} = 13$ TeV, at forward rapidity, fitted with the hadronic cocktail.

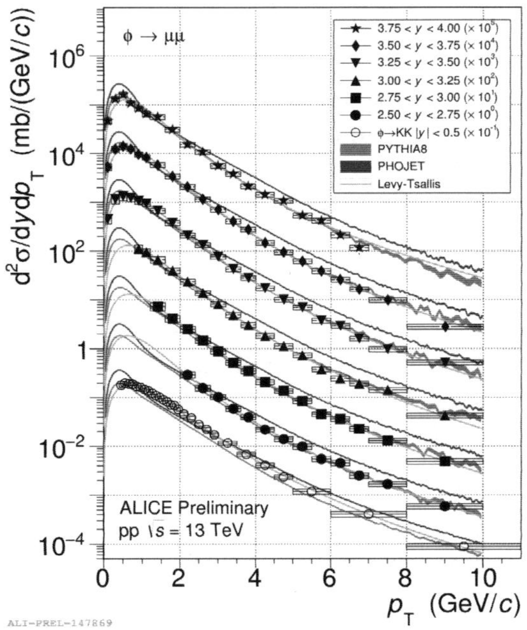

Figure 2. Differential ϕ-meson production cross section as a function of p_T in pp collisions at $\sqrt{s} = 13$ TeV, shown in several rapidity intervals, and compared with PYTHIA8 and PHOJET models and with measurements at mid-rapidity.

The p_T-differential ϕ-meson production cross section is shown in Fig. 2 in different rapidity intervals, compared with two model calculations, PYTHIA8 [10] and PHOJET [11], and previous measurements at mid-rapidity [12,13]. Cross sections are fitted with a Levy-Tsallis function. In general, a softening of the p_T spectra with increasing rapidity is observed. PYTHIA8 provides a quite good description of the different p_T distributions, in all rapidity ranges while PHOJET is in agreement with the data only for $1 < p_T < 2$ GeV/c. The energy dependence of the ϕ-meson production cross section has been also studied, as it can be seen in Fig. 3, where the ϕ-meson production cross section integrated in $1.5 < p_T < 5$ GeV/c is plotted as a function of $\sqrt{s}$. PHOJET describes quite well the energy dependence, while PYTHIA significantly underestimates the data. A hardening of the p_T spectra with increasing energy is observed [14].

Measurements of ϕ-meson production in Pb–Pb collisions have been performed at $\sqrt{s_{NN}} =$ 2.76 TeV [15] and more recently at 5.02 TeV. Also in this case, results at forward and mid-rapidity have been compared. Their compatibility hints to similarities in the interaction of

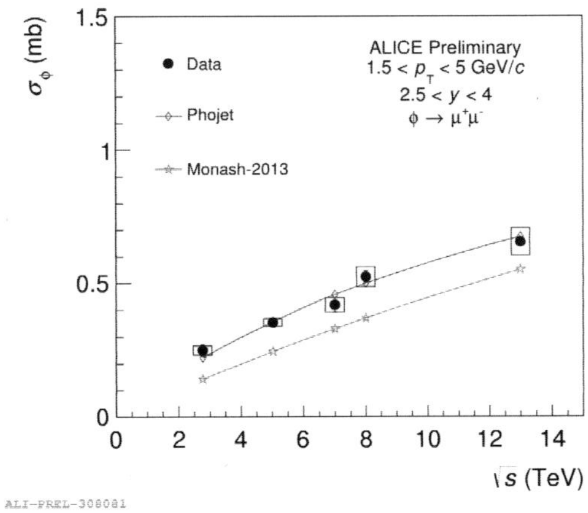

Figure 3. ϕ meson-production cross section for $1.5 < p_T < 5$ GeV/c and $2.5 < y < 4$, in pp collisions as a function of the centre-of-mass energy, compared with PHOJET and PYTHIA8 models.

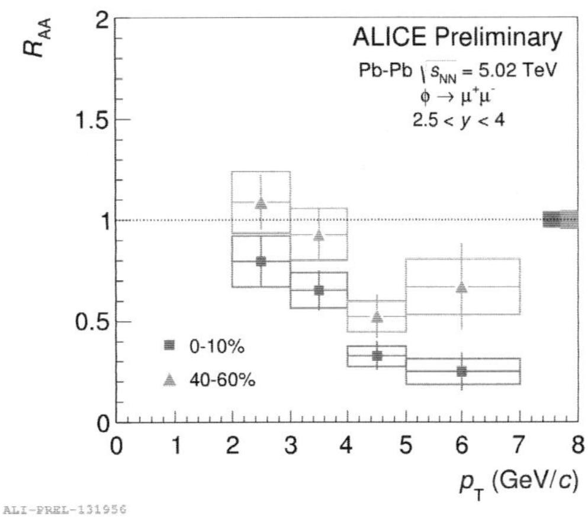

Figure 4. Nuclear modification factor R_{AA} of ϕ mesons, measured in Pb–Pb collisions at $\sqrt{s_{NN}}$= 5.02 TeV at forward rapidity, in 0–10% and 40–60% centrality intervals.

the ϕ meson with the medium in the two rapidity ranges at intermediate p_T. Figure 4 shows the nuclear modification factor R_{AA} measured in two different centrality classes. In peripheral collisions, the R_{AA} is compatible with unity within uncertainties, as expected in the case that these collisions behave as a superposition of incoherent pp collisions. In most central collisions, R_{AA} at forward rapidity is reduced to about 0.65, showing a clear suppression with respect to the pp reference in the intermediate p_T region.

5. Dielectron production in pp and p-Pb collisions

Recently, dielectron production has been measured in pp and in p-Pb collisions at $\sqrt{s_{NN}}$ = 5.02 TeV [16]. In Fig. 5 the dielectron production cross section in pp collisions is shown, as a function of the invariant mass m_{ee} (left panel) and transverse momentum of the pair (right panel). Charm and beauty cross sections are estimated by fitting the m_{ee} and $p_{T,ee}$ distributions in the IMR with templates obtained with two different event generetors, the leading-order parton shower generator PYTHIA [17] and the next-to-leading order generator POWHEG [18], for which hadronization of c- and b-quarks and decays of heavy-flavour hadrons are still implemented with PYTHIA. The two generators differ in the implementation of heavy-quark production mechanisms. Cocktails obtained with both calculations reproduce quite well the data, although POWHEG provides a slightly better description of the measurement at low m_{ee} .

The charm- and beauty-production cross sections evaluated with the two event generators are in agreement with previous measurements, performed through the reconstruction of single heavy-flavour hadron decays [19]. The difference in the cross section values obtained with the two models is compatible with the difference already observed in previous measurements, performed in pp collisions at 7 TeV [20] and 13 TeV [21]. As already observed in these previous publications, this difference reflects the sensitivity of the dielectron measurement to production mechanisms implemented in the Monte Carlo generators. A summary plot of the measurements of charm and beauty cross sections performed at different centre-of-mass energies is shown in Fig. 6.

The nuclear modification factor R_{pPb} is shown in Fig. 7 as a function of m_{ee} and for $p_{T,ee} < 8$ GeV/c. Data are compared with the R_{pPb} of the hadronic cocktail. Two additional models for the cocktail are taken into account and compared with the data. One includes cold-

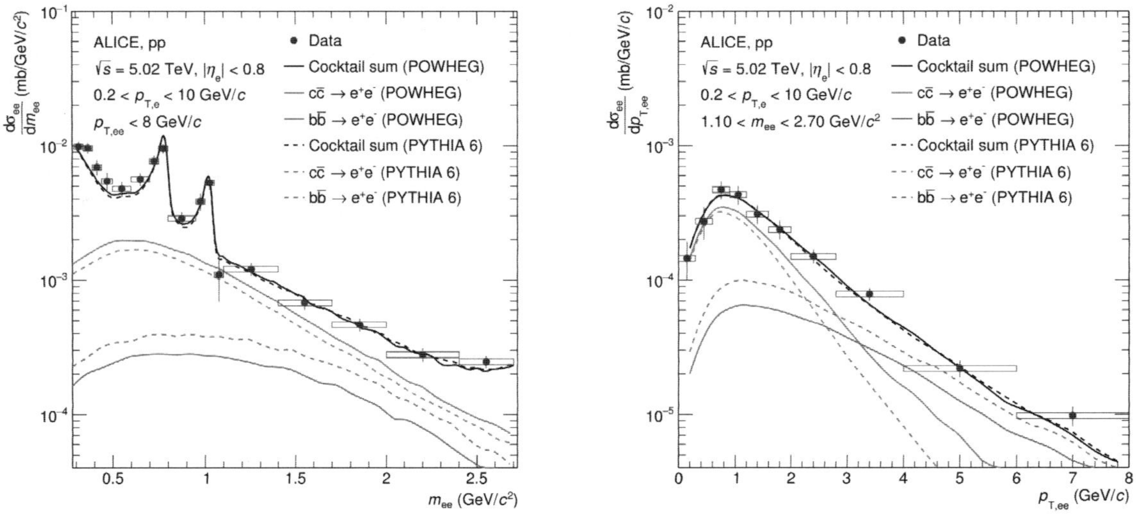

Figure 5. Dielectron cross section as a function of m_{ee} (left) and $p_{\mathrm{T,ee}}$ (right). Data in the IMR are fitted, in m_{ee} and $p_{\mathrm{T,ee}}$, with two different templates for open charm and beauty contributions, obtained using, respectively, POWHEG and PYTHIA event generators.

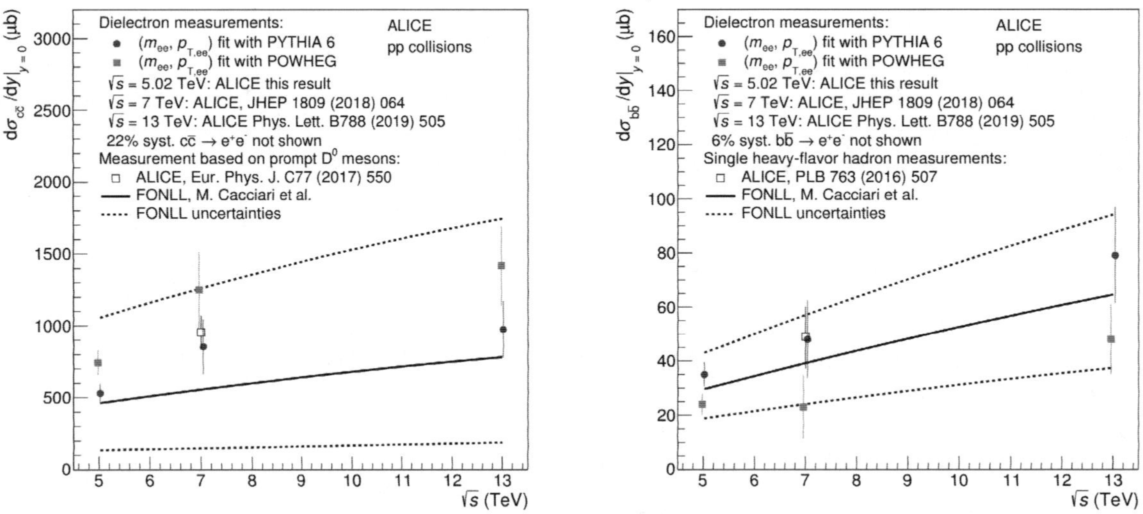

Figure 6. Charm and beauty production cross sections measured at midrapidity, as a function of $\sqrt{s}$ in pp collisions.

nuclear-matter (CNM) effects on the production of dielectrons from semileptonic decays of charm hadrons. CNM effects are estimated using EPS09 nuclear parton distribution functions (nPDF) [22]. The second model includes thermal radiation effects [23, 33]. The measured R_{pPb} stays below the theory expectations for $m_{\mathrm{ee}} < 1.1$ GeV$/c^2$, where the fraction of dielectrons from light-flavour hadron decays is not negligible. The R_{pPb} is instead consistent with unity in the IMR, and the behavior is reproduced by the hadronic cocktail, where no CNM effects are taken into account. The calculation with included CNM effects seems to describe better the data for $m_{\mathrm{ee}} < 1.1$ GeV$/c^2$, while, for the IMR, data are on the upper edge of the systematic uncertainties of the calculation, suggesting negligible CNM effects. Calculations including thermal radiation from hadronic and partonic phases are quite compatible with data within uncertainties, meaning that this source cannot be excluded.

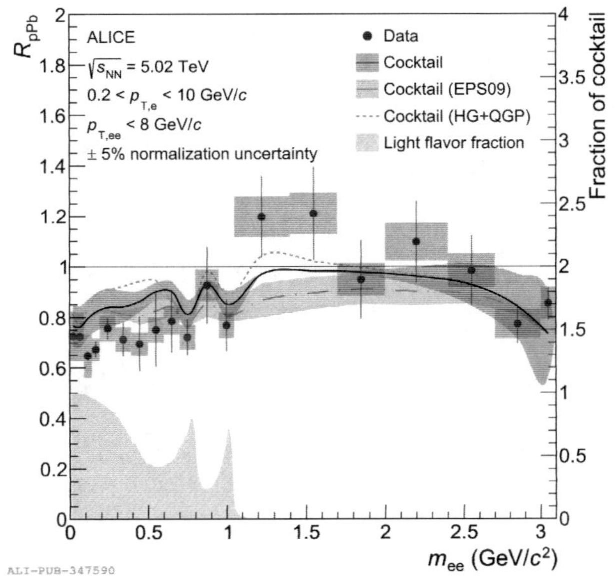

Figure 7. Dielectron nuclear modification factor R_{pPb} as a function of m_{ee} measured in pp and p–Pb collisions at $\sqrt{s}$ = 5.02 TeV. The measurement is compared with two cocktails, one including thermal radiation and the other one including CNM effects on the heavy-flavour production.

5.1. *Soft-dielectron excess in* pp *collisions at* $\sqrt{s}$= *13 TeV*

Figure 8 shows the dielectron cross section in pp collisions at $\sqrt{s}$ = 13 TeV as a function of m_{ee} (on the left) and as a function of $p_{\text{T,ee}}$ (on the right), at very low p_{T} and compared with the hadronic cocktail [25]. The η meson gives the dominant contribution to the hadronic cocktail in the very low p_{T} region under study. In order to estimate this contribution, ALICE measurements of η/π^0 performed in pp and p–Pb collisions have been parametrized and extended to low p_{T} using measurements from CERES/TAPS [26] and assuming that this ratio does not depend on $\sqrt{s}$. In the mass region $0.15 < m_{\text{ee}} < 0.6$ GeV/c^2 and for $p_{\text{T,ee}} < 0.4$ GeV/c, an enhancement of a factor 1.69 ± 0.14 (stat.) ± (syst., data) ± (syst., cockt.) with respect to the hadronic cocktail is observed.

The multiplicity dependence of the observed excess has been investigated, as shown in Fig. 9, where the dielectron yield and the ratio data/cocktail are plotted as a function of the charged-particle multiplicity. In all the multiplicity intervals, the dielectron yield for $0.15 < m_{\text{ee}} < 0.6$ GeV/c^2 is always above the cocktail, however no conclusions can be drawn on the multiplicity trend of the excess. The m_{ee} and $p_{\text{T,ee}}$ distributions of the excess, corrected for the detector acceptance, are shown in Fig. 10 compared with theoretical predictions. Calculations of bremsstrahlung from initial- and final-state hadrons are not able to describe data, as well as a calculation of thermal dielectron yield with a hadronic many-body model [23, 33]. This important result needs confirmation with more precise measurements, which will be possible with the upgraded ALICE detector [32]. If confirmed by future measurements, this result could help to solve the "soft-photon puzzle", i.e. the question of an anomalous production of soft

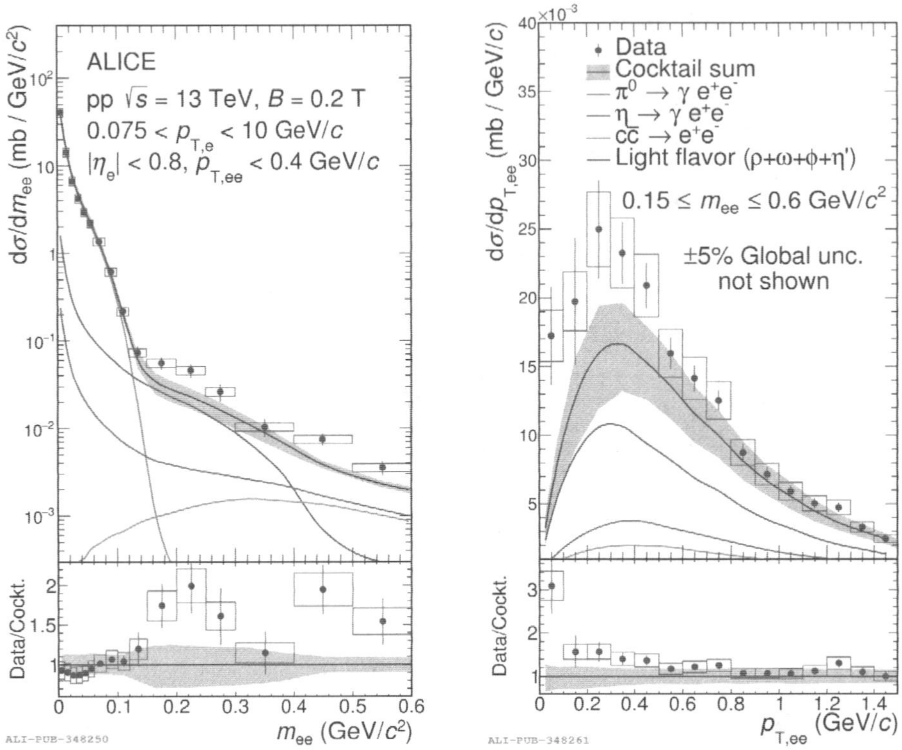

Figure 8. Differential dielectron production cross section as a function of m_{ee} (left) and $p_{\text{T,ee}}$ (right), measured in pp collisions at $\sqrt{s}=$ 13 TeV, with low magnetic field configuration, compared with the hadronic cocktail of the expected sources.

photons (and leptons) in hadronic collisions observed by several experiments since decades and still awaiting for a clear explanation [28–31].

6. Dielectron production measurements in Pb-Pb collisions

Dielectron production has been measured in Pb–Pb collisions at $\sqrt{s_{\text{NN}}} = 2.76$ TeV [2] and 5.02 TeV. The left panel of Fig. 11 shows the dielectron invariant mass distribution measured in central Pb–Pb collisions at $\sqrt{s_{\text{NN}}} = 5.02$ TeV, compared with the hadronic cocktail. There is a hint of enhancement at low m_{ee} (< 0.5 GeV/c^2). This enhancement could be explained as a contribution from thermal radiation from the QGP and hadron gas. This effect looks more pronounced in central collisions, where the thermal radiation contribution is expected to be larger, with respect to peripheral collisions. Thermal radiation calculations by R. Rapp [23, 33] are found to be compatible with the measurement. In the IMR, which is dominated by dielectrons from semileptonic heavy-flavour hadron decays, data are systematically below the hadronic cocktail, as can be observed in the left panel of Fig. 11. In the calculations for the cocktail, effects of open-charm hadron production modification due to cold-nuclear-matter effects are not taken into account. In the right panel of Fig. 11, EPPS16 nuclear parton distribution functions [24] are included in the calculations, without considering any hot-nuclear matter effects. As a consequence, the suppression is reduced and the data are better described by the hadronic cocktail. However, the statistical and systematic uncertainties of the measurement are still too large to allow us to constrain models and quantify the impact of CNM effects on the production of dielectrons from heavy-flavour hadron semileptonic decays. The Pb–Pb data sample collected

Journal of Physics: Conference Series **1602** (2020) 012016 doi:10.1088/1742-6596/1602/1/012016

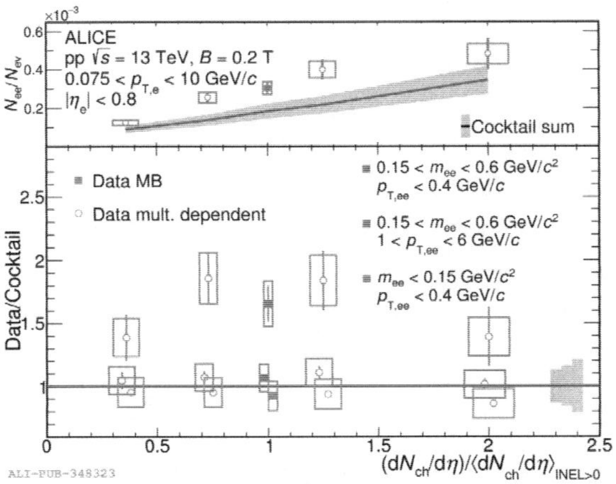

Figure 9. Dielectron yield per event in the excess region compared with the hadronic cocktail (top panel) and data/cocktail ratio in three m_{ee} intervals (bottom panel), as a function of the charged-particle multiplicity.

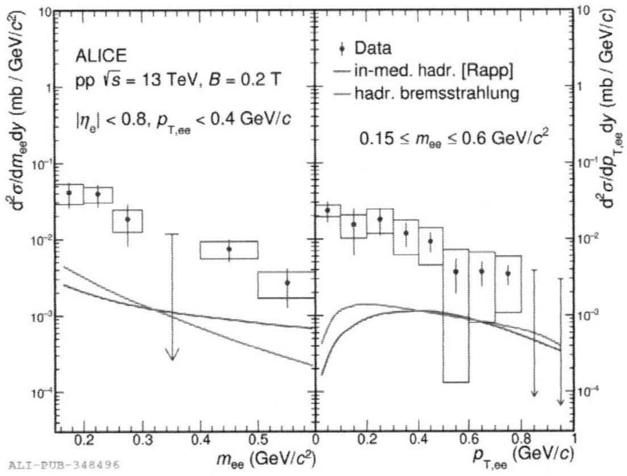

Figure 10. Dielectron excess as a function of m_{ee} (left panel) and $p_{\text{T,ee}}$ (right panel), after subtraction of the hadronic cocktail. Calculations of bremsstrahlung and thermal dielectron production are also shown.

in 2018 will help to improve the precision of the measurements.

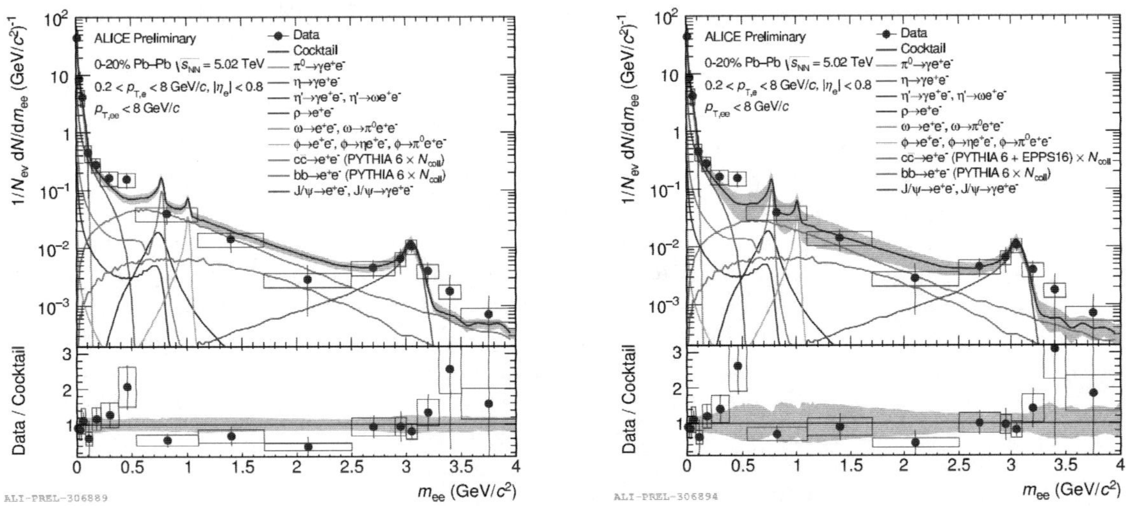

Figure 11. Dielectron invariant mass spectrum in Pb–Pb collisions at $\sqrt{s_{\text{NN}}}$= 5.02 TeV in 0–20% centrality class, compared with the hadronic cocktail. In the right panel, nPDF(EPPS16) are included in the charm contribution to the hadroni cocktail.

6.1. Photo-production in Pb-Pb collisions at $\sqrt{s_{\text{NN}}}$= 5.02 TeV

Photo-production of dielectrons has been studied in the IMR, which does not contain vector mesons and it is dominated by charm and beauty semileptonic decay contributions [34]. In Fig. 12 the dielectron $p_{\text{T,ee}}$ spectra in Pb–Pb collisions in two different centrality intervals, 0–40% (left panel) and 70–90% (right panel) are shown.

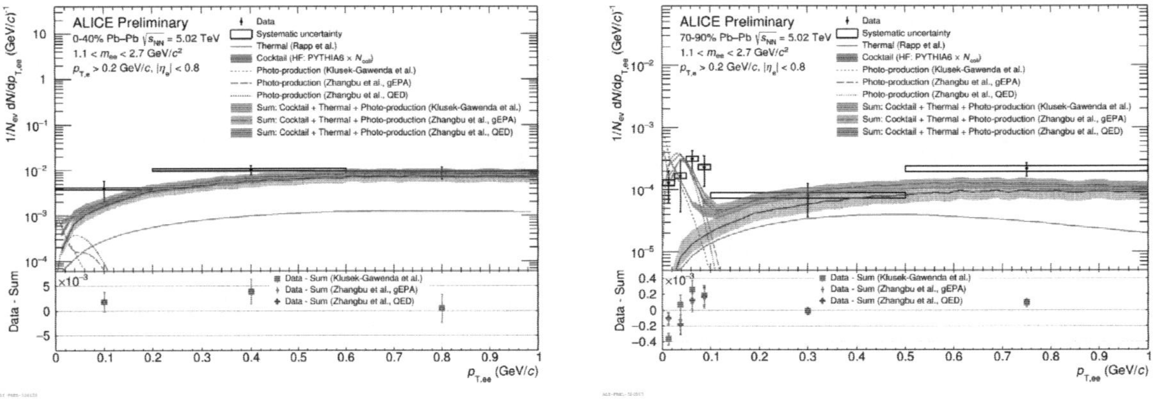

Figure 12. Dielectron spectrum as a function of $p_{T,ee}$ in the centrality ranges 0–40% (left) and 70–90% (right), compared with models of thermal radiation and photo-production.

Three different photo-production models [35, 36] are compared to data, together to the hadronic cocktail and with a thermal radiation model by R. Rapp [37]. Effects of tracking resolution are not taken into account in the photo-production models, however they should play a minor role in the comparison with data. In the 0-40% centrality class the spectrum is in agreement with the cocktail, and neither photo-production nor thermal effects can be inferred, due to the limited precision of the measurement. In peripheral collisions (70–90%) photo-production is expected to contribute stronger to the dielectron production with respect to central collisions. In this centrality range, an excess of dielectrons with respect to the cocktail is observed for $p_{T,ee} < 0.1$ GeV/c^2. All the photo-production calculations are overall compatible with data.

7. Outlook

Future dilepton studies are expected to reach much higher precision than the current measurements, thanks to the upgrade of the ALICE detector [32]. The new ITS, with improved resolution in the determination of secondary vertices, and reduced material budget, will allow a better separation of prompt sources (thermal radiation) and electrons from semileptonic decays of correlated heavy-flavour hadrons. The improved read-out capability will increase the expected number of events in the central barrel detector by a factor of 100. This will allow us to measure precisely the QGP temperature, and study the in-medium modification of the ρ meson.

The New Muon Forward Tracker (MFT) [38] will add vertexing capability to the muon spectrometer, thus allowing us to study at forward rapidity new observable like prompt/displaced J/ψ and to separate the individual contribution of charm and beauty in the open heavy-flavour studies. Equipped with the MFT, the upgraded muon spectrometer will allow us to increase the precision of low- and intermediate-mass dimuon measurements via the improvement of both the invariant mass resolution and the signal over background ratio.

References

[1] F. Karsch and M. Lutgemeier, "Deconfinement and chiral symmetry restoration in an SU(3) gauge theory with adjoint fermions," Nucl. Phys. B 550 (1999), 449-464 doi:10.1016/S0550-3213(99)00129-7 [arXiv:hep-lat/9812023 [hep-lat]].

[2] S. Acharya *et al.* [ALICE], "Measurement of dielectron production in central Pb-Pb collisions at $\sqrt{s_{NN}} = 2.76$ TeV," Phys. Rev. C 99 (2019) no.2, 024002 doi:10.1103/PhysRevC.99.024002 [arXiv:1807.00923 [nucl-ex]].

[3] K. Aamodt *et al.* [ALICE], "The ALICE experiment at the CERN LHC," JINST 3 (2008), S08002 doi:10.1088/1748-0221/3/08/S08002

[4] B. B. Abelev *et al.* [ALICE], "Performance of the ALICE Experiment at the CERN LHC," Int. J. Mod. Phys. A 29 (2014), 1430044 doi:10.1142/S0217751X14300440 [arXiv:1402.4476 [nucl-ex]].
[5] K. Aamodt *et al.* [ALICE], "Alignment of the ALICE Inner Tracking System with cosmic-ray tracks," JINST 5 (2010), P03003 doi:10.1088/1748-0221/5/03/P03003 [arXiv:1001.0502 [physics.ins-det]].
[6] J. Alme *et al.*, 'The ALICE TPC, a large 3-dimensional tracking device with fast readout for ultra-high multiplicity events," Nucl. Instrum. Meth. A 622 (2010), 316-367 doi:10.1016/j.nima.2010.04.042 [arXiv:1001.1950 [physics.ins-det]].
[7] A. Akindinov *et al.*, "Performance of the ALICE Time-Of-Flight detector at the LHC," Eur. Phys. J. Plus 128 (2013), 44 doi:10.1140/epjp/i2013-13044-x
[8] C. Finck [ALICE Muon Spectrometer], "The muon spectrometer of the ALICE," J. Phys. Conf. Ser. 50 (2006), 397-401 doi:10.1088/1742-6596/50/1/056
[9] B. Abelev *et al.* [ALICE], "Light vector meson production in *pp* collisions at $\sqrt{s} = 7$ TeV," Phys. Lett. B 710 (2012), 557-568 doi:10.1016/j.physletb.2012.03.038 [arXiv:1112.2222 [nucl-ex]].
[10] P. Skands, S. Carrazza and J. Rojo, "Tuning PYTHIA 8.1: the Monash 2013 Tune," Eur. Phys. J. C 74 (2014) no.8, 3024 doi:10.1140/epjc/s10052-014-3024-y [arXiv:1404.5630 [hep-ph]].
[11] R. Engel and J. Ranft, "Hadronic photon-photon interactions at high-energies," Phys. Rev. D 54 (1996), 4244-4262 doi:10.1103/PhysRevD.54.4244 [arXiv:hep-ph/9509373 [hep-ph]]
[12] S. Acharya *et al.* [ALICE], "$\mathrm{K}^*(892)^0$ and $\phi(1020)$ production at midrapidity in pp collisions at $\sqrt{s} = 8$ TeV," [arXiv:1910.14410 [nucl-ex]].
[13] S. Acharya *et al.* [ALICE], "Multiplicity dependence of $\mathrm{K}^*(892)^0$ and $\phi(1020)$ production in pp collisions at $\sqrt{s} = 13$ TeV," [arXiv:1910.14397 [nucl-ex]].
[14] A. Chauvin, "Low-Mass dimuon measurements in pp collisions with ALICE at the LHC," Poster presented at the conference "Strangeness in Quark Matter 2019".
[15] S. Acharya *et al.* [ALICE], "ϕ meson production at forward rapidity in Pb-Pb collisions at $\sqrt{s_{\mathrm{NN}}} = 2.76$ TeV," Eur. Phys. J. C 78 (2018) no.7, 559 doi:10.1140/epjc/s10052-018-6034-3 [arXiv:1804.08906 [nucl-ex]].
[16] S. Acharya *et al.* [ALICE], "Dielectron production in proton-proton and proton-lead collisions at $\sqrt{s_{\mathrm{NN}}} = 5.02$ TeV," [arXiv:2005.11995 [nucl-ex]].
[17] T. Sjostrand, S. Mrenna and P. Z. Skands, "PYTHIA 6.4 Physics and Manual," JHEP 05 (2006), 026 doi:10.1088/1126-6708/2006/05/026 [arXiv:hep-ph/0603175 [hep-ph]].
[18] S. Alioli, P. Nason, C. Oleari and E. Re, "A general framework for implementing NLO calculations in shower Monte Carlo programs: the POWHEG BOX," JHEP 06 (2010), 043 doi:10.1007/JHEP06(2010)043 [arXiv:1002.2581 [hep-ph]].
[19] S. Acharya *et al.* [ALICE], "Measurement of D-meson production at mid-rapidity in pp collisions at $\sqrt{s} = 7$ TeV," Eur. Phys. J. C 77 (2017) no.8, 550 doi:10.1140/epjc/s10052-017-5090-4 [arXiv:1702.00766 [hep-ex]]
[20] S. Acharya *et al.* [ALICE], "Dielectron production in proton-proton collisions at $\sqrt{s} = 7$ TeV," JHEP 09 (2018), 064 doi:10.1007/JHEP09(2018)064 [arXiv:1805.04391 [hep-ex]].
[21] S. Acharya *et al.* [ALICE], "Dielectron and heavy-quark production in inelastic and high-multiplicity proton-proton collisions at $\sqrt{s_{NN}} = 13$TeV," Phys. Lett. B 788 (2019), 505-518 doi:10.1016/j.physletb.2018.11.009 [arXiv:1805.04407 [hep-ex]].
[22] K. Eskola, H. Paukkunen and C. Salgado, "EPS09: A New Generation of NLO and LO Nuclear Parton Distribution Functions," JHEP 04 (2009), 065 doi:10.1088/1126-6708/2009/04/065 [arXiv:0902.4154 [hep-ph]].
[23] R. Rapp, "Signatures of thermal dilepton radiation at RHIC," Phys. Rev. C 63 (2001), 054907 doi:10.1103/PhysRevC.63.054907 [arXiv:hep-ph/0010101 [hep-ph]].
[24] K. J. Eskola, P. Paakkinen, H. Paukkunen and C. A. Salgado, "EPPS16: Nuclear parton distributions with LHC data," Eur. Phys. J. C 77 (2017) no.3, 163 doi:10.1140/epjc/s10052-017-4725-9 [arXiv:1612.05741 [hep-ph]].
[25] S. Acharya *et al.* [ALICE], "Soft-dielectron excess in proton-proton collisions at $\sqrt{s} = 13$ TeV," [arXiv:2005.14522 [nucl-ex]].
[26] G. Agakichiev *et al.*, Eur. Phys. J. C 4 (1998), 249-257 doi:10.1007/s100529800804
[27] R. Rapp, "Dilepton Spectroscopy of QCD Matter at Collider Energies," Adv. High Energy Phys. **2013** (2013), 148253 doi:10.1155/2013/148253 [arXiv:1304.2309 [hep-ph]].
[28] A. Belogianni *et al.*, "Observation of a soft photon signal in excess of QED expectations in p p interactions," Phys. Lett. B 548 (2002), 129-139 doi:10.1016/S0370-2693(02)02837-X
[29] A. Belogianni *et al.*, "Further analysis of a direct soft photon excess in pi- p interactions at 280-GeV/c," Phys. Lett. B 548 (2002), 122-128 doi:10.1016/S0370-2693(02)02836-8
[30] J. Antos *et al.*, "Soft photon production in 400-GeV/c p - Be collisions," Z. Phys. C 59 (1993), 547-554 doi:10.1007/BF01562546
[31] V. Balek, N. Pisutova and J. Pisut, "The Puzzle of Very Soft Photon Production in Hadronic Interactions,"

Acta Phys. Polon. B 21 (1990), 149 HU-TFT-89-33.
[32] Z. Citron *et al.* CERN Yellow Rep. Monogr. 7 (2019), 1159-1410 doi:10.23731/CYRM-2019-007.1159 [arXiv:1812.06772 [hep-ph]].
[33] R. Rapp, "Dilepton Spectroscopy of QCD Matter at Collider Energies," Adv. High Energy Phys. 2013 (2013), 148253 doi:10.1155/2013/148253 [arXiv:1304.2309 [hep-ph]].
[34] S. Lehner [ALICE], "Dielectron production at low transverse momentum in Pb-Pb collisions at $\sqrt{s_{\rm NN}} = 5.02$ TeV with ALICE," PoS LHCP2019 (2019), 164 doi:10.22323/1.350.0164 [arXiv:1909.02508 [nucl-ex]].
[35] M. Kusek-Gawenda, R. Rapp, W. Schfer and A. Szczurek, "Dilepton Radiation in Heavy-Ion Collisions at Small Transverse Momentum," Phys. Lett. B 790 (2019), 339-344 doi:10.1016/j.physletb.2019.01.035 [arXiv:1809.07049 [nucl-th]].
[36] W. Zha, J. D. Brandenburg, Z. Tang and Z. Xu, "Initial transverse-momentum broadening of Breit-Wheeler process in relativistic heavy-ion collisions," Phys. Lett. B 800 (2020), 135089 doi:10.1016/j.physletb.2019.135089 [arXiv:1812.02820 [nucl-th]]
[37] H. van Hees and R. Rapp, "Dilepton Radiation at the CERN Super Proton Synchrotron," Nucl. Phys. A 806 (2008), 339-387 doi:10.1016/j.nuclphysa.2008.03.009 [arXiv:0711.3444 [hep-ph]].
[38] B. Abelev *et al.* [ALICE], "Technical Design Report for the Muon Forward Tracker," CERN-LHCC-2015-001. ALICE-TDR-018

Journal of Physics: Conference Series **1602** (2020) 012017 doi:10.1088/1742-6596/1602/1/012017

Far From Equilibrium Hydrodynamics and the Beam Energy Scan

Travis Dore
University of Illinois at Urbana-Champaign, Urbana, IL 61801, USA

Emma McLaughlin
Department of Physics, Columbia University, 538 West 120th Street, New York, NY 10027, USA

Jacquelyn Noronha-Hostler
University of Illinois at Urbana-Champaign, Urbana, IL 61801, USA

Abstract. The existence of hydrodynamic attractors in rapidly expanding relativistic systems has shed light on the success of relativistic hydrodynamics in describing heavy-ion collisions at zero chemical potential. As the search for the QCD critical point continues, it is important to investigate how out of equilibrium effects influence the trajectories on the QCD phase diagram. In this proceedings, we study a Bjorken expanding hydrodynamic system based on DMNR equations of motion with initial out of equilibrium effects and finite chemical potential in a system with a critical point. We find that the initial conditions are not unique for a specific freeze-out point, but rather the system can evolve to the same final state freeze-out point with a wide range of initial baryon chemical potential, μ_B. For the same initial energy density and baryon density, depending on how far out of equilibrium the system begins, the initial μ_B can vary by $\Delta\mu_B \sim 350$ MeV. Our results indicate that knowledge of the out-of-equilibrium effects in the initial state provide vital information that influences the search for the QCD critical point.

1. Introduction

Determining the initial conditions for heavy-ion collisions is an ongoing problem in the field, and no less so as the Beam Energy Scan program at RHIC continues its search for the QCD critical point [1]. Studies of hydrodynamic attractors [2, 3, 4, 5, 6, 7, 6, 8, 9, 10, 11, 12, 13, 14, 15, 16, 17, 18, 19, 20, 21, 22, 23, 24] provide evidence that the effectiveness of hydrodynamics in describing heavy-ion data may come from the fact that the system quickly forgets about the initial conditions and collapses on to a universal attractor on short time scales. At LHC energies, where $\mu_B \sim 0$, the story is simpler as one does not need to worry about a multi-dimensional phase diagram. However, at lower beam energies where baryon stopping is relevant, as the system evolves it will traverse a path in the $T-\mu_B$ plane. If the system were to evolve entirely in equilibrium, it would simply follow isentropes (entropy over baryon number would be conserved i.e. $S/N_B = const$) in the phase diagram (see a number of recent studies relying on isentropes [25, 26, 27, 28, 29, 30]) However, the system evolves out of equilibrium, and depending on how far the system begins away from equilibrium, one would expect the deviations from the isentropic trajectories to be large.

Here, we study the effects of initializing the full $T^{\mu\nu}$, specifically the initial shear stress tensor $\pi^{\mu\nu}$ and initial bulk pressure Π, in an out of equilibrium state on the trajectories through the phase diagram. Previous studies of the full $T^{\mu\nu}$ on collective flow observables were only performed only at $\mu_B = 0$ [31, 32, 33] and, in this context, new causality conditions derived in [34] will be useful to constrain the far from equilibrium behavior. Here we run different initial conditions covering a wide range of initial shear and bulk viscous effects with a non-conformal equation of state that has a parametrized critical point. We find that far-from-equilibrium initial conditions can lead to significant deviations from isentropes, which implies connecting the initial state and final state can be significantly complicated at the beam energy scan. We also study whether or not hydrodynamic attractors persist in the presence of shear and bulk viscous terms that depend on both temperature and baryon chemical potential.

2. Hydrodynamic Model

Relativistic viscous hydrodynamics has enjoyed much success in predicting many observables in the low transverse momentum sector of heavy-ion collisions. As the Beam Energy Scan program at RHIC continues, a number of new developments concerning the hydrodynamic modeling of heavy-ion collisions have been performed to study a baryon-rich environment [35, 36, 37, 38] and transport coefficients that are also dependent on $\{T, \mu_B\}$ [39, 40, 41, 42, 43, 43, 44, 45]. However, in order to find the critical point in the QCD phase diagram (if it exists) it will be necessary to have a thorough understanding of how a dynamical evolution past the critical point can influence final state observables [46]. Moreover, an understanding of how out of equilibrium effects may influence criticality as well as system dynamics given two thermodynamic degrees of freedom, T and μ_B, is essential.

For the purpose of this proceedings, we will discuss only results obtained using DNMR hydrodynamic equations of motion [47]. As a starting point for future work, we use a simplified Bjorken expanding model [48] to obtain qualitative results. To our knowledge, this is the first work to study out of equilibrium effects on trajectories in the phase diagram, especially in regards to their impact on the search for the critical point. To that end, a simplified Bjorken picture is ideal for obtaining a basic understanding of the sought after effects. For a Bjorken expanding fluid (where $\tau = \sqrt{t^2 - z^2}$ is the propertime) the DNMR equations of motion become

$$\dot{\epsilon} = -\frac{1}{\tau}\left[e + p + \Pi - \pi^\eta_\eta\right] \quad (1)$$

$$\tau_\pi \dot{\pi}^\eta_\eta + \pi^\eta_\eta = \frac{1}{\tau}\left[\frac{4\eta}{3} - \pi^\eta_\eta\left(\delta_{\pi\pi} + \tau_{\pi\pi}\right) + \lambda_{\pi\Pi}\Pi\right] \quad (2)$$

$$\tau_\Pi \dot{\Pi} + \Pi = -\frac{1}{\tau}\left(\zeta + \delta_{\Pi\Pi}\Pi + \frac{2}{3}\lambda_{\Pi\pi}\pi^\eta_\eta\right) \quad (3)$$

$$\dot{\rho} = \frac{\rho_0}{\tau} \quad (4)$$

where ϵ is the energy density (with $\dot{\epsilon} = d\epsilon/d\tau$), p is the equilibrium pressure, π^η_η is the nonzero component of the shear stress tensor, Π is the bulk pressure, ρ is the baryon density (ρ_0 is the initial baryon density), η is the shear viscosity, ζ is the bulk viscosity, and the remaining

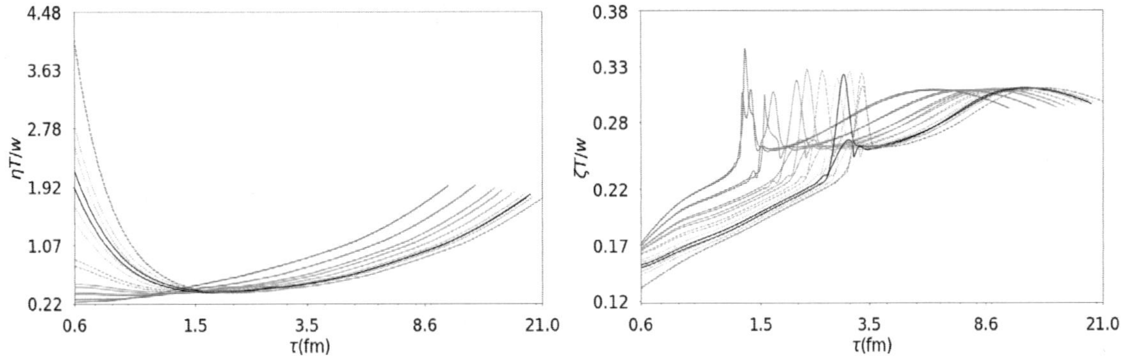

(a) (b)

Figure 1: Fig (a) shows the time evolution of $\eta T/w$ for hydrodynamic runs with different initial conditions. Fig (b) shows the time evolution of $\zeta T/w$ for hydrodynamic runs with different initial conditions having some critical point sensitivity.

transport coefficients are given by [36, 49]

$$\tau_\pi = \frac{5\,\eta}{\epsilon + p} \tag{5}$$

$$\tau_\Pi = \frac{\zeta}{15(e+p)\left(\frac{1}{3} - c_s^2\right)^2} \tag{6}$$

$$\lambda_{\pi\Pi} = \frac{6}{5}\tau_\pi \tag{7}$$

$$\delta_{\pi\pi} = \frac{4}{3}\tau_\pi \tag{8}$$

$$\tau_{\pi\pi} = \frac{10}{7}\tau_\pi \tag{9}$$

$$\lambda_{\Pi\pi} = \frac{8}{5}\left(\frac{1}{3} - c_s^2\right)\tau_\Pi \tag{10}$$

$$\delta_{\Pi\Pi} = \frac{2}{3}\tau_\Pi \tag{11}$$

where c_s is the isentropic speed of sound. We use a temperature dependent shear viscosity ($\eta T/w$ where $w = e + p$ is the enthalpy), given by an excluded volume calculation below the transition temperature [50] (using the PDG16+ [51]) that is matched onto a QCD motivated parameterization in the deconfined phase [52, 53]. The transition between the hadron gas phase and the QGP phase is matched using a tanh function in order to ensure a smooth matching. Additionally, we ensure here that the transition line matches that in the equation of state (EOS). At $\mu_B = 0$ we ensure that the shear viscosity to entropy density ratio has a minimum of 0.08, which occurs at $T_{\eta/s,min} = 196$ MeV, the μ_B dependence of the minimum $\eta T/w$ is determined by the change in $\eta T/w$ with μ_B within the excluded volume HRG model. The minimum of $\eta T/w$ converges to the transition line from the EOS at the critical point. Further work will extend this viscosity to the full 4 dimensional phase diagram of $\{T, \mu_B, \mu_S, \mu_Q\}$ taking into account strangeness and electric charge chemical potentials.

For the bulk viscosity, we use the parameterization

$$\frac{\zeta T}{w} = \frac{9}{2\pi}\left(\frac{1}{3} - c_s^2\right) \tag{12}$$

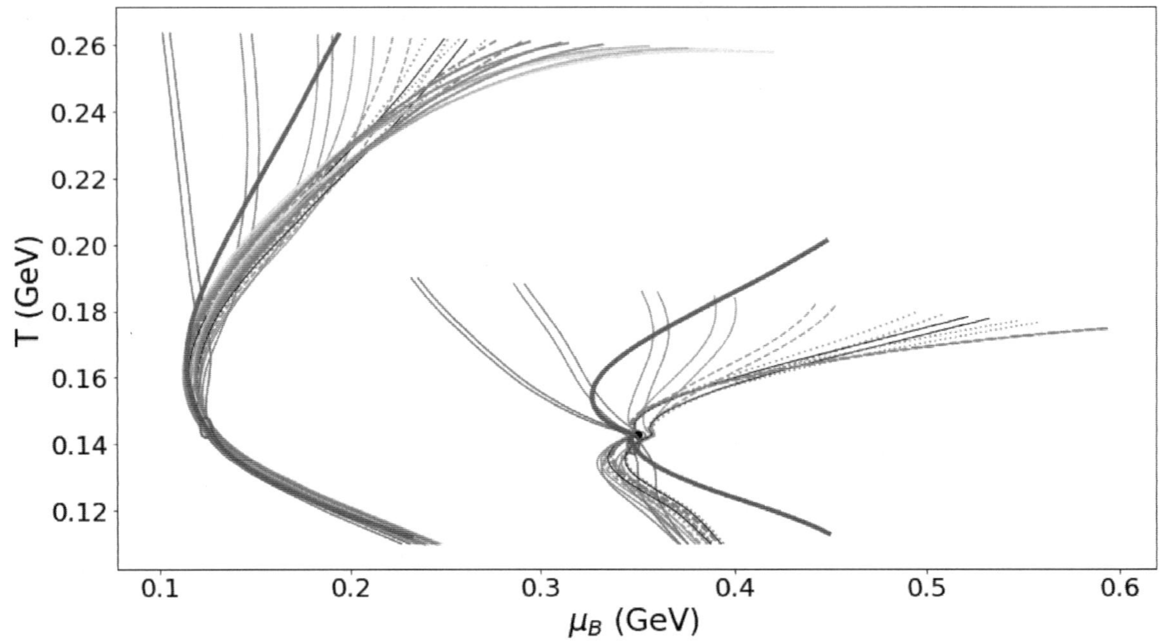

Figure 2: The $T-\mu_B$ trajectories of multiple hydrodynamic runs with different initial conditions, which all go through the same freeze-out region (green circle). The green line is the isentrope, for reference. For the isentrope away from the critical point, $S/N_B = 65$, and for the other $S/N_B = 21$.

which has a maximum $\frac{\zeta T}{w} \sim 0.3$ similar to that found in certain hydrodynamic simulations [54]. The inclusion of dependence on c_s^2 allows the bulk viscosity to have some sensitivity to critical dynamics, as it is expected to drop considerably at the critical point [27] while the bulk viscosity is expected to diverge [42].

In this work we use an equation of state that captures effects of criticality at finite baryon chemical potential. To this end, we use the most up to date equation of state reconstructed from Lattice QCD results, which are then mapped to a parameterized 3D Ising Model [27]. Here we place a critical point at $(T, \mu_B) = (143, 350)$ MeV, with parameters $\alpha_1 = 3$, $\alpha_2 = 93$, $\omega = 1$, and $\rho = 2$, which was a default EOS setting from [27]. We note that the current version of this EOS is only reliable up to $\mu_B \sim 400$ MeV. We, however, extend this out to $\mu_B \sim 600$ MeV in the high temperature region. That being said, we find limitations due to the maximum $\mu_B \sim 600$ MeV when studying the critical point. Finally, at this time we assume $\mu_S = \mu_Q = 0$ but in future studies plan to incorporate the full 4D EOS in $\{T, \mu_B, \mu_S, \mu_Q\}$ from [28] but we note that the 3D Ising critical point has not yet been implemented in the 4D EOS.

3. Trajectories in the QCD Phase Diagram

In Fig. 2 we display the thermodynamic trajectories generated by hydrodynamics for a range of different initial conditions where the ratio between the dissipative stresses and the enthalpy

$$\chi \equiv \pi_\eta^\eta/(\epsilon + p) \quad (13)$$

$$\Omega \equiv \Pi/(\epsilon + p) \quad (14)$$

are varied between ± 0.5. The initial energy density for the trajectories away from the critical point is 7.0 GeV/fm^3 and 1.5 GeV/fm^3 for those that pass through the critical region. We place a constraint on the trajectories produced by imposing that all trajectories must pass through a freeze-out region defined along an isentrope close to where one would expect the system to freeze-out at given thermal fits [55, 56, 57, 28, 58, 59, 60]. The initial baryon density is varied until trajectories are found that pass through either side of the freeze-out region. The region was chosen sufficiently small to mimic the uncertainty in the freeze-out temperature and is represented in Fig. 2 by a small green circle which trajectories pass through. If it were true that the system evolved isentropically, then one could uniquely determine the path traversed by the system given its freeze-out temperature and chemical potential. This path would correspond to the solid and thicker, green line (also shown). In this work, we argue that when considering viscous corrections, the unique correspondence between a freeze-out state and a trajectory in the QCD phase diagram no longer exists.

It can be seen in Eq. 3 that shear and bulk out of equilibrium effects both contribute linearly for determining the dynamic behavior of ϵ, albeit with opposite sign (see Table 3 for the different initial conditions used). Directly, one can notice that positive shear and negative bulk contributions work to slow down the rate at which energy density decreases. With this in mind, when looking at Fig. 2, it is consistent that those trajectories with the largest positive shear and negative bulk initial contributions start out the flattest in temperature and therefore must start at larger chemical potential in order to make it to the same freeze-out point as the other trajectories. We note that limitations in the EOS for producing data at large chemical potential ($>$ 600 MeV) prevented us from being able to explore the full range of initial positive shear and negative bulk, as was done for trajectories away from the critical point at low μ_B.

Again in Fig. 2, one can see a spread in initial μ_B of about 350 MeV and a spread in initial temperature of about 10 MeV. In order to better constrain the trajectory path, it is important to be able to calculate the initial conditions for the system (which include the initial values for the dissipative stresses) precisely. Currently there exists some models for hydrodynamic initial conditions at both zero [61, 62, 32] and finite chemical potential [63, 62, 64]. However, the initial condition models at finite μ_B do not incorporate a fully initialized $T^{\mu\nu}$ such that we have no way to estimate how wide of fluctuations to anticipate in $\pi^{\mu\nu}$ and Π on an event-by-event basis.

4. Hydrodynamic Attractor

In recent studies, it has been suggested that much of the success of relativistic viscous hydrodynamics in describing the system with such little knowledge of the initial conditions can be attributed to the existence of an attracting solution in the equations of motion of the viscous fields. In this work, we expand on this by studying a non-conformal hydrodynamic system with a finite chemical potential.

A natural question to ask when using more realistic transport coefficients that depend on temperature and chemical potential is whether or not the hydrodynamic attractor persists. Typically, one would expect that when scaling the time evolution of these quantities by their respective relaxation times of τ_π and τ_Π, these quantities would collapse onto a non-zero and non-trivial attractor and then eventually evolve towards zero (i.e., equilibrium). However, past work has primarily only focused on shear viscosity, which then only has one characteristic time scale. Here we have two and a nontrivial equation of state. In this work, we vary the initial values χ_0 and Ω_0 between ± 0.5.

In Figs. 4a, 4b, and 6a one can see that the presence of an attracting solution is likely, even though the system never fully reaches equilibrium (since χ approaches a constant $\neq 0$ at late times). The added help of the arrows pointing in the direction of the χ trajectory at the final time in Fig. 4a allows one to see that while the curves have not collapsed onto each other they do appear to be converging to a single line. At infinitely large times, it is likely that the χ trajectories under this rescaling would also collapse on to each other. Interesting to note, is the presence of weak attraction in the unscaled evolution, Fig. 6a. There has not yet been much

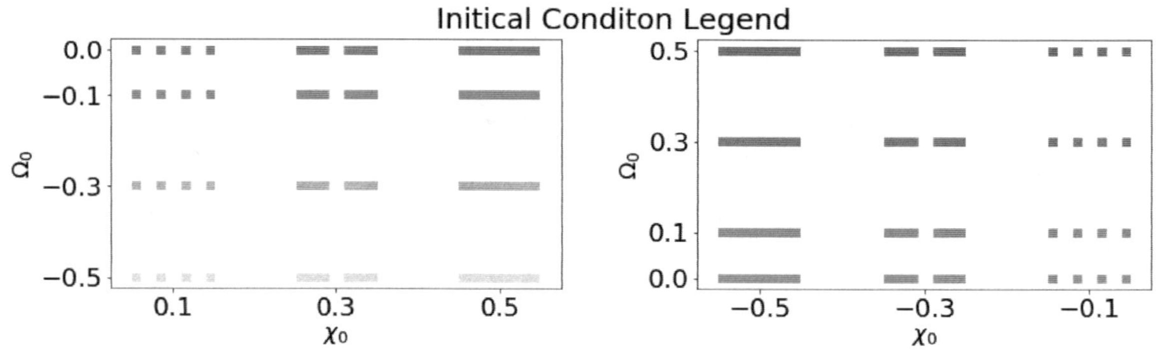

Figure 3: Reference for identifying which lines correspond to which initial conditions. Not shown here is the legend for $\chi_0 = \Omega_0 = 0$, which is represented by solid black curves in other figures. Note that not all combinations of χ_0 and Ω_0 are used, especially not in those trajectories that pass through the critical region.

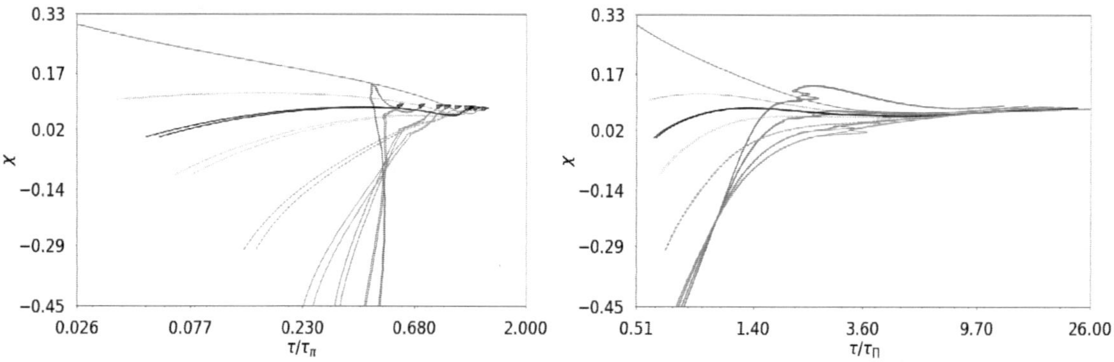

(a) (b)

Figure 4: Fig. (a) shows time evolution of χ rescaled by its relaxation time. The end arrows point along the instantaneous directional derivative. Fig. (b) shows time evolution of χ rescaled by the bulk relaxation time.

evidence to support the existence of an attractor in the unscaled time evolution. One should also note that the weak attraction seemingly present in Fig. 4b may simply be due to the fact that τ_Π is of $\mathcal{O}(1)$.

This explanation seems consistent with the fact that the attracting behavior for both χ and Ω in their unscaled evolution appears stronger than when their evolution is scaled by the bulk relaxation time, τ_Π. This seems especially noticeable in comparison between Figs. 5b and 6b. However, this notion goes against the intuition that the scaled time should be controlling the attracting behavior for its relevant quantity. Here we offer no solution to this slight paradox, but only point out its existence.

Either way, Fig. 6b shows a clear non-trivial and non-zero attractor for Ω. Even as the system goes through the critical point, which the bulk viscosity is sensitive to via c_s^2, it is about an order of magnitude smaller than the shear viscosity. The implications are interesting given that the system exhibits memory effects of the initial state and never actually relaxes to equilibrium. While we are still missing critical fluctuations in this approach, it is interesting to see attractor-like behavior in the bulk viscosity near the critical point. Given that there is no reason to believe initial viscous effects in heavy-ion collisions at BES energies should be small, theoretical models

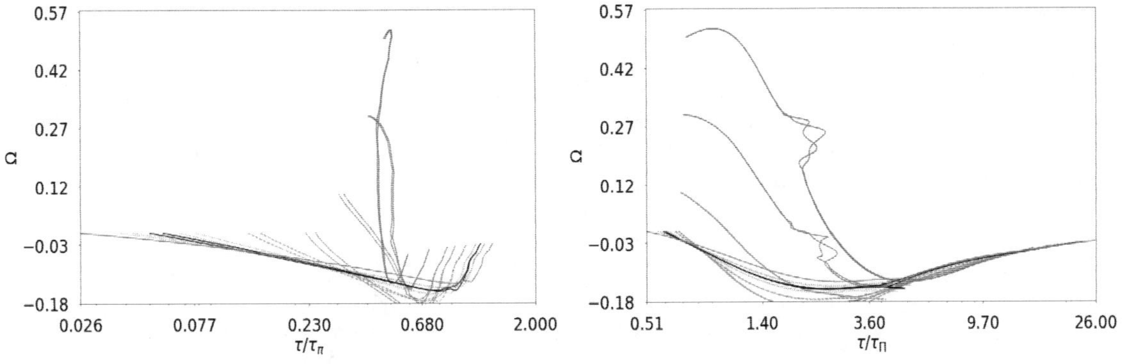

(a) (b)

Figure 5: Fig. (a) shows time evolution of Ω rescaled by the shear relaxation time. Fig (b) shows time evolution of Ω rescaled by its relaxation time.

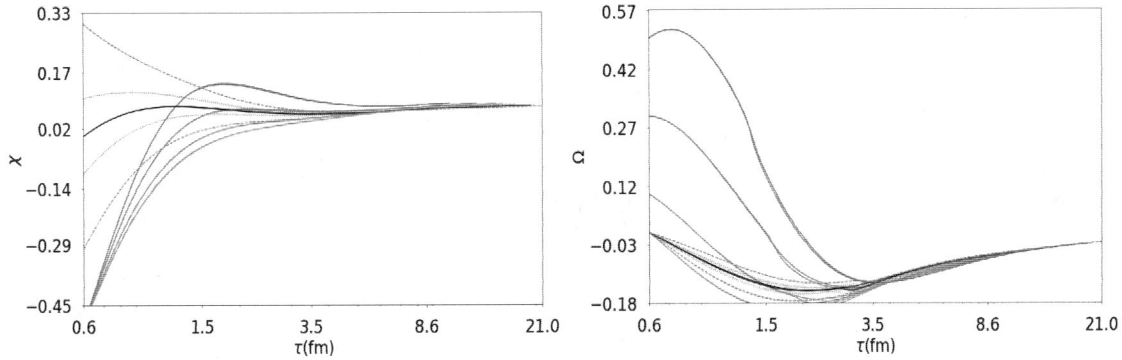

(a) (b)

Figure 6: Fig (a) shows time evolution of χ without rescaling. Fig (b) shows time evolution of Ω without rescaling

should take far-from-equilibrium initial conditions into consideration.

5. Conclusions

Opening up of the new degree of freedom, μ_B, in conjunction with exploring hydrodynamic non-equilibrium effects, lead to system dynamics that must be more thoroughly understood. In connection with the Beam Energy Scan program at RHIC, there are many important implications. For instance, given an initial ϵ and ρ, widely different trajectories may be seen throughout the phase diagram depending on how far-from-equilibrium the system begins. For these reasons, it is crucial to begin studying more deeply how to properly initialize the full $T_{\mu\nu}$ as input to hydrodynamics, compatible with event-by-event fluctuations, non-conformal system dynamics, and a finite chemical potential. Ongoing efforts such as [65] make it possible to study diffusive dynamics in the more heavily studied $\mu_B = 0$ regime by taking into consideration local fluctuations of the conserved charges. Moving forward, this seems to be a promising framework to begin asking the same kinds of questions asked here, but in a more realistic scenario.

Acknowledgements
J.N.H. acknowledges support from the US-DOE Nuclear Science Grant No. DE-SC0019175, the Alfred P. Sloan Foundation, and the Illinois Campus Cluster, a computing resource that is operated by the Illinois Campus Cluster Program (ICCP) in conjunction with the National Center for Supercomputing Applications (NCSA) and which is supported by funds from the University of Illinois at Urbana-Champaign.

References
[1] Bzdak A, Esumi S, Koch V, Liao J, Stephanov M and Xu N 2020 *Phys. Rept.* **853** 1–87 (*Preprint* 1906.00936)
[2] Heller M P and Spalinski M 2015 *Phys. Rev. Lett.* **115** 072501 (*Preprint* 1503.07514)
[3] Buchel A, Heller M P and Noronha J 2016 *Phys. Rev.* **D94** 106011 (*Preprint* 1603.05344)
[4] Heller M P, Kurkela A, Spaliński M and Svensson V 2018 *Phys. Rev.* **D97** 091503 (*Preprint* 1609.04803)
[5] Spaliński M 2018 *Phys. Lett.* **B776** 468–472 (*Preprint* 1708.01921)
[6] Romatschke P 2017 *JHEP* **12** 079 (*Preprint* 1710.03234)
[7] Romatschke P 2018 *Phys. Rev. Lett.* **120** 012301 (*Preprint* 1704.08699)
[8] Behtash A, Cruz-Camacho C N and Martinez M 2018 *Phys. Rev.* **D97** 044041 (*Preprint* 1711.01745)
[9] Strickland M, Noronha J and Denicol G 2018 *Phys. Rev.* **D97** 036020 (*Preprint* 1709.06644)
[10] Denicol G S and Noronha J 2018 *Phys. Rev.* **D97** 056021 (*Preprint* 1711.01657)
[11] Blaizot J P and Yan L 2018 *Phys. Lett.* **B780** 283–286 (*Preprint* 1712.03856)
[12] Casalderrey-Solana J, Gushterov N I and Meiring B 2018 *JHEP* **04** 042 (*Preprint* 1712.02772)
[13] Heller M P and Svensson V 2018 *Phys. Rev.* **D98** 054016 (*Preprint* 1802.08225)
[14] Rougemont R, Critelli R and Noronha J 2018 *Phys. Rev.* **D98** 034028 (*Preprint* 1804.00189)
[15] Denicol G S and Noronha J 2019 *Phys. Rev.* **D99** 116004 (*Preprint* 1804.04771)
[16] Almaalol D and Strickland M 2018 *Phys. Rev.* **C97** 044911 (*Preprint* 1801.10173)
[17] Casalderrey-Solana J, Herzog C P and Meiring B 2019 *JHEP* **01** 181 (*Preprint* 1810.02314)
[18] Behtash A, Cruz-Camacho C N, Kamata S and Martinez M 2019 *Phys. Lett.* **B797** 134914 (*Preprint* 1805.07881)
[19] Behtash A, Kamata S, Martinez M and Shi H 2019 *Phys. Rev.* **D99** 116012 (*Preprint* 1901.08632)
[20] Strickland M 2018 *JHEP* **12** 128 (*Preprint* 1809.01200)
[21] Strickland M and Tantary U 2019 *JHEP* **10** 069 (*Preprint* 1903.03145)
[22] Kurkela A, van der Schee W, Wiedemann U A and Wu B 2019 (*Preprint* 1907.08101)
[23] Jaiswal S, Chattopadhyay C, Jaiswal A, Pal S and Heinz U 2019 *Phys. Rev.* **C100** 034901 (*Preprint* 1907.07965)
[24] Denicol G S and Noronha J 2019 (*Preprint* 1908.09957)
[25] Günther J, Bellwied R, Borsanyi S, Fodor Z, Katz S D, Pasztor A and Ratti C 2017 *EPJ Web Conf.* **137** 07008
[26] Bellwied R, Noronha-Hostler J, Parotto P, Portillo Vazquez I, Ratti C and Stafford J M 2019 *Phys. Rev.* **C99** 034912 (*Preprint* 1805.00088)
[27] Parotto P, Bluhm M, Mroczek D, Nahrgang M, Noronha-Hostler J, Rajagopal K, Ratti C, Schäfer T and Stephanov M 2018 (*Preprint* 1805.05249)
[28] Noronha-Hostler J, Parotto P, Ratti C and Stafford J M 2019 *Phys. Rev.* **C100** 064910 (*Preprint* 1902.06723)
[29] Monnai A, Schenke B and Shen C 2019 *Phys. Rev.* **C100** 024907 (*Preprint* 1902.05095)
[30] Stafford J M, Alba P, Bellwied R, Mantovani-Sarti V, Noronha-Hostler J, Parotto P, Portillo-Vazquez I and Ratti C 2019 Determination of Chemical Freeze-out Parameters from Net-kaon Fluctuations at RHIC *18th International Conference on Strangeness in Quark Matter (SQM 2019) Bari, Italy, June 10-15, 2019* (*Preprint* 1912.12968)
[31] Liu J, Shen C and Heinz U 2015 *Phys. Rev. C* **91** 064906 [Erratum: Phys.Rev.C 92, 049904 (2015)] (*Preprint* 1504.02160)
[32] Kurkela A, Mazeliauskas A, Paquet J F, Schlichting S and Teaney D 2019 *Phys. Rev. Lett.* **122** 122302 (*Preprint* 1805.01604)
[33] Schenke B, Shen C and Tribedy P 2020 *Phys. Lett. B* **803** 135322 (*Preprint* 1908.06212)
[34] Bemfica F S, Disconzi M M, Hoang V, Noronha J and Radosz M 2020 (*Preprint* 2005.11632)
[35] Du L and Heinz U 2019 (*Preprint* 1906.11181)
[36] Denicol G S, Gale C, Jeon S, Monnai A, Schenke B and Shen C 2018 *Phys. Rev.* **C98** 034916 (*Preprint* 1804.10557)
[37] Batyuk P, Blaschke D, Bleicher M, Ivanov Yu B, Karpenko I, Malinina L, Merts S, Nahrgang M, Petersen H and Rogachevsky O 2018 *EPJ Web Conf.* **182** 02056 (*Preprint* 1711.07959)
[38] Fotakis J A, Greif M, Denicol G, Niemi H and Greiner C 2019 (*Preprint* 1912.09103)

[39] Demir N and Bass S A 2009 *Phys. Rev. Lett.* **102** 172302 (*Preprint* 0812.2422)
[40] Denicol G S, Gale C, Jeon S and Noronha J 2013 *Phys. Rev.* **C88** 064901 (*Preprint* 1308.1923)
[41] Kadam G P and Mishra H 2014 *Nucl. Phys.* **A934** 133–147 (*Preprint* 1408.6329)
[42] Monnai A, Mukherjee S and Yin Y 2017 *Phys. Rev.* **C95** 034902 (*Preprint* 1606.00771)
[43] Rougemont R, Critelli R, Noronha-Hostler J, Noronha J and Ratti C 2017 *Phys. Rev.* **D96** 014032 (*Preprint* 1704.05558)
[44] Auvinen J, Bernhard J E, Bass S A and Karpenko I 2018 *Phys. Rev.* **C97** 044905 (*Preprint* 1706.03666)
[45] Martinez M, Schäfer T and Skokov V 2019 *Phys. Rev.* **D100** 074017 (*Preprint* 1906.11306)
[46] Son D and Stephanov M 2004 *Phys. Rev. D* **70** 056001 (*Preprint* hep-ph/0401052)
[47] Denicol G S, Niemi H, Molnar E and Rischke D H 2012 *Phys. Rev.* **D85** 114047 [Erratum: Phys. Rev.D91,no.3,039902(2015)] (*Preprint* 1202.4551)
[48] Bjorken J 1983 *Phys. Rev. D* **27** 140–151
[49] Bazow D, Heinz U W and Strickland M 2018 *Comput. Phys. Commun.* **225** 92–113 (*Preprint* 1608.06577)
[50] Noronha-Hostler J, Noronha J and Greiner C 2012 *Phys. Rev.* **C86** 024913 (*Preprint* 1206.5138)
[51] Alba P *et al.* 2017 *Phys. Rev. D* **96** 034517 (*Preprint* 1702.01113)
[52] Christiansen N, Haas M, Pawlowski J M and Strodthoff N 2015 *Phys. Rev. Lett.* **115** 112002 (*Preprint* 1411.7986)
[53] Dubla A, Masciocchi S, Pawlowski J M, Schenke B, Shen C and Stachel J 2018 *Nucl. Phys.* **A979** 251–264 (*Preprint* 1805.02985)
[54] Ryu S, Paquet J F, Shen C, Denicol G S, Schenke B, Jeon S and Gale C 2015 *Phys. Rev. Lett.* **115** 132301 (*Preprint* 1502.01675)
[55] Ejiri S, Karsch F, Laermann E and Schmidt C 2006 *Phys. Rev.* **D73** 054506 (*Preprint* hep-lat/0512040)
[56] Schmid C (COL-NOTE = RBC, HotQCD) 2008 Lattice QCD thermodynamic results with improved staggered fermions *Proceedings, 3rd International Conference on Hard and Electromagnetic Probes of High-Energy Nuclear Collisions (Hard Probes 2008): Illa da Toxa, Spain, June 8-14, 2008* (*Preprint* 0810.0374)
[57] Bellwied R, Borsanyi S, Fodor Z, Gunther J, Katz S D, Pasztor A, Ratti C and Szabo K K 2016 *Nucl. Phys.* **A956** 797–800 (*Preprint* 1601.00466)
[58] Motornenko A, Steinheimer J, Vovchenko V, Schramm S and Stoecker H 2020 *Phys. Rev.* **C101** 034904 (*Preprint* 1905.00866)
[59] Bellwied R, Borsanyi S, Fodor Z, Guenther J N, Noronha-Hostler J, Parotto P, Pasztor A, Ratti C and Stafford J M 2020 *Phys. Rev.* **D101** 034506 (*Preprint* 1910.14592)
[60] Alba P, Sarti V M, Noronha-Hostler J, Parotto P, Portillo-Vazquez I, Ratti C and Stafford J M 2020 (*Preprint* 2002.12395)
[61] Gale C, Jeon S, Schenke B, Tribedy P and Venugopalan R 2013 *Phys. Rev. Lett.* **110** 012302 (*Preprint* 1209.6330)
[62] Werner K 1993 *Strings, pomerons, and the venus model of hadronic interactions at ultrarelativistic energies* Other thesis
[63] Weil J *et al.* 2016 *Phys. Rev. C* **94** 054905 (*Preprint* 1606.06642)
[64] Shen C and Schenke B 2018 *Phys. Rev.* **C97** 024907 (*Preprint* 1710.00881)
[65] Martinez M, Sievert M D, Wertepny D E and Noronha-Hostler J 2019 (*Preprint* 1911.10272)

Journal of Physics: Conference Series **1602** (2020) 012018 doi:10.1088/1742-6596/1602/1/012018

Pseudothermalization of the quark-gluon plasma

Michael Strickland

Department of Physics, Kent State University, Kent, OH 44242, United States

E-mail: mstrick6@kent.edu

Abstract. In this proceedings contribution I review recent work in kinetic theory which demonstrates that, for system undergoing Bjorken expansion, there exists an attractor in all moments of the one-particle distribution function. I discuss how this attractor emerges in both exact solutions obtained in relaxation time approximation (RTA) and the effective kinetic theory approach to high-temperature quantum chromodynamics (QCD). The QCD effective kinetic theory collisional kernel used includes both elastic ($2 \leftrightarrow 2$) and LPM-resummed inelastic ($2 \leftrightarrow 1$) contributions. The results obtained indicate that a pseudothermal attractor exists in both RTA and QCD kinetic theory and that their respective attractors can be extended to early times when the system is far from equilibrium. Finally, I discuss how knowledge of the QCD effective kinetic theory attractor can be used to assess different hydrodynamic freeze-out prescriptions used in heavy-ion phenomenology. The results obtained show that improved freeze-out prescriptions such as anisotropic hydrodynamics perform better in conditions corresponding to those generated in high-multiplicity pA and pp collisions, e.g. short lifetime and high inverse Reynolds number.

1. Introduction

In recent years there has been intense interest in the degree to which the dynamics of non-equilibrium quantum field theories can be reliably described by relativistic viscous hydrodynamics. This has impacts on our understanding on the quark-gluon plasma (QGP) generated in relativistic heavy-ion collisions [1–3], condensed matter systems in which a hydrodynamical description appears to be successful [4, 5], and potentially studies of neutron star mergers [6–9]. In the context of heavy-ion collisions (AA) the fundamental question is how well can quarks and gluons produced in the first fractions of a fm/c be described by relativistic hydrodynamics and whether or not such a description can be extended to collisions of small systems such as pA and pp. In the context of AA collisions first indications for the effectiveness of relativistic viscous hydrodynamics came from the AdS/CFT studies of Chesler and Yaffe [10] wherein it was shown that, despite rather large early-time pressure anisotropies in the local rest frame of the system, the dynamics of the energy-momentum tensor could be well-described by viscous hydrodynamics. This implied that the system does not rapidly thermalize and instead rapidly *hydrodynamizes*.

Since the work of Chesler and Yaffe there have been many papers addressing the phenomenon of hydrodynamization which have found that, in a variety of contexts, there exists a "non-equilibrium attractor" for the components of the energy-momentum tensor which can be well-described by relativistic dissipative hydrodynamics after a short amount of time in the center of the fireball ($\lesssim 1$ fm/c) [11–33]. In the majority of these studies, the authors concentrated on the evolution of the energy-momentum tensor, which in a kinetic theory language translates into a particular set of moments of the one-particle distribution function. A natural question which arises is whether or not a non-equilibrium attractor can be observed in higher moments of the distribution function or, even more generally, the distribution function itself. The first attempts to address this question was presented in Refs. [26, 29], where it was shown that exact solutions to relaxation time approximation (RTA) Boltzmann equation in a Bjorken-expanding background exhibited an attractor in all computed moments of the distribution function.

In this proceedings contribution, I review past work using exact solutions to the RTA kinetic theory in systems undergoing Bjorken expansion. I then present similar findings reported recently in Ref. [34] using the effective kinetic theory (EKT) approach to the dynamics of a non-equilibrium quark-gluon plasma [12, 35, 36]. The EKT evolution used includes both elastic and inelastic contributions to the collisional kernel allowing for a numerical realization of the bottom-up thermalization scenario [37].

2. Exact RTA attractor and dynamics

For a system subject to boost-invariant and transversally homogenous Bjorken flow and subject to a RTA collisional kernel, the underlying kinetic equation is simple

$$p^\mu \partial_\mu f(x,p) = \frac{p\cdot u}{\tau_{\rm eq}(\tau)}\left(f_{\rm eq} - f\right), \tag{1}$$

where $\tau_{\rm eq}(\tau) = 5\bar{\eta}(\tau)/T(\tau)$ is the relaxation time with $\bar{\eta}(\tau) = \eta(\tau)/s(\tau)$ being the shear viscosity to entropy density ratio and $T(\tau)$ being the local effective temperature. Eq. (1) can be cast into simpler form by writing it in terms of manifestly boost-invariant variables [38, 39]. The resulting simpler equation can easily be shown to have a general solution given by [40–42]

$$f(\tau,w,p_T) = D(\tau,\tau_0)f_0(w,p_T) + \int_{\tau_0}^{\tau}\frac{d\tau'}{\tau_{\rm eq}(\tau')}\,D(\tau,\tau')\,f_{\rm eq}(\tau',w,p_T)\,, \tag{2}$$

where $w = tp_L - zE$ and D is the damping function

$$D(\tau_2,\tau_1) = \exp\left[-\int_{\tau_1}^{\tau_2}\frac{d\tau''}{\tau_{\rm eq}(\tau'')}\right]. \tag{3}$$

Equation (2) can be turned into an infinite tower of equations for moments of the one-particle distribution function

$$\mathcal{M}^{nm}[f] \equiv \int dP\,(p\cdot u)^n\,(p\cdot z)^{2m}\,f(\tau,w,p_T)\,. \tag{4}$$

Assuming classical statistics, one obtains [26]

$$\begin{aligned}\mathcal{M}^{nm}(\tau) \;=\; & \frac{\Gamma(n+2m+2)}{(2\pi)^2}\Bigg[D(\tau,\tau_0)T_0^{n+2m+2}\frac{\mathcal{H}^{nm}\left(\frac{\alpha_0\tau_0}{\tau}\right)}{[\mathcal{H}^{20}(\alpha_0)/2]^{(n+2m+2)/4}} \\ & + \int_{\tau_0}^{\tau}\frac{d\tau'}{\tau_{\rm eq}(\tau')}\,D(\tau,\tau')\,T^{n+2m+2}(\tau')\mathcal{H}^{nm}\!\left(\tfrac{\tau'}{\tau}\right)\Bigg],\end{aligned} \tag{5}$$

with

$$\mathcal{H}^{nm}(y) = \tfrac{2y^{2m+1}}{2m+1}\,{}_2F_1\!\left(\tfrac{1}{2}+m,\tfrac{1-n}{2};\tfrac{3}{2}+m;1-y^2\right). \tag{6}$$

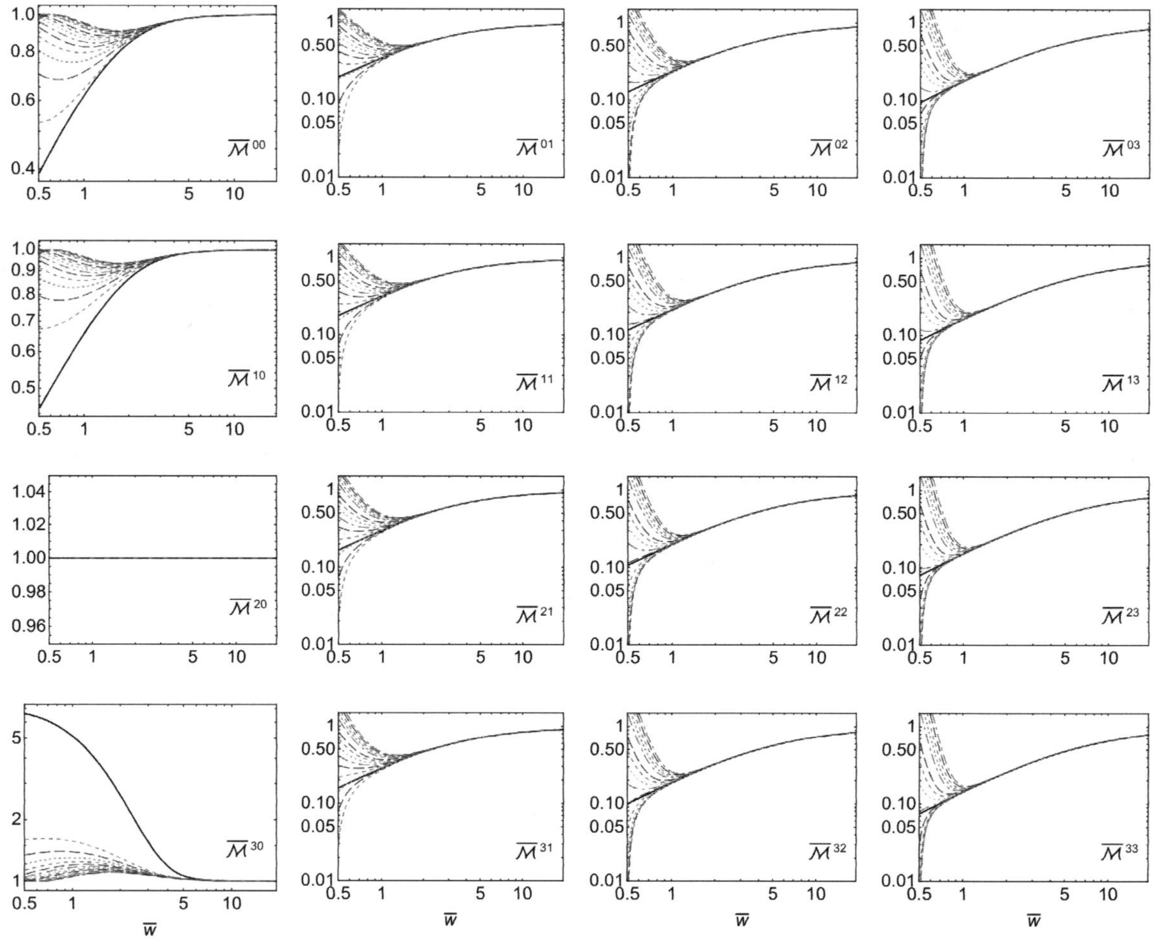

Figure 1. Scaled moments $\overline{\mathcal{M}}^{nm}$ obtained from the exact RTA attractor solution (solid black line) compared to a set of exact RTA solutions (various colored dotted and dashed lines) initialized at $\tau = 0.1$ fm/c with varying initial pressure anisotropy. The horizontal axis is $\overline{w} \equiv \tau/\tau_{\rm eq} = \tau T/5\bar{\eta}$. Panels show a grid in n and m.

Note that certain moments map to familiar hydrodynamics variables, e.g. $n = \mathcal{M}^{10}$, $\varepsilon = \mathcal{M}^{20}$, and $P_L = \mathcal{M}^{01}$.

One can obtain a closed integral equation for $T(\tau)$ by considering the integral equation obeyed by $\mathcal{M}^{20} = \varepsilon = \varepsilon_{\rm eq}$ which simplifies to

$$T^4(\tau) = D(\tau,\tau_0) T_0^4 \frac{\mathcal{H}\left(\frac{\alpha_0 \tau_0}{\tau}\right)}{\mathcal{H}(\alpha_0)} + \int_{\tau_0}^{\tau} \frac{d\tau'}{2\tau_{\rm eq}(\tau')} D(\tau,\tau')\, T^4(\tau')\mathcal{H}\!\left(\frac{\tau'}{\tau}\right).$$

This equation can be numerically solved iteratively [41, 42]. Once the solution for $T(\tau)$ is obtained, one can use this to solve for all other moments $\mathcal{M}^{nm}(\tau)$ using Eq. (5) and the full distribution function itself using Eq. (2).

In Fig. 1 I present the results obtained originally in Ref. [26]. Each panel shows the dependence of a scaled moment of the distributions function, $\overline{\mathcal{M}}^{nm} \equiv \mathcal{M}^{nm}/\mathcal{M}^{nm}_{\rm eq}$. The black line in each panel is the exact attractor for that moment of the distribution function and the colored dashed lines are solutions with particular momentum-space anisotropic initial

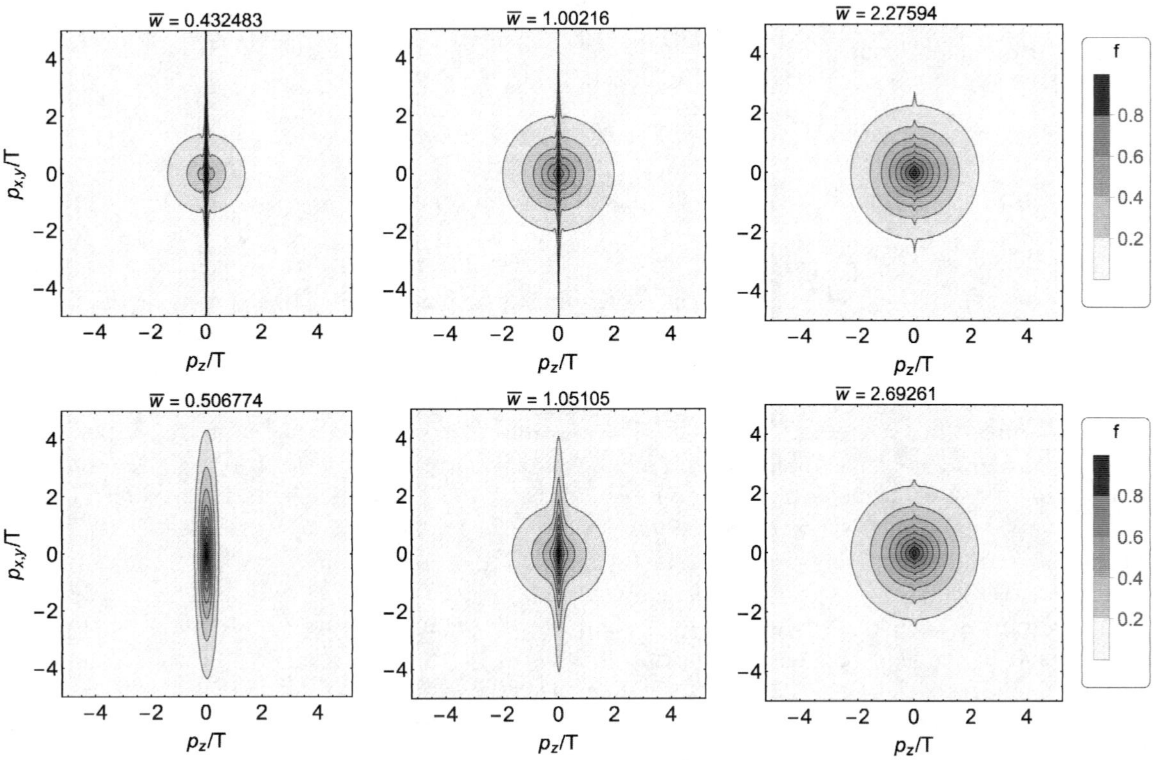

Figure 2. (Top) Visualization of the one-particle distribution function associated with the RTA attractor. (Bottom) Visualization of the one-particle distribution function obtained using a typical (non-attractor) anisotropic initial condition and an RTA collisional kernel.

conditions. As can be seen from these panels, for all moments with $n > 1$ the solutions converge to the attractor solution by approximately $\overline{w} = \tau/\tau_{\rm eq} = 2$. Closer analysis reveals that higher order moments converge to their respective attractors more quickly [26, 29]. The convergence of moments with $n = 0$ is slower and can be traced to the free streaming contribution to the solution [26,29]. Overall the results indicate that there is an attractor for the entire distribution function.

In Fig. 2 I present contour plots of the corresponding one-particle distribution functions obtained using Eq. (2). The top panels show the distribution function obtained along the exact RTA attractor and the bottom panels show a typical evolution obtained with a spheroidal initial condition. The columns from left to right correspond to different snapshots in rescaled time $\overline{w}$. As can be seen from this Figure, the bottom panels approach the form given by the attractor solution presented in the top panels, converging to one another at rather short rescaled time. For more details concerning the nature of the attractor for the full one-particle distribution function see Ref. [26].

3. Effective kinetic theory approach to QCD

RTA is a nice toy model in which it is possible to obtain the exact solutions presented above, however, it would be better to ask the same questions in QCD. In Ref. [34] we made use of a numerical implementation of the Arnold, Moore, and Yaffe effective kinetic theory (EKT) [35] which allows for the description of both thermally occupied and over-occupied gluonic plasmas

Journal of Physics: Conference Series **1602** (2020) 012018 doi:10.1088/1742-6596/1602/1/012018

[36,43]. In practice, for a system undergoing boost-invariant Bjorken expansion, the code solves an EKT Boltzmann equation of the form

$$-\frac{df(\mathbf{p})}{d\tau} = \mathcal{C}_{1\leftrightarrow 2}[f(\mathbf{p})] + \mathcal{C}_{2\leftrightarrow 2}[f(\mathbf{p})] + \mathcal{C}_{\rm exp}[f(\mathbf{p})]\,, \tag{7}$$

where $f(\mathbf{p})$ is the gluonic one-particle distribution function. We discretize $\mathbf{p}$ in spherical coordinates, $p = |\mathbf{p}|$, $x \equiv \cos\theta$, and ϕ. The effect of longitudinal expansion is included through $\mathcal{C}_{\rm exp}[f(\mathbf{p})] = -\frac{p_z}{\tau}\frac{\partial}{\partial p_z} f(\mathbf{p})$ [44]. The elastic scattering term $\mathcal{C}_{2\leftrightarrow 2}$ and the effective inelastic term $\mathcal{C}_{1\leftrightarrow 2}$ include physics of dynamical screening and Landau-Pomeranchuck-Migdal (LPM) suppression and, in order to find the form of the collision kernels, self-energy and ladder resummations are required. For details, see Refs. [35,36,43].

For the numerical solution of Eq. (7), we discretized $n(\mathbf{p}) = p^2 f(\mathbf{p})$ on a three-dimensional grid in momentum space and used Monte Carlo sampling to compute the integrals appearing in the elastic and inelastic collisional kernels. The algorithm used is based on Ref. [43] and exactly conserves energy while also exactly accounting for the particle number violation originating from the inelastic contributions to the collisional kernel. Due to the azimuthal symmetry of Bjorken flow, we used an effectively two-dimensional grid: 250×2000 points in the p and $x = \cos\theta$ directions, respectively.

We computed the time evolution of a complete set of integral moments characterizing the momentum dependence of the distribution function (4). As mentioned previously, the energy density is given by $\varepsilon = \nu\mathcal{M}^{20}$, longitudinal pressure by $P_L = \nu\mathcal{M}^{01}$, and number density by $n = \nu\mathcal{M}^{10}$ for ν degrees of freedom ($\nu = 2d_A$ for d_A adjoint colors of gluons). The other moments do not have an interpretation in terms of the usual hydrodynamic moments considered in the literature.[1] As with the RTA, these moments will be scaled by their corresponding equilibrium values with $\overline{\mathcal{M}}^{nm}(\tau) \equiv \mathcal{M}^{nm}(\tau)/\mathcal{M}^{nm}_{\rm eq}(\tau)$, where, using a Bose distribution, one obtains

$$\mathcal{M}^{nm}_{\rm eq} = \frac{T^{n+2m+2}\Gamma(n+2m+2)\zeta(n+2m+2)}{2\pi^2(2m+1)}\,. \tag{8}$$

The temperature T here corresponds to the temperature of an equilibrium system with the same energy density, given by $T = (30\varepsilon/\nu\pi^2)^{1/4}$. Note that the different moments are sensitive to different momentum regions of the distribution function and for future comparisons, we note that, in equilibrium, the typical momentum contributing to a given moment is $\langle p\rangle^{nm}_{\rm eq} = \mathcal{M}^{n+1,m}_{\rm eq}/\mathcal{M}^{nm}_{\rm eq}$, giving, *e.g.*, $\langle p\rangle^{10}_{\rm eq} \simeq 2.7\,T$, $\langle p\rangle^{01}_{\rm eq} = \langle p\rangle^{20}_{\rm eq} \simeq 3.83\,T$, $\langle p\rangle^{21}_{\rm eq} \simeq 5.95\,T$, and $\langle p\rangle^{33}_{\rm eq} \simeq 11\,T$.

In Fig. 3 I present a panel of results for the scaled moments. The horizontal (time) axis is scaled by the instantaneous interaction time of the system $\tau/\tau_R(\tau)$, which measures the age of the system in units of the instantaneous interaction time, $\tau_R(\tau) \equiv 4\pi\bar{\eta}/T(\tau)$. In all panels the black dotted and dashed lines correspond to the EKT results obtained using two different types of initial conditions: (i) spheroidally deformed thermal (RS) and (ii) over-occupied CGC-type initial conditions. For details of the initial conditions used see Ref. [34]. For both types of initial conditions, we varied both the initial momentum-space anisotropy and initialization time. In all panels we also show the attractors corresponding to RTA, viscous hydrodynamics (vHydro) with a quadratic freeze-out form (i), vHydro with a LPM-modified freeze-out form (ii), and anisotropic hydrodynamics [22].

As can be seen from Fig. 3, one finds that all EKT solutions afpproach a universal EKT attractor irrespective of the initial conditions assumed. In particular, we observe that, as the initialization time is decreased, the solutions approach this attractor more rapidly.

[1] The $m = 0$ modes are simply related to the effective temperatures introduced in Refs. [45,46].

Journal of Physics: Conference Series **1602** (2020) 012018 doi:10.1088/1742-6596/1602/1/012018

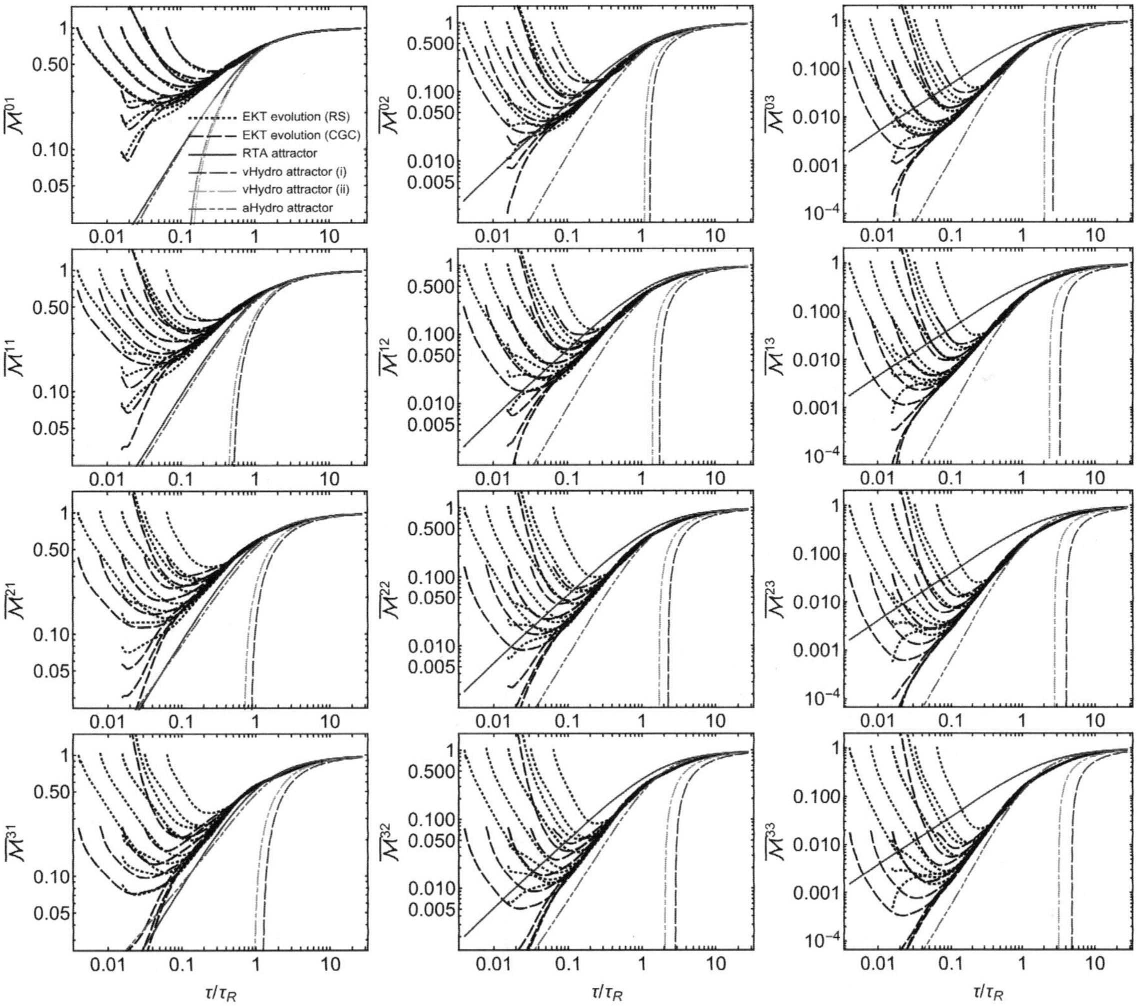

Figure 3. Evolution of the scaled moments $\overline{\mathcal{M}}^{nm}$ with $n \in \{0,1,2\}$ and $m \in \{0,1,2,3\}$. Black dotted and dashed lines show EKT evolution with RS amd CGC initial conditions, respectively. The purple solid line is the exact RTA attractor, the blue long-dashed line is the DNMR vHydro attractor using δf parameterization (i), the green dot-dashed line is the DNMR vHydro attractor using δf parameterization (ii), and the red dot-dot-dashed line is the aHydro attractor.

This establishes the existence of an early-time or pull-back [28] attractor in EKT QCD. We additionally observe that, while all approximate attractors (RTA, vHydro, and aHydro) do a reasonable job in reproducing low-order moments, e.g. $P_L/P_L^{\rm eq} = \overline{\mathcal{M}}^{01}$, the different schemes do not accurately describe the EKT QCD attractor. Of the set of approximations, aHydro is the best in reproducing the EKT QCD attractor.

Having determined the EKT QCD attractor, one can use this to assess the ability of different viscous hydrodynamics freeze-out prescriptions in reproducing the behavior observed in the high-order moments. For this purpose one can determine the shear viscous correction, Π or ξ for vHydro and aHydro, respectively, that allows one to exactly reproduce the EKT QCD attractor for $P_L/P_L^{\rm eq} = \overline{\mathcal{M}}^{01}$. Using the extracted Π or ξ one can than make predictions for

Journal of Physics: Conference Series **1602** (2020) 012018 doi:10.1088/1742-6596/1602/1/012018

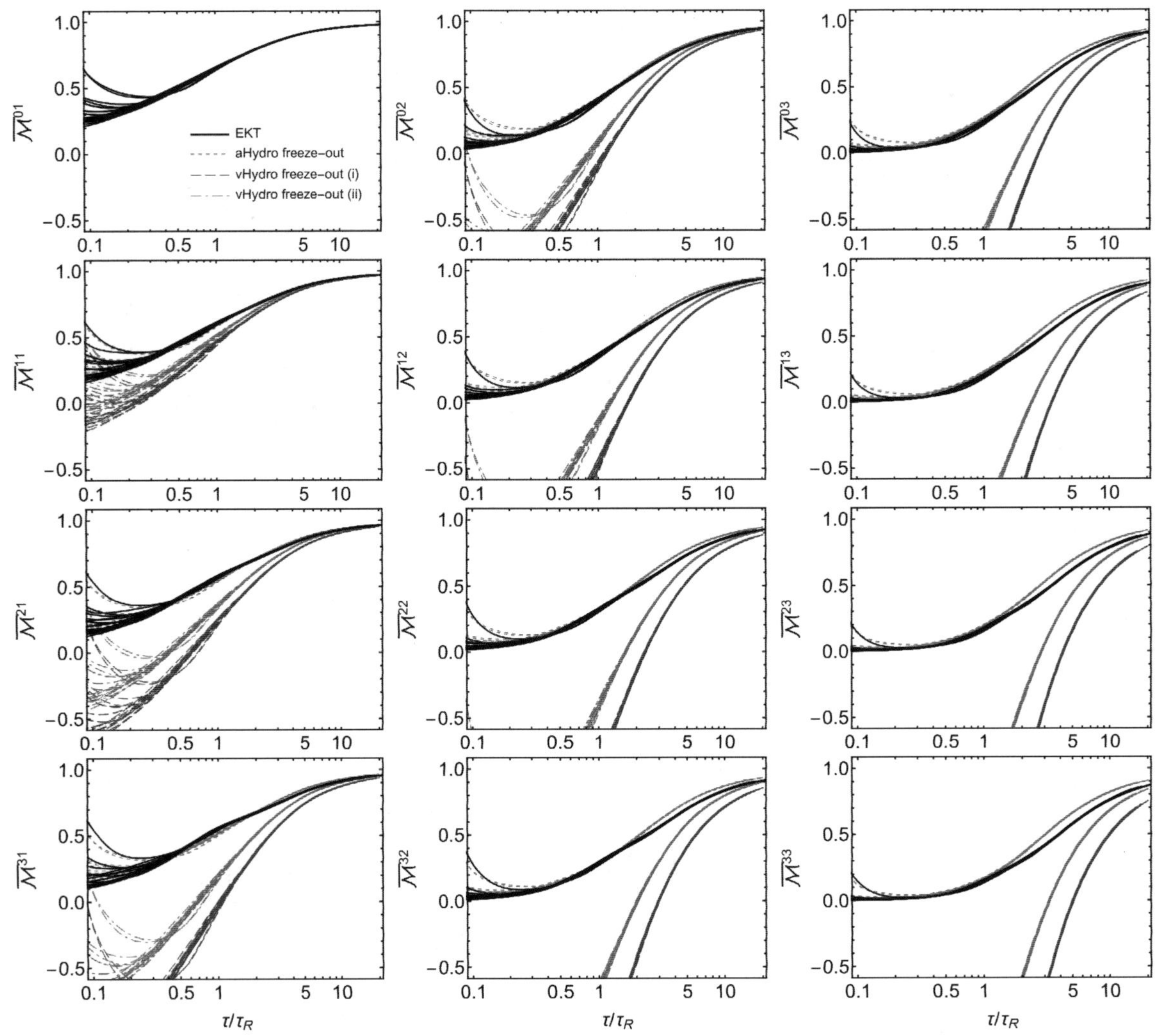

Figure 4. Evolution of the scaled moments $\overline{\mathcal{M}}^{nm}$ with $n \in \{0,1,2\}$ and $m \in \{0,1,2,3\}$ for all runs (CGC +RS) shown as black lines. Other lines – red-dashed, blue long-dashed, green dot-dashed – correspond to results obtained in three freeze-out scenarios, aHydro, vHydro (i), and vHydro (ii), respectively. Note that all scenarios agree on $\overline{\mathcal{M}}^{01}$ by construction (P_L-matching). The other panels are predictions based on this matching.

high-order moments.[2] I present the results of this exercise in Fig. 4. As this figure demonstrates, the aHydro freeze-out prescription does the best in reproducing the behavior of all moments. One sees a modest improvement in the agreement of vHydro when using the LPM-improved prescription, however, it still shows poor agreement at early times where its predictions become negative and unphysical. This provides some motivation to implement aHydro-type freeze-out for the phenomenological analysis of small collisions systems at RHIC and LHC.

[2] For details concerning the freeze-out prescriptions used, see Ref. [34].

4. Conclusions

In this proceedings contribution, I have presented evidence for the existence of a non-equilibrium attractor in both RTA and EKT QCD kinetic theories. I demonstrated that in both theories on can identify an attractor in *all* moments, that can be extended to early times. In both cases, the non-equilibrium kinetic-theory attractor smoothly connects to the late-time hydrodynamic attractor. In addition to this fundamental finding, I also discussed how one can use the determined EKT QCD attractor to test different hydrodynamic freeze-out prescriptions. This addresses an important open question in the phenomenological analysis of nuclear collisions which is to determine the best prescription for converting hydrodynamical fields into particle distributions. Currently, the quadratic ansatz (i) is widely used. This ansat assumes linear deviations from thermal equilibrium, which is in stark contrast to the far-from-equilibrium conditions in which fluid-dynamical modeling is practiced in current phenomenological applications, in particular in modeling of small systems (see e.g. Refs. [47–52]). To address whether these linearized procedures remain quantitatively predictive far from equilibrium, in Ref. [34] we confronted them with far-from-equilibrium simulations of QCD effective kinetic theory. The results of Ref. [34] show that the aHydro freeze-out ansatz performed better in reconstructing moments of the distribution function compared to linearized ansatze in far-from-equilibrium systems. Future directions include inclusion of quarks and implementation of aHydro freeze-out into widely used viscous hydrodynamics codes.

Acknowlegements
I thank my collaborators D. Almaalol, A. Kurkela, and U. Tantary. The work reported here was supported by the U.S. Department of Energy, Office of Science, Office of Nuclear Physics Award No. DE-SC0013470.

References
[1] Averbeck R, Harris J W and Schenke B 2015 *Heavy-Ion Physics at the LHC* (Springer, Cham) pp 355–420
[2] Jeon S and Heinz U 2016 *Quark-Gluon Plasma 5* ed Wang X N (World Scientific) pp 131–187
[3] Romatschke P and Romatschke U 2019 *Relativistic Fluid Dynamics In and Out of Equilibrium* Cambridge Monographs on Mathematical Physics (Cambridge University Press) ISBN 9781108483681, 9781108750028 (*Preprint* 1712.05815)
[4] Sachdev S 2010 [Lect. Notes Phys.828,273(2011)] (*Preprint* 1002.2947)
[5] Nastase H 2017 *String Theory Methods for Condensed Matter Physics*
[6] Duez M D, Liu Y T, Shapiro S L and Stephens B C 2004 *Phys. Rev.* **D69** 104030 (*Preprint* astro-ph/0402502)
[7] Shibata M and Kiuchi K 2017 *Phys. Rev.* **D95** 123003 (*Preprint* 1705.06142)
[8] Most E R, Papenfort L J, Dexheimer V, Hanauske M, Schramm S, Stöcker H and Rezzolla L 2019 *Phys. Rev. Lett.* **122** 061101 (*Preprint* 1807.03684)
[9] Alford M, Harutyunyan A and Sedrakian A 2019 *Phys. Rev.* **D100** 103021 (*Preprint* 1907.04192)
[10] Chesler P M and Yaffe L G 2010 *Phys. Rev.* **D82** 026006 (*Preprint* 0906.4426)
[11] Heller M P and Spalinski M 2015 *Phys. Rev. Lett.* **115** 072501 (*Preprint* 1503.07514)
[12] Keegan L, Kurkela A, Romatschke P, van der Schee W and Zhu Y 2016 *JHEP* **04** 031 (*Preprint* 1512.05347)
[13] Heller M P, Kurkela A, Spaliński M and Svensson V 2018 *Phys. Rev.* **D97** 091503 (*Preprint* 1609.04803)
[14] Florkowski W, Heller M P and Spalinski M 2018 *Rept. Prog. Phys.* **81** 046001 (*Preprint* 1707.02282)
[15] Romatschke P 2018 *Phys. Rev. Lett.* **120** 012301 (*Preprint* 1704.08699)
[16] Bemfica F S, Disconzi M M and Noronha J 2018 *Phys. Rev.* **D98** 104064 (*Preprint* 1708.06255)
[17] Spaliński M 2018 *Phys. Lett.* **B776** 468–472 (*Preprint* 1708.01921)
[18] Romatschke P 2017 *JHEP* **12** 079 (*Preprint* 1710.03234)
[19] Behtash A, Cruz-Camacho C N and Martinez M 2018 *Phys. Rev.* **D97** 044041 (*Preprint* 1711.01745)
[20] Florkowski W, Maksymiuk E and Ryblewski R 2018 *Phys. Rev.* **C97** 024915 (*Preprint* 1710.07095)
[21] Florkowski W, Maksymiuk E and Ryblewski R 2018 *Phys. Rev.* **C97** 014904 (*Preprint* 1711.03872)
[22] Strickland M, Noronha J and Denicol G 2018 *Phys. Rev.* **D97** 036020 (*Preprint* 1709.06644)
[23] Almaalol D and Strickland M 2018 *Phys. Rev.* **C97** 044911 (*Preprint* 1801.10173)

[24] Denicol G S and Noronha J 2019 *Phys. Rev.* **D99** 116004 (*Preprint* 1804.04771)
[25] Behtash A, Cruz-Camacho C N, Kamata S and Martinez M 2019 *Phys. Lett.* **B797** 134914 (*Preprint* 1805.07881)
[26] Strickland M 2018 *JHEP* **12** 128 (*Preprint* 1809.01200)
[27] Heller M P and Svensson V 2018 *Phys. Rev.* **D98** 054016 (*Preprint* 1802.08225)
[28] Behtash A, Kamata S, Martinez M and Shi H 2019 (*Preprint* 1911.06406)
[29] Strickland M and Tantary U 2019 *JHEP* **10** 069 (*Preprint* 1903.03145)
[30] Jaiswal S, Chattopadhyay C, Jaiswal A, Pal S and Heinz U 2019 *Phys. Rev.* **C100** 034901 (*Preprint* 1907.07965)
[31] Kurkela A, van der Schee W, Wiedemann U A and Wu B 2020 *Phys. Rev. Lett.* **124** 102301 (*Preprint* 1907.08101)
[32] Chattopadhyay C and Heinz U W 2020 *Phys. Lett. B* **801** 135158 (*Preprint* 1911.07765)
[33] Brewer J, Yan L and Yin Y 2019 (*Preprint* 1910.00021)
[34] Almaalol D, Kurkela A and Strickland M 2020 (*Preprint* 2004.05195)
[35] Arnold P B, Moore G D and Yaffe L G 2003 *JHEP* **01** 030 (*Preprint* hep-ph/0209353)
[36] Kurkela A and Zhu Y 2015 *Phys. Rev. Lett.* **115** 182301 (*Preprint* 1506.06647)
[37] Baier R, Mueller A H, Schiff D and Son D T 2001 *Phys. Lett.* **B502** 51–58 (*Preprint* hep-ph/0009237)
[38] Białas A and Czyż W 1984 *Phys. Rev. D* **30**(11) 2371–2378
[39] Białas A and Czyż W 1988 *Nuclear Physics B* **296** 611 – 624 ISSN 0550-3213
[40] Baym G 1984 *Phys. Lett.* **138B** 18–22
[41] Florkowski W, Ryblewski R and Strickland M 2013 *Nucl. Phys.* **A916** 249–259 (*Preprint* 1304.0665)
[42] Florkowski W, Ryblewski R and Strickland M 2013 *Phys. Rev.* **C88** 024903 (*Preprint* 1305.7234)
[43] Abraao York M C, Kurkela A, Lu E and Moore G D 2014 *Phys. Rev.* **D89** 074036 (*Preprint* 1401.3751)
[44] Mueller A H 2000 *Phys. Lett.* **B475** 220–224 (*Preprint* hep-ph/9909388)
[45] Kurkela A and Mazeliauskas A 2019 *Phys. Rev. Lett.* **122** 142301 (*Preprint* 1811.03040)
[46] Kurkela A and Mazeliauskas A 2019 *Phys. Rev.* **D99** 054018 (*Preprint* 1811.03068)
[47] Bozek P and Broniowski W 2013 *Phys. Rev. C* **88** 014903 (*Preprint* 1304.3044)
[48] Shen C, Paquet J F, Denicol G S, Jeon S and Gale C 2017 *Phys. Rev. C* **95** 014906 (*Preprint* 1609.02590)
[49] Alqahtani M, Nopoush M and Strickland M 2017 *Phys. Rev.* **C95** 034906 (*Preprint* 1605.02101)
[50] Weller R D and Romatschke P 2017 *Phys. Lett. B* **774** 351–356 (*Preprint* 1701.07145)
[51] Mäntysaari H, Schenke B, Shen C and Tribedy P 2017 *Phys. Lett. B* **772** 681–686 (*Preprint* 1705.03177)
[52] Strickland M 2019 *Nucl. Phys.* **A982** 92–98 (*Preprint* 1807.07191)

Heavy flavor dynamics across system size at the LHC

Roland Katz
SUBATECH, Université de Nantes, EMN, IN2P3/CNRS, 44307 Nantes, France

Caio A. G. Prado
Institute of Particle Physics, Central China Normal University (CCNU), Wuhan, Hubei 430079, China

Jacquelyn Noronha-Hostler
University of Illinois at Urbana-Champaign, Urbana, IL 61801, USA
E-mail: jnorhos@illinois.edu

Alexandre A. P. Suaide
Instituto de Física, Universidade de São Paulo, C.P. 66318, 05315-970 São Paulo, SP, Brazil

Abstract. One of the fundamental signatures of the Quark Gluon Plasma has been the suppression of heavy flavor (specifically D mesons), which has been measured via the nuclear modification factor, R_{AA} and azimuthal anisotropies, v_n, in large systems. However, multiple competing models can reproduce the same data for R_{AA} to v_n. In this talk we break down the competing effects that conspire together to successfully reproduce R_{AA} and v_n in experimental data using Trento+v-USPhydro+DAB-MOD. Then using our best fit model we make predictions for R_{AA} and v_n across system size for ^{208}PbPb, ^{129}XeXe, ^{40}ArAr, and ^{16}OO collisions. We find that 0–10% centrality has a non-trivial interplay between the system size and eccentricities such that system size effects are masked in v_2 whereas in 30–50% centrality the eccentricities are approximately constant across system size and, therefore, is a better centrality class to study D meson dynamics across system size.

1. Introduction

The suppression of heavy probes has been a fundamental signature used to constrain the microscopic properties of the Quark Gluon Plasma (QGP). The soft sector of the QGP consists of a nearly perfect fluid composed of light quarks (up, down, and strange) and gluons. The collective flow observables in the soft sector are well described through event-by-event relativistic viscous hydrodynamics [1, 2, 3, 4, 5, 6, 7, 8, 9, 10, 11, 12, 13, 14, 15, 16, 17, 18]. Due to the much larger mass of charm quarks, they are created at very short time scales and are very unlikely to be fully thermalized within the soft sector of the QGP. Therefore, open heavy flavor mesons, like D mesons can provide orthogonal information to that gathered in the soft sector. Studying the nuclear modification factor, R_{AA}, which is approximately 1 when no suppression occurs and much less than 1 in large AA collisions where heavy flavor quarks loose energy by being bumped around by the strongly interacting QGP, can provide insight into the microscopic dynamics of the QGP. Additionally, due to the path length dependence of the heavy quarks passing through the azimuthally anisotropic medium, a large v_2 is measured for D mesons. In more recent years it has been shown that there is a strong correlation between the soft and hard/heavy sectors when it comes to azimuthal anisotropies [19, 20, 21, 22, 23, 24, 25]

In light of the recent small system results from the LHC and RHIC wherein collective flow and strangeness enhancement [26, 27, 28, 29, 30, 31, 32, 33, 34, 35, 36, 37, 38, 39, 40, 41, 42, 43, 44, 45, 46, 47] have been measured in small systems that can be reasonably well described by theoretical hydrodynamic models [48, 49, 50, 4, 51, 52, 53, 54, 15, 53], physicists have been trying to reconcile these results with the lack of hard or heavy flavor suppression (i.e. $R_{pA} \sim 1$ [55, 56]). For an overview and discussion on future measurements see [57]. A further piece in the puzzle is that a large v_2 has now been measured for D mesons in pPb collisions at the LHC [58]. Currently, there are no theoretical models that can simultaneously reproduce $R_{pA} \sim 1$ with a large v_2 prediction, rather $R_{pA} \sim 0.8$ [59] or no v_2 calculations have been attempted [60, 61]. In order to shed light on this mystery we study the system size dependence of R_{AA} and v_2 in the potential upcoming system size run of symmetric AA collisions (first proposed at the LHC in [62]) using our best fit models in PbPb collisions: trento+v-USPhydro+DAB-MOD.

2. Model

For our heavy flavor description we have created the modular D and B meson Monte Carlo simulation package DAB-MOD [22, 63, 25] that can run both Langevin and energy loss models (and easily incorporate different components of the modeling such as energy loss fluctuations, coalescence, varying the diffusion transport coefficient, and path length dependence of the energy loss). The initial heavy flavor quarks are sampled from pQCD FONLL calculations [64, 65] distribution. After they are evolved dynamically on top of an event-by-event viscous hydrodynamical evolving medium (Trento [66]+v-USPhydro [67, 68, 69]) using either a Langevin description or energy loss, they are hadronized at the decoupling temperature T_d via a hybrid fragmentation/coalescence model from which the final nuclear modification factor can be reconstructed. Here we use our "best fit" model tuned at PbPb 5.02 TeV collisions [63] where the purely collisional spatial diffusion coefficient model [70] $D_s(2\pi T) = 2.23$ reasonably reproduces experimental data at low $p_T \lesssim 5$–6 GeV, while an energy loss model (with the same best fit description as in [71]) works best for for the high $p_T \gtrsim 5$–6 GeV. While a number of lessons can be learned from PbPb collisions, it is not clear across system size in what p_T range Langevin versus energy loss descriptions will dominate. Therefore, in the plots shown here we include both in an overlapping region.

Because we also plot the azimuthal anisotropies, we point out that this is inherently a soft hard correlation (between one soft particle and one hard particle, see [20, 21, 22, 23, 24] for further details on the implications). Therefore, we pay special attention to a reasonable hydrodynamic description of the soft sector. Here we used Trento initial conditions with the parameters $p = 0$, $k = 1.6$, and $\sigma = 0.51$ fm determined from a Bayesian analysis [8] and that works reasonably well compared to multiparticle cumulants [72, 73]. The hydrodynamical description fixed the parameters $\tau_0 = 0.6$ fm, $\eta/s = 0.047$, $T_{\rm FO} = 150$ MeV by comparing to experimental data from the LHC [11, 13, 74] and uses the state-of-the-art Lattice QCD equation of state [75, 11] matched to the particle data list PDG16+ [76, 77]. Additionally, due to the sensitivity of the soft sector to a deformed ^{129}Xe nucleus [13] we also compare a spherical versus prolate xenon nucleus.

Finally, we would be remiss not to point out that there is a tuning parameter for $R_{\rm AA}$ that adjusts its overall magnitude (generally tuned in most central collisions). Because these are predictions for XeXe, ArAr, and OO i.e. there is no experimental data available for $R_{\rm AA}$ in these collisional systems, there is inherently an amount of uncertainty due to this parameter. Furthermore, our hydrodynamic calculations have a similar parameter that is typically tuned to the dN/dy in central collisions. Here we assume in both cases that these constants do not vary

Journal of Physics: Conference Series **1602** (2020) 012019 doi:10.1088/1742-6596/1602/1/012019

with system size.

3. Results

Within the heavy flavor sector there are many competing models that can reasonable well reproduce experimental data (see for these comparisons in [78]). However, in [79] it was found that once all models are required to use the same hydrodynamical background that these similarities no longer hold. In this work, we constrain our background to fit multiple observables in the soft sector and then study what contributes to produce a reasonable $R_{\rm AA}$ and v_2.

In Fig. 1 we compare the R_{AA} for PbPb collisions at 5.02 TeV for our two different "best fit" models: the Langevin description and energy loss (see [63] for further comparisons of energy loss fluctuations, initial state models, diffusion coefficients, and energy loss parameterizations). Furthermore, we compare both models with and without coalescence. It is clear that at low p_T the need for coalescence is much more obvious whereas above $p_T > 5$ GeV, the need for coalescence is minimal.

When it comes to $v_2\{2\}$ the influence of coalescence appears to play a role even in the energy loss results up to roughly $p_T \sim 10$ GeV. For both Langevin and energy loss the effect of coalescence shifts the peak in the v_2 curve to the right (higher p_T). Additionally, the v_2 results make it quite clear that energy loss is needed above $p_T > 5$ GeV because the Langevin results significantly under-predict the data whereas the energy loss model can reproduce the experimental data perfectly.

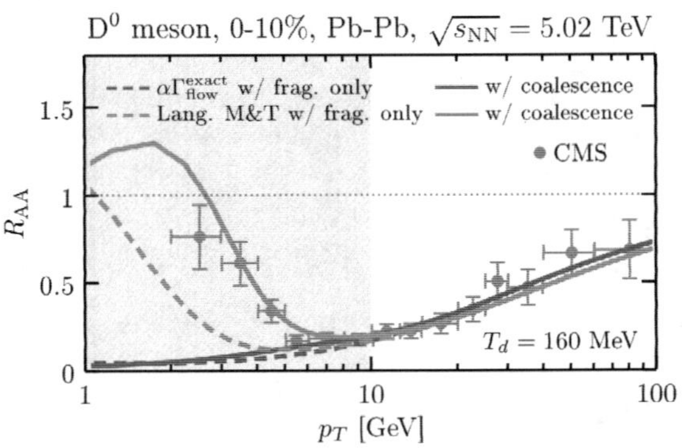

Figure 1. Comparison of R_{AA} for both our best fit Langevin and energy loss models both with and without coalescence.

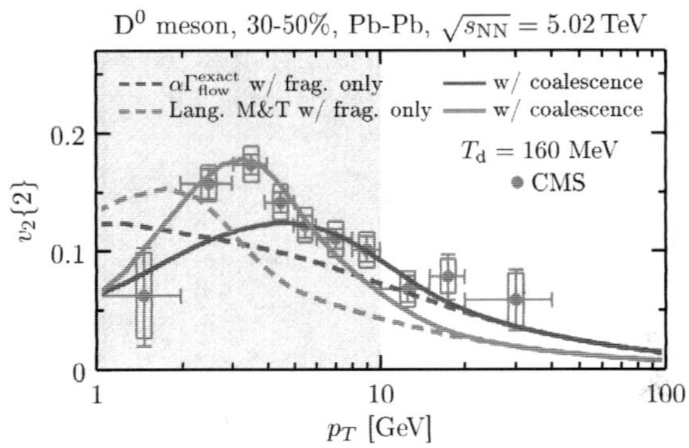

Figure 2. Comparison of $v_2\{2\}$ for both our best fit Langevin and energy loss models both with and without coalescence.

Using then our best fit models (always including coalescence) we then make predictions across system size in Fig. 3 for the nuclear modification factor. We find that the smaller systems, indeed, bring $R_{\rm AA}$ closer to 1 but still have rather significant deviations from 1 at $p_T \sim 10$ GeV where a minimum is seen. Generally $30-50\%$ centrality class can get an $R_{\rm AA} \sim 0.8$ at its minimum whereas $0-10\%$ centrality class has a minimum closer to 0.5 for our smallest system size of OO collisions. We note that OO collisions are about double the radius of pPb collisions. Finally, we see that Langevin and energy loss descriptions are nearly identical in $R_{\rm AA}$ at high p_T but between $p_T \sim 5-10$ GeV the energy loss model predicts significantly more heavy flavor suppression than the Langevin model. This point specifically is interesting because it may be that in smaller systems a Langevin description, which predicts less suppression, is preferred in contrast to large systems where Langevin matches data only at low p_T.

Because we study v_2, it is important to study the shape of the initial state through the eccentricities. It has been shown that the initial eccentricities are very strongly correlated

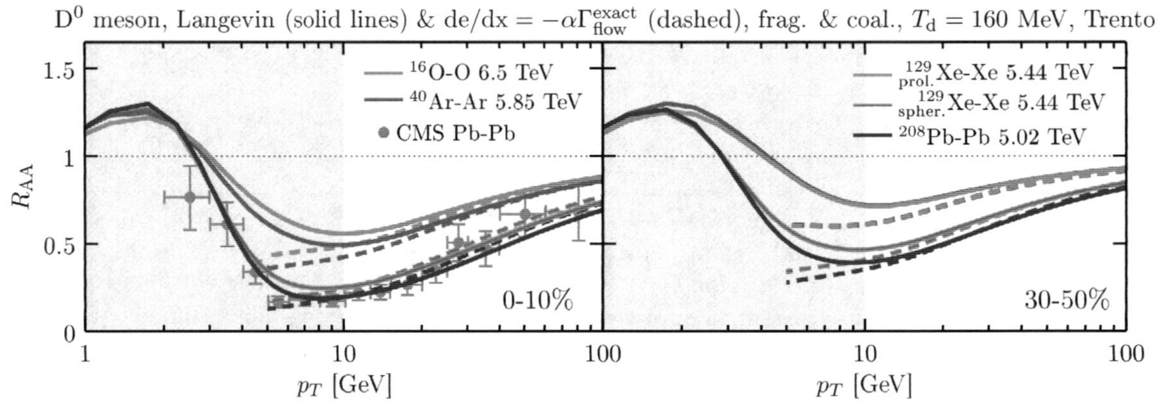

Figure 3. Plot of R_{AA} across different collisional systems (i.e. system sizes), comparing two different centrality classes: $0-10\%$ and $30-50\%$.

with the final v_2 in both the soft [80, 81, 82, 83, 84, 85] and the hard/heavy sector [20, 21, 22, 23, 24, 63, 25]. Thus, studying the eccentricities across system size can shed light on the anticipated final v_2 of D mesons. In Fig. 4 we perform precisely this comparison and find quite different results depending on the centrality class. In centrality collisions of $0-10\%$ there is a strong system size dependence in ε_2 such that as the system size decreases, the eccentricities increase. In contrast, in midcentral collisions of $30-50\%$ we find that the eccentricities are nearly constant and only the system size changes. Thus, $30-50\%$ is the best centrality class to actually observe system size effects (to reduce the competing effect of variation in eccentricities).

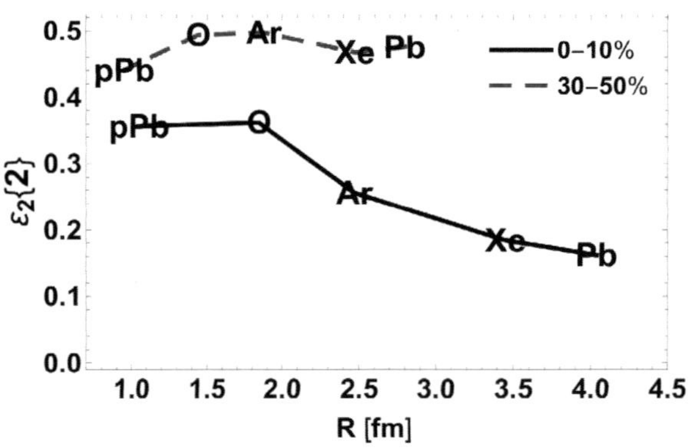

Figure 4. Plot of elliptical eccentricities versus average radius for different collisional systems, comparing two different centrality classes: $0-10\%$ and $30-50\%$.

Then in Fig. 5 we make the full comparison of the v_2 calculations. We find that as anticipated from the eccentricities that $0-10\%$ paints a complicated picture where there is very little system size dependence. This is because the increase in ε_2 in small systems would have the effect of increasing v_2 whereas the system size decrease (demonstrated through the average radius, R) would suppress v_2. Overall, these two competing effects lead to a nearly identical v_2 across system size. In contrast, the $30-50\%$ centrality class sees a very clear suppression of v_2 as one decrease the system size.

Additionally, in Fig. 5 we see that a prolate nucleus leads to a larger v_2 for D mesons. This is somewhat surprising because typically the effect of a deformation only shows up in very central collisions. Additionally, one would likely expect that D mesons are less sensitive to such small deformation, however, it appears that they indeed can see these differences.

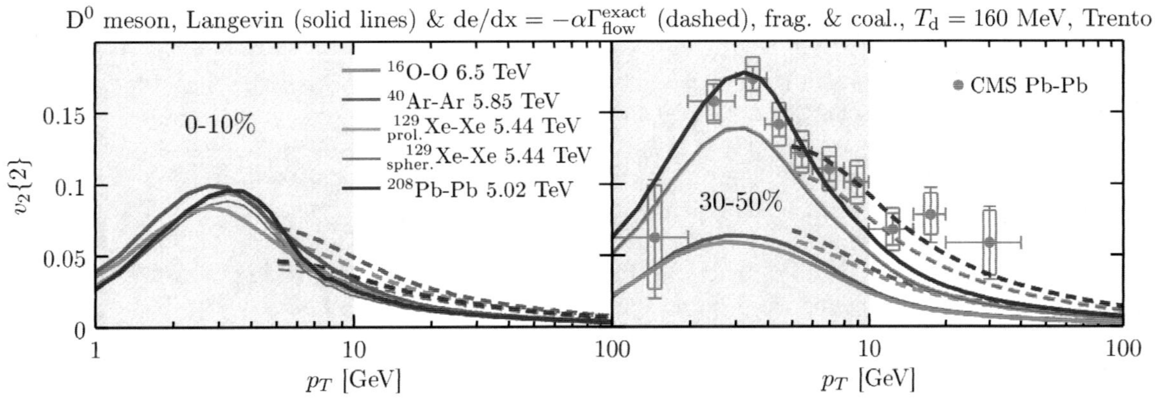

Figure 5. Plot of $v_2\,\{2\}$ across different collisional systems (i.e. system sizes), comparing two different centrality classes: $0-10\%$ and $30-50\%$.

4. Conclusions

Similar to what was found in the soft sector in [74], different behaviors can appear when scaling either by multiplicity (see also [86]) or centrality. In this talk, we argue that when scaling by centrality the v_2 results have a complex interplay between eccentricities and shrinking system size such that the D meson v_2 is nearly identical across system size in most central collisions. However, we argue that comparing $30-50\%$ centrality allows a focus only on system size effects while keeping the eccentricities nearly identical. In this case we predict that D meson v_2 will be significantly suppressed in smaller systems in $30-50\%$ centrality, this result occurs both in Langevin and energy loss descriptions.

One surprising feature is that we do find a sensitive of D meson v_2 to a deformed ^{129}Xe nucleus in $0-10\%$ collisions in the range of $p_T = 3-5$ GeV, in light of these results it would be quite interesting to revisit RHIC UU results in central collisions in order to study D meson v_2 or to also explore effects of deformed nuclei in the future sPHENIX experiment.

Finally, we study the system size dependence of a heavy flavor Langevin description versus energy loss description. While the overage magnitude shifts somewhat between the two descriptions, we find that they are both influenced by system size in roughly equivalent manners. In the case of R_{AA} they both approach 1 as the system size is decreased (however, a Langevin description is slightly closer to 1 compared to the energy loss description). However, both still are quite a bit farther away from the $R_{pPb} \sim 1$ result. The v_2 results for energy loss tend to produce more v_2 for $p_T > 5$ GeV (bringing it closer to experimental data in PbPb collisions than the Langevin models at high p_T). In future work we hope to extent our model to even smaller systems such as pPb and pp to compare to recent results from ATLAS and CMS. Additionally, we expect a similar effect to occur at RHIC (as discussed in [87]). Furthermore, it will be interesting to explore further soft-heavy correlations such as [88] or study the initialization time of heavy flavor dynamics more carefully [24]. Furthermore, it would be interesting to explore these ideas further in proposed measurements such as polarized beams [89] and ultracentral deformed ion-ion collisions [90].

5. Acknowledgments.
The authors thank Fundação de Amparo à Pesquisa do Estado de São Paulo (FAPESP) and Conselho Nacional de Desenvolvimento Científico e Tecnológico (CNPq) for support. R.K. is supported by the Region Pays de la Loire (France) under contract No. 2015-08473. C.A.G.P. is supported by the NSFC under grant No. 11521064, MOST of China under Project No. 2014CB845404. J.N.H. acknowledges the support of the Alfred P. Sloan Foundation; support from the US-DOE Nuclear Science Grant No. DE-SC0019175; the Illinois Campus Cluster, a computing resource that is operated by the Illinois Campus Cluster Program (ICCP) in conjunction with the National Center for Supercomputing Applications (NCSA) and which is supported by funds from the University of Illinois at Urbana-Champaign; and the Office of Advanced Research Computing (OARC) at Rutgers, The State University of New Jersey for providing access to the Amarel cluster and associated research computing resources that have contributed to the results reported here.

References
[1] Song H, Bass S A, Heinz U, Hirano T and Shen C 2011 *Phys. Rev. Lett.* **106** 192301 [Erratum: Phys.Rev.Lett. 109, 139904 (2012)] (*Preprint* 1011.2783)
[2] Bozek P and Wyskiel-Piekarska I 2012 *Phys. Rev. C* **85** 064915 (*Preprint* 1203.6513)
[3] Gardim F G, Grassi F, Luzum M and Ollitrault J Y 2012 *Phys. Rev. Lett.* **109** 202302 (*Preprint* 1203.2882)
[4] Bozek P and Broniowski W 2013 *Phys. Rev.* **C88** 014903 (*Preprint* 1304.3044)
[5] Niemi H, Eskola K J and Paatelainen R 2016 *Phys. Rev.* **C93** 024907 (*Preprint* 1505.02677)
[6] Ryu S, Paquet J F, Shen C, Denicol G S, Schenke B, Jeon S and Gale C 2015 *Phys. Rev. Lett.* **115** 132301 (*Preprint* 1502.01675)
[7] McDonald S, Shen C, Fillion-Gourdeau F, Jeon S and Gale C 2017 *Phys. Rev.* **C95** 064913 (*Preprint* 1609.02958)
[8] Bernhard J E, Moreland J S, Bass S A, Liu J and Heinz U 2016 *Phys. Rev.* **C94** 024907 (*Preprint* 1605.03954)
[9] Gardim F G, Grassi F, Luzum M and Noronha-Hostler J 2017 *Phys. Rev.* **C95** 034901 (*Preprint* 1608.02982)
[10] Giacalone G, Yan L, Noronha-Hostler J and Ollitrault J Y 2016 *Phys. Rev. C* **94** 014906 (*Preprint* 1605.08303)
[11] Alba P, Mantovani Sarti V, Noronha J, Noronha-Hostler J, Parotto P, Portillo Vazquez I and Ratti C 2018 *Phys. Rev.* **C98** 034909 (*Preprint* 1711.05207)
[12] Gardim F G, Grassi F, Ishida P, Luzum M, Magalhães P S and Noronha-Hostler J 2018 *Phys. Rev.* **C97** 064919 (*Preprint* 1712.03912)
[13] Giacalone G, Noronha-Hostler J, Luzum M and Ollitrault J Y 2018 *Phys. Rev.* **C97** 034904 (*Preprint* 1711.08499)
[14] Eskola K J, Niemi H, Paatelainen R and Tuominen K 2018 *Phys. Rev.* **C97** 034911 (*Preprint* 1711.09803)
[15] Weller R D and Romatschke P 2017 *Phys. Lett.* **B774** 351–356 (*Preprint* 1701.07145)
[16] Schenke B, Shen C and Tribedy P 2019 (*Preprint* 1901.04378)
[17] Giacalone G, Gardim F G, Noronha-Hostler J and Ollitrault J Y 2020 (*Preprint* 2004.09799)
[18] Giacalone G, Gardim F G, Noronha-Hostler J and Ollitrault J Y 2020 (*Preprint* 2004.01765)
[19] Nahrgang M, Aichelin J, Bass S, Gossiaux P B and Werner K 2015 *Phys. Rev.* **C91** 014904 (*Preprint* 1410.5396)
[20] Betz B, Gyulassy M, Luzum M, Noronha J, Noronha-Hostler J, Portillo I and Ratti C 2017 *Phys. Rev.* **C95** 044901 (*Preprint* 1609.05171)
[21] Noronha-Hostler J, Betz B, Noronha J and Gyulassy M 2016 *Phys. Rev. Lett.* **116** 252301 (*Preprint* 1602.03788)
[22] Prado C A G, Noronha-Hostler J, Katz R, Suaide A A P, Noronha J, Munhoz M G and Cosentino M R 2017 *Phys. Rev.* **C96** 064903 (*Preprint* 1611.02965)
[23] Sirunyan A M *et al.* (CMS) 2018 *Phys. Lett.* **B776** 195–216 (*Preprint* 1702.00630)
[24] Andres C, Armesto N, Niemi H, Paatelainen R and Salgado C A 2019 (*Preprint* 1902.03231)
[25] Katz R, Prado C A, Noronha-Hostler J and Suaide A A 2019 (*Preprint* 1907.03308)
[26] Chatrchyan S *et al.* (CMS) 2013 *Phys. Lett.* **B724** 213–240 (*Preprint* 1305.0609)
[27] Aaboud M *et al.* (ATLAS) 2017 *Eur. Phys. J.* **C77** 428 (*Preprint* 1705.04176)
[28] Aaboud M *et al.* (ATLAS) 2018 *Phys. Rev.* **C97** 024904 (*Preprint* 1708.03559)
[29] Aad G *et al.* (ATLAS) 2013 *Phys. Lett.* **B725** 60–78 (*Preprint* 1303.2084)
[30] Sirunyan A M *et al.* (CMS) 2018 *Phys. Rev. Lett.* **121** 082301 (*Preprint* 1804.09767)

[31] Khachatryan V *et al.* (CMS) 2015 *Phys. Lett.* **B742** 200–224 (*Preprint* 1409.3392)
[32] Khachatryan V *et al.* (CMS) 2015 *Phys. Rev. Lett.* **115** 012301 (*Preprint* 1502.05382)
[33] Khachatryan V *et al.* (CMS) 2015 *Phys. Rev.* **C92** 034911 (*Preprint* 1503.01692)
[34] Sirunyan A M *et al.* (CMS) 2018 *Phys. Rev. Lett.* **120** 092301 (*Preprint* 1709.09189)
[35] Abelev B B *et al.* (ALICE) 2013 *Phys. Lett.* **B726** 164–177 (*Preprint* 1307.3237)
[36] Abelev B B *et al.* (ALICE) 2014 *Phys. Rev.* **C90** 054901 (*Preprint* 1406.2474)
[37] Adare A *et al.* (PHENIX) 2013 *Phys. Rev. Lett.* **111** 212301 (*Preprint* 1303.1794)
[38] Adare A *et al.* (PHENIX) 2015 *Phys. Rev. Lett.* **114** 192301 (*Preprint* 1404.7461)
[39] Aidala C *et al.* (PHENIX) 2018 (*Preprint* 1805.02973)
[40] Adare A *et al.* (PHENIX) 2018 (*Preprint* 1807.11928)
[41] Adare A *et al.* (PHENIX) 2015 *Phys. Rev. Lett.* **115** 142301 (*Preprint* 1507.06273)
[42] Aidala C *et al.* 2017 *Phys. Rev.* **C95** 034910 (*Preprint* 1609.02894)
[43] Adare A *et al.* (PHENIX) 2018 *Phys. Rev.* **C97** 064904 (*Preprint* 1710.09736)
[44] Adare A *et al.* (PHENIX) 2018 *Phys. Rev.* **C98** 014912 (*Preprint* 1711.09003)
[45] Aidala C *et al.* (PHENIX) 2017 *Phys. Rev.* **C96** 064905 (*Preprint* 1708.06983)
[46] Aidala C *et al.* (PHENIX) 2018 *Phys. Rev. Lett.* **120** 062302 (*Preprint* 1707.06108)
[47] Adam J *et al.* (ALICE) 2017 *Nature Phys.* **13** 535–539 (*Preprint* 1606.07424)
[48] Bozek P 2012 *Phys. Rev.* **C85** 014911 (*Preprint* 1112.0915)
[49] Bozek P and Broniowski W 2013 *Phys. Lett.* **B718** 1557–1561 (*Preprint* 1211.0845)
[50] Bozek P, Broniowski W and Torrieri G 2013 *Phys. Rev. Lett.* **111** 172303 (*Preprint* 1307.5060)
[51] Kozlov I, Luzum M, Denicol G, Jeon S and Gale C 2014 (*Preprint* 1405.3976)
[52] Zhou Y, Zhu X, Li P and Song H 2015 *Phys. Rev.* **C91** 064908 (*Preprint* 1503.06986)
[53] Zhao W, Zhou Y, Xu H, Deng W and Song H 2018 *Phys. Lett.* **B780** 495–500 (*Preprint* 1801.00271)
[54] Mäntysaari H, Schenke B, Shen C and Tribedy P 2017 *Phys. Lett.* **B772** 681–686 (*Preprint* 1705.03177)
[55] Adam J *et al.* (ALICE) 2016 *Phys. Lett. B* **754** 81–93 (*Preprint* 1509.07491)
[56] Acharya S *et al.* (ALICE) 2019 *JHEP* **12** 092 (*Preprint* 1906.03425)
[57] Perepelitsa D V 2020 (*Preprint* 2005.05981)
[58] Zhang C, Marquet C, Qin G Y, Wei S Y and Xiao B W 2019 *Phys. Rev. Lett.* **122** 172302 (*Preprint* 1901.10320)
[59] Xu Y, Cao S, Qin G Y, Ke W, Nahrgang M, Auvinen J and Bass S A 2016 *Nucl. Part. Phys. Proc.* **276-278** 225–228 (*Preprint* 1510.07520)
[60] Kang Z B, Vitev I, Wang E, Xing H and Zhang C 2015 *Phys. Lett.* **B740** 23–29 (*Preprint* 1409.2494)
[61] Sharma R, Vitev I and Zhang B W 2009 *Phys. Rev.* **C80** 054902 (*Preprint* 0904.0032)
[62] Citron Z *et al.* 2018 (*Preprint* 1812.06772)
[63] Katz R, Prado C A G, Noronha-Hostler J, Noronha J and Suaide A A P 2019 (*Preprint* 1906.10768)
[64] Cacciari M, Greco M and Nason P 1998 *JHEP* **05** 007 (*Preprint* hep-ph/9803400)
[65] Cacciari M, Frixione S and Nason P 2001 *JHEP* **03** 006 (*Preprint* hep-ph/0102134)
[66] Moreland J S, Bernhard J E and Bass S A 2015 *Phys. Rev.* **C92** 011901 (*Preprint* 1412.4708)
[67] Noronha-Hostler J, Noronha J and Grassi F 2014 *Phys. Rev.* **C90** 034907 (*Preprint* 1406.3333)
[68] Noronha-Hostler J, Denicol G S, Noronha J, Andrade R P G and Grassi F 2013 *Phys. Rev.* **C88** 044916 (*Preprint* 1305.1981)
[69] Noronha-Hostler J, Noronha J and Gyulassy M 2016 *Phys. Rev. C* **93** 024909 (*Preprint* 1508.02455)
[70] Moore G D and Teaney D 2005 *Phys. Rev.* **C71** 064904 (*Preprint* hep-ph/0412346)
[71] Aaboud M *et al.* (ATLAS) 2018 *Phys. Rev.* **C98** 044905 (*Preprint* 1805.05220)
[72] Giacalone G, Noronha-Hostler J and Ollitrault J Y 2017 *Phys. Rev.* **C95** 054910 (*Preprint* 1702.01730)
[73] Rao S, Sievert M and Noronha-Hostler J 2019 (*Preprint* 1910.03677)
[74] Sievert M D and Noronha-Hostler J 2019 (*Preprint* 1901.01319)
[75] Borsanyi S, Fodor Z, Hoelbling C, Katz S D, Krieg S and Szabo K K 2014 *Phys. Lett.* **B730** 99–104 (*Preprint* 1309.5258)
[76] Alba P *et al.* 2017 *Phys. Rev.* **D96** 034517 (*Preprint* 1702.01113)
[77] Alba P, Sarti V M, Noronha-Hostler J, Parotto P, Portillo-Vazquez I, Ratti C and Stafford J 2020 (*Preprint* 2002.12395)
[78] Acharya S *et al.* (ALICE) 2018 *Phys. Rev. Lett.* **120** 102301 (*Preprint* 1707.01005)
[79] Xu Y *et al.* 2019 *Phys. Rev.* **C99** 014902 (*Preprint* 1809.10734)
[80] Teaney D and Yan L 2011 *Phys. Rev.* **C83** 064904 (*Preprint* 1010.1876)
[81] Gardim F G, Grassi F, Luzum M and Ollitrault J Y 2012 *Phys. Rev.* **C85** 024908 (*Preprint* 1111.6538)
[82] Niemi H, Denicol G S, Holopainen H and Huovinen P 2013 *Phys. Rev.* **C87** 054901 (*Preprint* 1212.1008)
[83] Teaney D and Yan L 2012 *Phys. Rev.* **C86** 044908 (*Preprint* 1206.1905)
[84] Qiu Z and Heinz U W 2011 *Phys. Rev.* **C84** 024911 (*Preprint* 1104.0650)

[85] Gardim F G, Noronha-Hostler J, Luzum M and Grassi F 2015 *Phys. Rev.* **C91** 034902 (*Preprint* 1411.2574)
[86] Acharya S *et al.* (ALICE) 2019 *Phys. Rev. Lett.* **123** 142301 (*Preprint* 1903.01790)
[87] Huang S, Chen Z, Jia J and Li W 2019 (*Preprint* 1904.10415)
[88] Plumari S, Coci G, Das S K, Minissale V and Greco V 2019 *Nucl. Phys.* **A982** 655–658 (*Preprint* 1901.07815)
[89] Bozek P and Broniowski W 2018 *Phys. Rev. Lett.* **121** 202301 (*Preprint* 1808.09840)
[90] Noronha-Hostler J, Paladino N, Rao S, Sievert M D and Wertepny D E 2019 (*Preprint* 1905.13323)

Journal of Physics: Conference Series **1602** (2020) 012020 doi:10.1088/1742-6596/1602/1/012020

High p_T identified hadrons in large and small systems measured by PHENIX

Balazs Ujvari (for the PHENIX Collaboration)

University of Debrecen, H-4032 Debrecen, Egyetem ter 1, Hungary

E-mail: `balazs.ujvari@science.unideb.hu`

Abstract. The PHENIX experiment at RHIC measured identified hadrons in several collision systems and centrality classes. Measurements were performed in different years and independent decay modes. The mesons show universal high-p_T supression in the different collision systems which suggest no dependency on the size of the projectile. Yields of π^0 , η, K_S, K^* and Φ mesons measured in large systems show similar suppression pattern at $\sqrt{s_{NN}}$ = 200 GeV for similar numbers of participant nucleons. In the intermediate p_T region, between 2 and 5 GeV/c, a significant enhancement of baryon-to-meson ratios compared to those measured in pp collisions is observed. In small systems an ordering ($R_{\mathrm{pA}} > R_{\mathrm{dA}} > R_{\mathrm{HeA}}$) for the mesons ($\pi^0$ and Φ) at mid-p_T was observed.

Journal of Physics: Conference Series **1602** (2020) 012020 doi:10.1088/1742-6596/1602/1/012020

1. Introduction

One of the most important tools to study the properties of the Quark Gluon Plasma (QGP), formed in relativistic heavy ion collisions [1, 2, 3, 4], is the production of high p_{T} hadrons which as a rule are leading fragments of jets from hard scattered partons at the earliest stage of the collision [5]. Except for possible nuclear modifications of the parton distribution functions, the initial spectrum of the hard scattered partons is expected to be the same irrespective whether the scattering occurred in pp or nucleus-nucleus (NN) collision. However, if a hot, dense, colored medium is formed in NN, the scattered parton will traverse it and interact with it, losing energy [6]. Therefore, when it fragments into final state colorless particles, their average momentum, including that of the leading particle, will usually be smaller compared to the momenta of final state particles from a pp collision with similar hard scattering. As a result, the spectra of high p_{T} particles in NN will be shifted down with respect to the pp spectra scaled by the number of binary nucleon-nucleon collisions. The simplest observable to quantify this shift is the nuclear modification factor R_{AB} defined as

$$R_{AB}^{cent}(p_T) = \frac{1}{T_{AB}^{cent}} \frac{dN_{AB}{}^{cent}/dp_T}{d\sigma_{pp}/dp_T}$$

where A, B are the two nuclei, the upper index ^{cent} is the centrality of the collision as derived from the Glauber-model [7, 8], dN_{AB}^{cent} is the invariant yield of the observable (high p_T hadron), $d\sigma_{pp}/dp_T$ is the cross-section for production of the hadron in pp collisions, while T_{AB} is the nuclear overlap function, proportional to the the number of binary nucleon-nucleon collisions at a given centrality ($N_{coll} = T_{AB} * \sigma_{NN}^{inel}$). The volume of the overlap region is characterized by the number of "participating" nucleons N_{part} [9].

The energy loss of the hard scattered parton depends, among others, on the path-length in the medium, which in turn can be controlled to some extent with the collision geometry. The idea and its limitations are illustrated in Figure 1. Partons scattered near the surface of the overlap volume (color shaded area) often have little medium to traverse before they exit in the vacuum and fragment, on the other hand, less frequently, have unusually high path in the medium. (Note that this "surface bias" is a well-known phenomenon, that can be remedied studying the asymmetry of back-to-back dijets, but here we study single inclusive spectra.) The bias clearly increases with the surface to volume ratio, which in turn, for any fixed overlap volume (i.e. N_{part}) is maximum if the two colliding nuclei are of the same size (Au+Au in Figure 1). If one is much smaller than the other (Cu+Au), the same size (N_{part}) overlap volume is closer to spherical, decreasing the surface bias. Therefore, it is very interesting to compare R_{AB} as a function of N_{part} for same size (Au+Au) and different size (Cu+Au) nuclei. Note that both nuclei are spherical which makes determination of the overlap geometry (volume shape) relatively straightforward. The same is not true for U+U collisions; the nuclei are oblong, and their relative orientation can not be clearly determined event-by-event, also, N_{part} is no longer a reliable measure of the overlap volume. Fully aligned "tip-to-tip" or "body-to-body" collisions have the same N_{part}, but much smaller volume and higher energy density in case of "tip-to-tip", and vice versa for "body-to-body" collisions. Nonetheless, in extreme cases they provide the highest possible (at RHIC) energy density QGP, so measuring their R_{AA} is very important.

Another interesting question is to study whether the flavor content of the leading (highest p_{T}) hadron makes any difference in the R_{AB}. The results presented include R_{AB} of mesons with light quarks only, with hidden strangeness and with open strangeness.

Contrary to mesons – where R_{AB} is everywhere at or below unity – baryons are typically enhanced at 2-3 GeV/c in p_{T}. This observation helps to shed light on the physics of hadronization. While at higher p_{T} virtually the sole possible source of hadrons is jet fragmentation, at lower p_{T} this is not the case. If a thermalized QGP was formed, up to a certain baryon p_{T} recombination of three quarks with momenta at the higher end of the thermal

Journal of Physics: Conference Series **1602** (2020) 012020 doi:10.1088/1742-6596/1602/1/012020

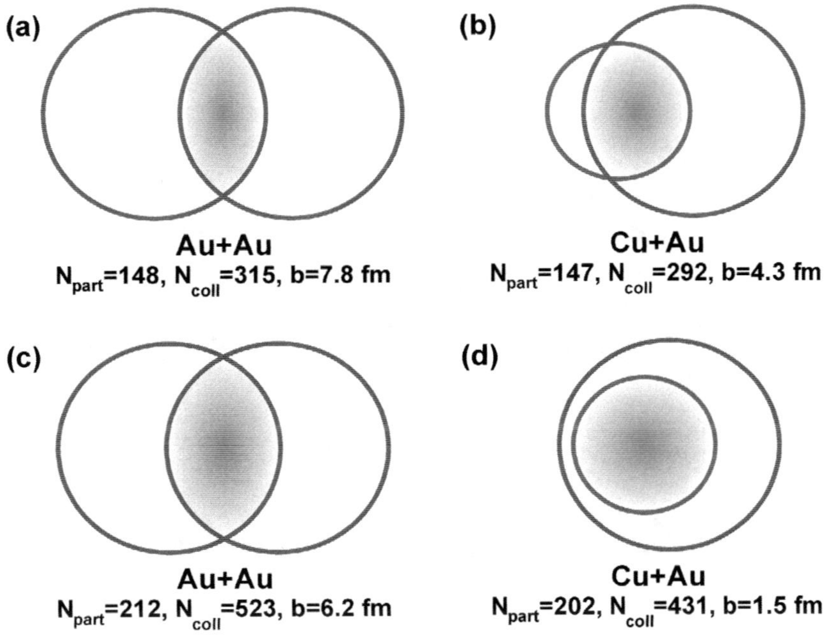

Figure 1. Cartoon showing Au+Au and Cu+Au collisions with comparable $N_{\rm part}$. On panel (b) the overlap area is asymmetric, and part of the dilute surfaces of Au and Cu (corona) overlap.

distribution is a plausible source of moderate $p_{\rm T}$ baryons. Therefore, the observed "baryon enhancement" is a powerful constraint on the QGP equation of state and the recombination model.

A fourth line of inquiry is collisions of very small and large ions (p+Au, d+Au, ^{3}He+Au). Azimuthal asymmetries in particle emission at mid-rapidity indicated that formation of QGP even in such small systems is a plausible possibility. In large colliding systems such asymmetries always came hand in hand with strong suppression of high $p_{\rm T}$ leading hadrons (small $R_{\rm AB}$). Therefore, it is paramount to study $R_{\rm AB}$ in small-on-large systems, too, although we should emphasize, that *lack of suppression in p/d/^{3}He+Au in itself does not disprove the creation of QGP in those systems.* Parton energy loss depends on path length in the medium, and even if droplets of QGP are formed, they might be too small to induce any measurable energy loss. – Also, at intermediate $p_{\rm T}$ (2-5 GeV/c) it is important to see whether a "Cronin-type" enhancement can be observed or not [10].

2. Experimental Details

All results presented are from data taken and analyzed by the PHENIX experiment (see Fig. 2) at the Relativistic Heavy Ion Collider (RHIC). The Cu+Au data were collected in 2012 [11], Au+Au in 2004 [12] and 2007 [13], Cu+Cu in 2005 [14], U+U in 2012 [15], p+Au in 2015, d+Au in 2003 [16] and 2008 [17], and ^{3}He+Au in 2014.

For Year-2004 data collision centrality was determined from the correlation between the number of charged particles detected in the Beam-Beam Counters (BBC, $3.0 < |\eta| < 3.9$) and the energy measured in the Zero Degree Calorimeters (ZDC). For all other data the centrality classification was done solely with the total charge observed in the BBC. The two central spectrometer arms cover $|\eta| < 0.35$ and an azimuthal angle range of $\pi/2$ each. The

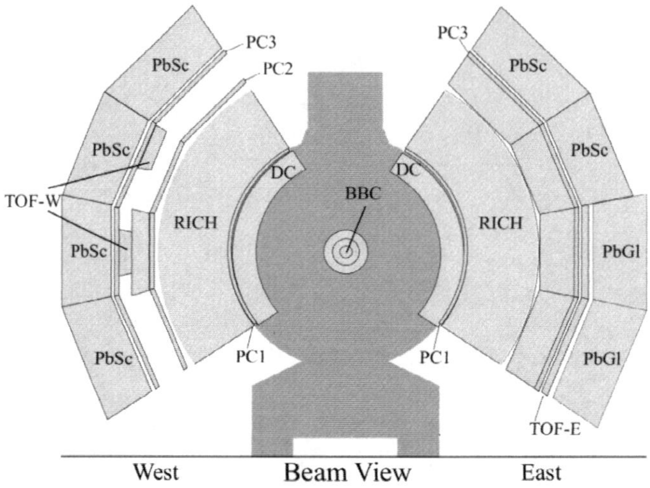

Figure 2. Layout of the PHENIX detector

electromagnetic calorimeter (EMCal) consists of eight sectors, two are composed of lead-glass (PbGl) towers in the bottom sectors of the east arm while the remaining six are composed of lead-scintillator (PbSc) modules. Requiring energy-momentum matching with an associated hit in the Ring Imaging Čerenkov Counter (RICH) provides a hadron rejection factor of better than 10^4 up to 4.9 GeV/c momentum, thus providing good electron identification. Each arm is instrumented with a drift chamber (DC) and pad chambers (PCs) that determine the trajectories, and together with a magnetic field, measure the momenta of charged particles. The Time-of-Flight (TOF) detector identifies charged hadrons, i.e. pions, kaons and protons. Neutral pions and η are measured via the $\pi^0(\eta) \rightarrow \gamma\gamma$ decay channel. The K^0_S meson is reconstructed via the $K^0_S \rightarrow \pi^0\pi^0 \rightarrow 4\gamma$ decay mode. The K^0 and $\overline{K^0}$ mesons are reconstructed via the $K^0 \rightarrow K^+\pi^-$ and $\overline{K^0} \rightarrow K^-\pi^+$ decay modes, respectively. Φ is reconstructed in the K^+K^- decay channel.

3. Results for large systems

In central Cu+Au collisions the Cu nucleus is fully submerged in the Au nucleus, which results in the reduction of nucleon-nucleon interactions in the corona region of the collision (see Fig. 1). In semi-central Cu+Au collisions an asymmetry of the nuclear overlap region is present along the axis connecting the centers of the interacting nuclei. These features make Cu+Au collision system an important part of the systematic study of the final-state effects in heavy-ion collisions.

Measurements of the production of different types of mesons (π^0, η, K, ϕ) allow a systematic study of jet quenching with respect to the fragmentation function and quantum numbers (mass, flavour, spin, etc.) of the final state hadrons. The nuclear modification factor R_{AB} of π^0, η, K* and Φ mesons as functions of p_{T} are shown in Fig. 3 for different Cu+Au centrality intervals. For both π^0 and η R_{AB} is consistent within uncertainties in the whole p_{T} range for every analyzed centrality interval of Cu+Au collisions. For open or pure strangeness mesons the R_{AB} is larger at low p_{T} but the difference disappears at higher p_{T}.

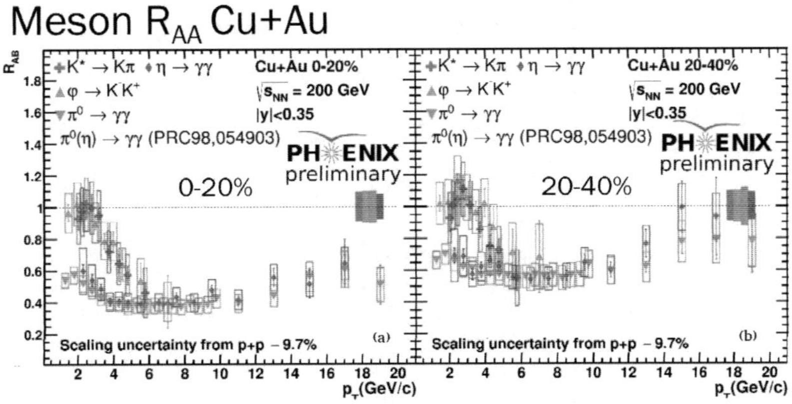

Figure 3. R_{AB} for K* and Φ meson > R_{AB} for π^0 and η at low p_T in central events, difference disappears at high p_T

The nuclear modification factor R_{AA} of π^0, η, K_S, K* and Φ mesons as functions of p_T is shown in Fig. 4 for different U+U centrality intervals. The absolute values are smaller than those in Cu+Au, but the pattern of separation of the π^0 and η from the open or pure strangeness mesons is similar, as is the p_T where the difference between them disappears.

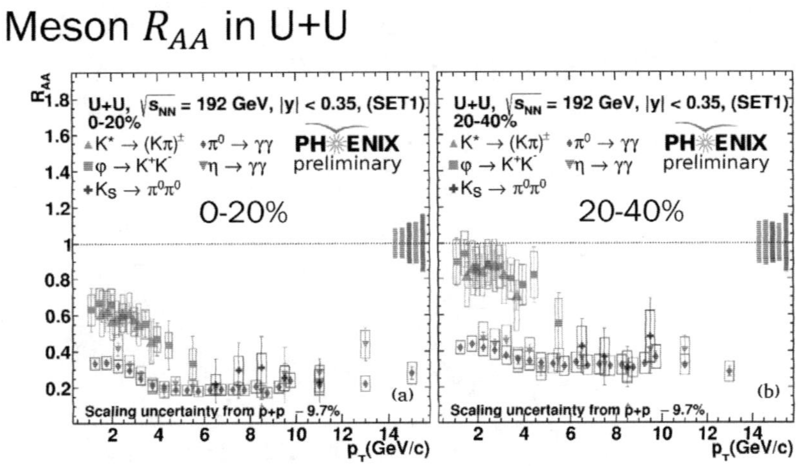

Figure 4. R_{AA} of π^0, η, K*, K_S and Φ mesons measured as a function of p_T in different centrality intervals of U+U collisions at $\sqrt{s_{NN}} = 192$ GeV.

Fig. 5 compares R_{AB} of π^0 mesons measured as a function of p_T in Cu+Au, Au+Au and Cu+Cu collisions at $\sqrt{s_{NN}} = 200$ GeV and at similar N_{part}. In central and semi-central Cu+Au collisions the π^0 R_{AB} is consistent with those measured in Au+Au and Cu+Cu, if applicable, which suggests that π^0 suppression mostly depends on the energy density and size of the produced medium. Since in the most central collisions the Cu ion is fully submerged in Au, without any "corona", but the suppression is the same as in Au+Au at comparable N_{part}, the corona-effect is either non-existent or very small compared to the uncertainties of the measurement.

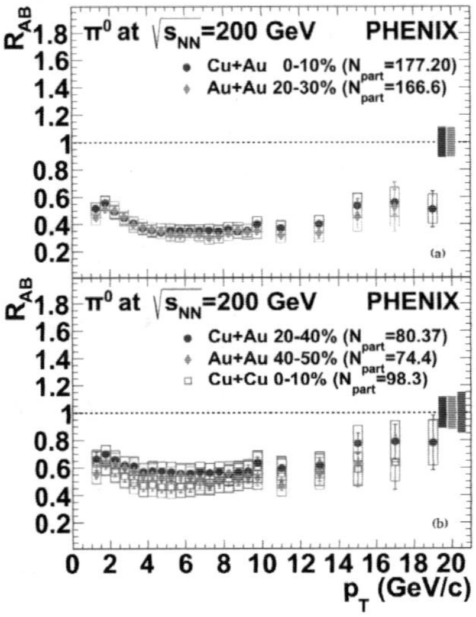

Figure 5. Comparison of π^0 R_{AB} measured in Cu+Au, Au+Au and Cu+Cu collisions.

As for baryons, Fig. 6 shows $R_{\rm AB}$ of $(p+\bar{p})/2$ as a function of $p_{\rm T}$ in different centrality classes in Cu+Au and Au+Au [18] at $\sqrt{s_{NN}}$ = 200 GeV and similar $N_{\rm part}$. Baryon enhancement is present in both systems, it reaches a maximum value above unity between 2 and 3 GeV, but above 3 GeV the $R_{\rm AB}$ values decrease and a suppression pattern emerges. The peripheral proton $R_{\rm AB}$– smaller to begin with – decreases more slowly than the central proton $R_{\rm AB}$ and they reach unity at about the same $p_{\rm T}$.

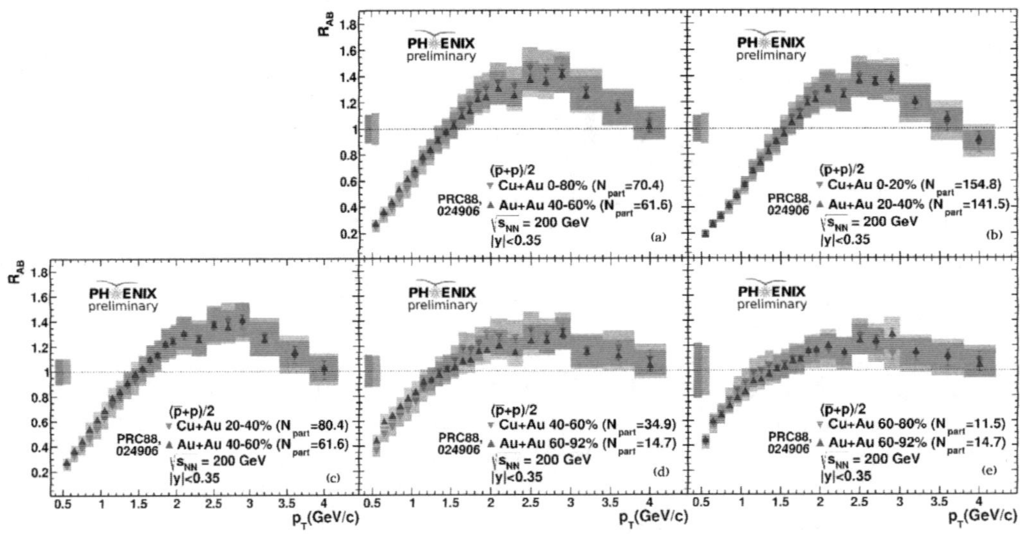

Figure 6. $R_{\rm AB}$ of $(p+\bar{p})/2$ consistent across Cu+Au and Au+Au [18] at similar N_{part}

Journal of Physics: Conference Series **1602** (2020) 012020 doi:10.1088/1742-6596/1602/1/012020

4. Results for small systems

Since π^0 R_{AB} has been widely investigated in all large collision systems and up to a high p_{T}, it is a natural first candidate to investigate the nuclear modification factor of leading hadrons of jets in small system collisions, too. PHENIX measured the R_{AB} of π^0 in in p+Au (preliminary), d+Au [19] and ^{3}He+Au (preliminary) collisions at $\sqrt{s_{NN}}$ = 200 GeV, shown in Fig. 7. In central collisions a clear enhancement is observed at p_{T} about 5 GeV/c in p+Au collisions, while it disappears in the d+Au and ^{3}He+Au collisions. Note that if the enhancement in p+Au is indeed a "Cronin-peak", it shows up at higher p_{T} than expected [10]. On the other hand at high p_{T} all three systems show a comparable suppression. The high-p_{T} nuclear modification is very comparable in the different collision systems: it appears to be independent on the projectile (p, d or ^{3}He).

Due to its composition $\Phi(s\bar{s})$ is a very sensitive probe of the formation and evolution of the QGP both when compared to non-strange mesons and double-strange baryons [20]. For central collisions the R_{AB} of Φ shows enhancement at p+Au and d+Au [17] at p_{T} about 5 GeV, in

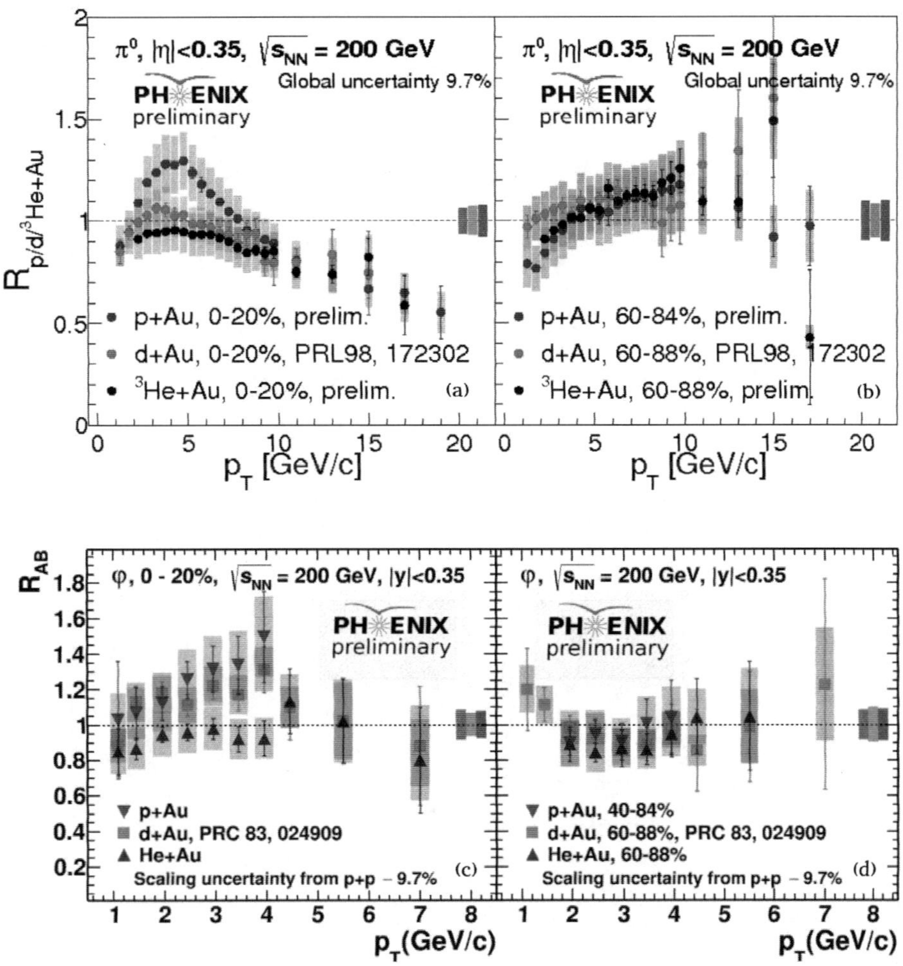

Figure 7. R_{AB} of π^0, Φ mesons measured as a function of p_{T} in small systems in central and peripheral collisions

Journal of Physics: Conference Series **1602** (2020) 012020 doi:10.1088/1742-6596/1602/1/012020

peripheral collisions the data are also consistent with unity within the experimental uncertainties.

5. Summary

In the Au+Au, Cu+Au, Cu+Cu and U+U collisions we observed an universal high p_T suppression for all measured mesons, which suggests that the QGP medium produced in these collisions either does not affect the jet fragmentation into light mesons or it affects them the same way. The nuclear modification factor of the mesons mostly depends on the energy density and size of the produced medium, characterized by N_{part}. At low p_T the strange mesons show an intermediate suppression between the more suppressed π^0 and the unsuppressed baryons. In the intermediate p_T region, between 2 and 5 GeV, a significant enhancement of baryon-to-meson ratios compared to those measured in *pp* collisions is observed. In small systems the data suggest an ordering ($R_{\text{pA}} > R_{\text{dA}} > R_{\text{HeA}}$) between the different collision systems for the mesons at mid-p_T. At high-p_T the nuclear modification factor is very similar between the different collision systems which suggest no dependency on the size of the projectile.

Acknowledgments

We thank the staff of the Collider-Accelerator and Physics Departments at Brookhaven National Laboratory and the staff of the other PHENIX participating institutions for their vital contributions. Thanks to the National Research, Development and Innovation Office (Hungary) OTKA-131991. The research was financed by the Thematic Excellence Programme of the Ministry for Innovation and Technology in Hungary (ED 18-1-2019-0028), within the framework of the Space Sciences thematic programme of the University of Debrecen.

References

[1] I. Arsene et al. "Quark gluon plasma and color glass condensate at RHIC? The Perspective from the BRAHMS experiment". In: *Nucl. Phys. A* 757 (2005), p. 1. DOI: 10.1016/j.nuclphysa.2005.02.130.

[2] B. B. Back et al. "The PHOBOS perspective on discoveries at RHIC". In: *Nucl. Phys. A* 757 (2005), p. 28. DOI: 10.1016/j.nuclphysa.2005.03.084.

[3] J. Adams et al. "Experimental and theoretical challenges in the search for the quark gluon plasma: The STAR Collaboration's critical assessment of the evidence from RHIC collisions". In: *Nucl. Phys. A* 757 (2005), p. 102. DOI: 10.1016/j.nuclphysa.2005.03.085.

[4] K. Adcox et al. "Formation of dense partonic matter in relativistic nucleus-nucleus collisions at RHIC: Experimental evaluation by the PHENIX collaboration". In: *Nucl. Phys. A* 757 (2005), p. 184. DOI: 10.1016/j.nuclphysa.2005.03.086.

[5] J. D. Bjorken. *Energy Loss of Energetic Partons in Quark - Gluon Plasma: Possible Extinction of High p(t) Jets in Hadron - Hadron Collisions.* FERMILAB-PUB-82-059-THY, FERMILAB-PUB-82-059-T. 1982.

[6] Guang-You Qin and Xin-Nian Wang. "Jet quenching in high-energy heavy-ion collisions". In: *Quark-Gluon Plasma 5.* Ed. by Xin-Nian Wang. Vol. 24. 11. Oct. 2015, p. 1530014. DOI: 10.1142/S0218301315300143. arXiv: 1511.00790 [hep-ph].

[7] R. J. Glauber. *Lectures in Theoretical Physics.* New York: Interscience, 1959, p. 315.

[8] Roy J. Glauber. "Quantum Optics and Heavy Ion Physics". In: *Nucl. Phys.* A774 (2006), pp. 3–13. DOI: 10.1016/j.nuclphysa.2006.06.009. arXiv: nucl-th/0604021 [nucl-th].

[9] Michael L. Miller et al. "Glauber modeling in high energy nuclear collisions". In: *Ann. Rev. Nucl. Part. Sci.* 57 (2007), pp. 205–243. DOI: 10.1146/annurev.nucl.57.090506.123020. arXiv: nucl-ex/0701025.

[10] J.W. Cronin et al. "Production of hadrons with large transverse momentum at 200, 300, and 400 GeV". In: *Phys. Rev. D* 11 (1975). Ed. by J.R. Smith, pp. 3105–3123. DOI: 10.1103/PhysRevD.11.3105.

[11] C. Aidala et al. "Production of π^0 and η mesons in Cu+Au collisions at $\sqrt{s_{NN}}$=200 GeV". In: *Phys. Rev. C* 98.5 (2018), p. 054903. DOI: 10.1103/PhysRevC.98.054903. arXiv: 1805.04389 [hep-ex].

[12] A. Adare et al. "Suppression pattern of neutral pions at high transverse momentum in Au+Au collisions at $\sqrt{s_{NN}}$ = 200 GeV and constraints on medium transport coefficients". In: *Phys. Rev. Lett.* 101 (2008), p. 232301. DOI: 10.1103/PhysRevLett.101.232301. arXiv: 0801.4020 [nucl-ex].

[13] A. Adare et al. "Neutral pion production with respect to centrality and reaction plane in Au+Au collisions at $\sqrt{s_{NN}}$=200 GeV". In: *Phys. Rev. C* 87.3 (2013), p. 034911. DOI: 10.1103/PhysRevC.87.034911. arXiv: 1208.2254 [nucl-ex].

[14] A. Adare et al. "Onset of π^0 Suppression Studied in Cu+Cu Collisions at $\sqrt{s_{NN}}$ =22.4, 62.4, and 200 GeV". In: *Phys. Rev. Lett.* 101 (2008), p. 162301. DOI: 10.1103/PhysRevLett.101.162301. arXiv: 0801.4555 [nucl-ex].

[15] U. Acharya et al. "Production of π^0, η, and K_S mesons in U+U collisions at $\sqrt{s_{NN}}$ = 192 GeV". In: (May 2020). arXiv: 2005.14686 [hep-ex].

[16] S.S. Adler et al. "Centrality dependence of pi0 and eta production at large transverse momentum in s(NN)**(1/2) = 200-GeV d+Au collisions". In: *Phys. Rev. Lett.* 98 (2007), p. 172302. DOI: 10.1103/PhysRevLett.98.172302. arXiv: nucl-ex/0610036.

Journal of Physics: Conference Series **1602** (2020) 012020 doi:10.1088/1742-6596/1602/1/012020

[17] Adare et al. "Nuclear modification factors of ϕ mesons in $d +$ Au, Cu + Cu, and Au + Au collisions at $\sqrt{s_{NN}} = 200$ GeV". In: *Phys. Rev. C* 83 (2 Feb. 2011), p. 024909. DOI: 10.1103/PhysRevC.83.024909. URL: https://link.aps.org/doi/10.1103/PhysRevC.83.024909.

[18] A. Adare et al. "Spectra and ratios of identified particles in Au+Au and d+Au collisions at $\sqrt{s_{NN}} = 200$ GeV". In: *Phys. Rev. C* 88 (2 Aug. 2013), p. 024906. DOI: 10.1103/PhysRevC.88.024906. URL: https://link.aps.org/doi/10.1103/PhysRevC.88.024906.

[19] S. Adler et al. "Centrality Dependence of π^0 and η Production at Large Transverse Momentum in $\sqrt{s_{NN}} = 200$GeV $d +$ Au Collisions". In: *Phys. Rev. Lett.* 98 (17 Apr. 2007), p. 172302. DOI: 10.1103/PhysRevLett.98.172302. URL: https://link.aps.org/doi/10.1103/PhysRevLett.98.172302.

[20] Peter Koch, Berndt Müller, and Johann Rafelski. "From strangeness enhancement to quark–gluon plasma discovery". In: *Int. J. Mod. Phys. A* 32.31 (2017), p. 1730024. DOI: 10.1142/S0217751X17300241. arXiv: 1708.08115 [nucl-th].

Journal of Physics: Conference Series **1602** (2020) 012021 doi:10.1088/1742-6596/1602/1/012021

Status of the MPD Experiment at JINR

Adam Kisiel for the MPD Collaboration

Joint Institute for Nuclear Research, Joliot-Curie 6, 141980 Dubna, Moscow Region, Russia
Warsaw University of Technology, Faculty of Physics, Koszykowa 75, 00-662 Warsaw, Poland

E-mail: Adam.Kisiel@jinr.ru

Abstract. The Mega-Science project NICA (Nuclotron-based Ion Collider fAcility) is under construction at the Joint Institute for Nuclear Research (JINR) in Dubna. The Multi-Purpose Detector (MPD) is part of this complex, as a main physics experiment operating there. It aims to study the phase diagram of QCD matter at maximum baryonic density, determine the nature of the phase transition between the deconfined and hadronic matter and perform a search for the conjectured critical point in the diagram. Status of the preparation of critical elements of the infrastructure will be discussed, including the construction of the MPD subsystems.

The designed physics performance of the detector components will be discussed. Spectra of identified hadrons, including hyperons and hypernuclei will be discussed, with emphasis on differential measurement and total yield extraction. The quality of directed and elliptic flow determination will be discussed, with comparison to model expectations. The sensitivity of event-by-event fluctuations and femtoscopic measurements to the nature of the phase transition and the presence of a critical point will be given. Performance of the electromagnetic calorimeter working in conjunction with the tracking system for the di-lepton measurements and the potential for identification of charmed mesons will be described. In summary, all the main components of the physics program of the MPD Collaboration will be presented.

1. Introduction

The investigation of the phase diagram of the strongly interacting matter has been the focus of many years of research at several experimental facilities, including the Super Proton Synchrotron (SPS) at CERN [1, 2, 3, 4], as well as the Beam Energy Scan program at the Relativistic Heavy-Ion Collider at Brookhaven National Laboratory (BNL) [5, 6, 7]. The existence of a deconfined phase, where quarks and gluons are no longer bound into hadrons, the so-called Quark Gluon Plasma has been confirmed in collisions at the energy frontier at RHIC and Large Hadron Collider (LHC) energies. In such collisions maximum energy densities are reached, while the net baryon density is close to zero. The transition to the ordinary hadronic matter is of the cross-over type there. The theoretical calculations on the lattice predict that at large net baryon densities the transition is of the first order. That implies the existence of the conjectured critical point in the phase diagram. Matter created in collisions at lower energies is reaching lower temperatures and baryon densities above 300 MeV, where the search for these features may be performed, and the determination of the Equation Of State (EOS) of nuclear matter may be achieved.

Recent discovery of a neutron star merger via the observation of the gravitational wave [8], as well as accompanying signal in the traditional electromagnetic observations [9] was the first example of the so-called multi-messenger astronomy. It is expected that such observations may become more frequent, giving unique experimental insight into the behavior of nuclear matter

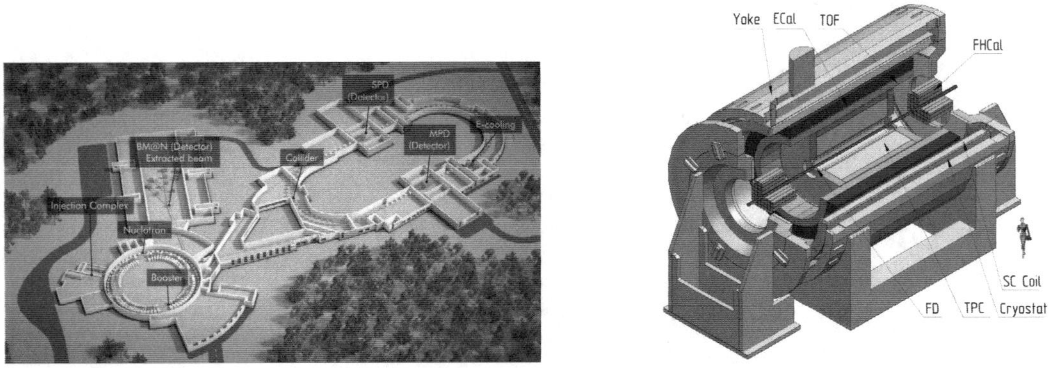

Figure 1. (left) The general overview of the NICA Accelerator Complex. (right) The schematic view of the MPD detector in Stage 1.

at extreme conditions. Simulations of neutron star collisions are driven by the EOS, which is investigated in heavy-ion collisions. In fact it is expected that in certain regions of the neutron star collision zone matter may reach temperatures and baryonic densities of the same order as in the collisions mentioned above. This presents an unique opportunity to study the same phenomena in fundamentally different experimental regimes, which provides crucial cross-checks for any theoretical understanding of the EOS.

The new experimental and accelerator complex: the Nuclotron-based Ion Collider fAcility (NICA) is being constructed at the Joint Institute for Nuclear Research (JINR) in Dubna. The design parameters of NICA in the collider mode are Au+Au collisions in the $\sqrt{s_{\rm NN}}$ range of 4-11 GeV per nucleon pair. The facility will also provide collisions of polarized protons and deuterons at the second stage of operation. The main detector dedicated to QCD physics at NICA is the Multi-Purpose Detector (MPD), which is being constructed at one of the two beam crossings at NICA. This work will shortly describe the status of MPD construction as well as main physics goals of the experiment.

2. NICA Accelerator Complex

The NICA Accelerator Complex, shown schematically in Fig. 1, is located in the Veksler-Baldin Laboratory for High Energy Physics (VBLHEP) of JINR. The Nuclotron accelerates (polarized) protons, (polarized) deuterons as well as heavy ions up to Bi, delivering maximum kinetic energy of 10.71 GeV/u for protons and 3.8 GeV/u for Au ions. It is in operation since 1993 and has recently undergone significant upgrade. The specific ion sources together with LU-20 Linac provide light ions (up to C) to Nuclotron. The beams can be then extracted to the Fixed Target Area, where the Baryonic Matter at Nuclotron (BM@N) experiment is currently operating. All systems mentioned above are fully commissioned and operational. The heavy ion beams originate in the KRION source and are first accelerated in the HILac linear accelerator. Both are fully commissioned and operational. Beams will then transfer to the Booster auxiliary accelerator. Its main purpose is the storage of ions with A/Z not larger than 3 and initial acceleration to 600 MeV/u, which allows full stripping of ions during the transfer to Nuclotron. Booster is currently in construction, commissioning of critical systems is expected in 2020 and 2021. Together, Booster and Nuclotron will provide heavy-ion beams. The final part of the complex is the NICA storage and acceleration with the racetrack shaped ring of 503,04 m circumference. It will provide collisions of Au (and other) ions in the range of 4 to 11 GeV center-of-mass energy with 1×10^{27} $cm^{-2}s^{-1}$ luminosity. Two collision points are located at two straight sections of NICA - the MPD experiment will be located around one of them. Civil construction of buildings

Journal of Physics: Conference Series **1602** (2020) 012021 doi:10.1088/1742-6596/1602/1/012021

for NICA and its transfer lines is significantly advanced, while installation and commissioning of the accelerator components is planned for 2021 and 2022. First beam in NICA is expected in 2022.

3. The MPD apparatus

The civil construction for MPD Hall, where MPD apparatus will be located is finished. The general schematic view of the apparatus is shown in the right panel of Fig. 1. The central-barrel detectors will be housed inside the large-aperture magnet. The outer iron yoke, serving also as structural support for the setup is already delivered to JINR and its assembly will take place in summer of 2020. The superconducting solenoid is manufactured and will be transported to JINR in late 2020 and assembled with the yoke. Together, they will provide a uniform magnetic field of up to 0.5 T along the beam axis. A dedicated carbon fiber support structure will be attached to the magnet structure, and will allow installation of the remaining MPD subsystems.

The main tracking detector of MPD will be the Time Projection Chamber, a gaseous detector, with cylindrical shape enclosing the beam pipe, 340 cm in length, with 140 cm and 25 cm outer and inner radii respectively. It will provide 3D tracking of charged particles, as well as measurement of specific ionization energy loss for particle identification for $|\eta| < 1.2$. Expected relative transverse momentum resolution of less than 2.5% below 1 GeV/c and primary vertex z-coordinate better than 200 μm for events with track multiplicities above 200 will be achieved. The main components of the TPC vessel have been manufactured. Mass production of the TPC read-out chambers is ongoing.

The TPC will be surrounded by cylindrical barrel of the Time-of-Flight (TOF) detector, based on the Multigap Resistive Plate Chambers (MRPC) detectors, with strip read-out. Total of 28 modules with 13440 channels are in production and commissioning. The TOF will provide the measurement of the particle's arrival time with the time resolution of the order of 50 ps, as well as the position of the track. By performing the TPC-TOF matching, particle identification will be possible allowing to distinguish pions from kaons up to 1.5 GeV/c and protons from pions and kaons up to 2.5 GeV/c.

The Electro-Magnetic Calorimeter (ECal) will consist of shashlyk-type towers of approximately 11 interaction lengths with lead absorber and plastic scintillators. It will be a cylinder surrounding the TOF, consisting of almost 40 000 towers arranged in projective geometry. Its main purpose it to measure the position and total energy of electrons and photons in heavy-ion collisions. The production of ECal modules has already started, with the installation of the limited number of finished ECal sectors expected for first data-taking.

In the forward direction two detectors will be installed. The forward Hadronic Calorimeter (FHCal) will provide centrality determination as well as reaction plane measurement. It is composed of 88 modules placed in two parts on the opposite side of the interaction point. Each module consist of 42 lead-scintillator sandwiches read out by WLS fibers. All FHCal modules are produced and ready for installation.

The Fast-Forward Detector (FFD) will provide fast triggering for nucleus-nucleus collisions, as well as a start time T0 for the TOF. The two Cherenkov modular arrays are placed at $2.7 < |\eta| < 4.1$, close to the beam pipe. All components of the FFD have been manufactured.

In summary, in the first stage MPD will be able to provide full tracking of charged particles in the central barrel, together with full particle identification, including the separation of electrons from hadrons as well as total energy measurement for photons. Centrality determination, as well as reaction plane measurement will be possible, as well as precise determination of primary and secondary vertices. With these capabilities the physics program of the MPD is expected to be rich and should enable direct comparison with results obtained at similar collision energies at RHIC and SPS.

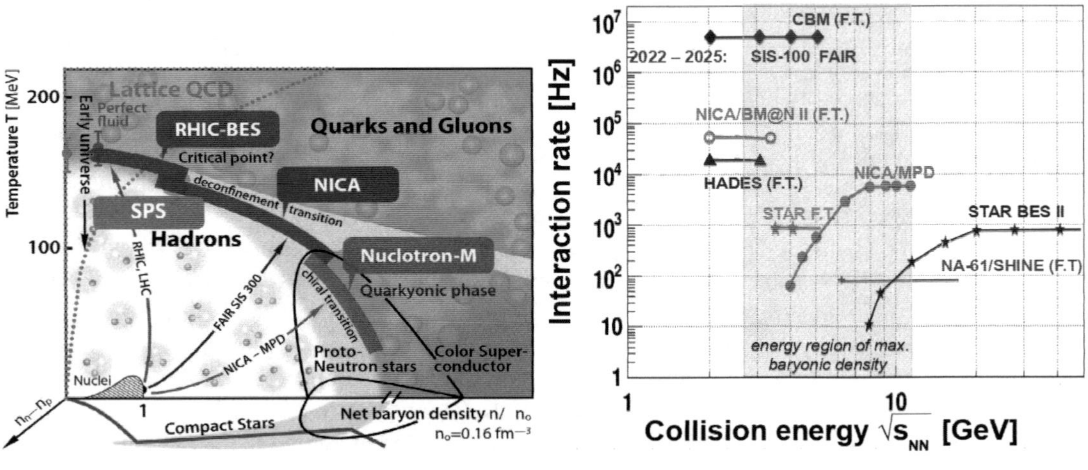

Figure 2. (left) The schematic view of the QCD Phase diagram. (right) The collision energies and expected interaction rates of currently ongoing and planned experimental programs.

4. The physics program of the MPD experiment

The main goal of the physics program of the MPD experiment is the characterization of the QCD phase diagram, shown schematically in Figure 2, in the less explored area of large temperatures and large baryochemical potential. The region of maximum temperatures and vanishing baryon density has been explored by the experiments at the LHC, as well as measurements at top RHIC energy. The production of QGP was confirmed at these conditions, and the transition back to hadronic matter has been discovered to be of the cross-over type. At larger baryon densities models consistently predict the existence of first-order phase transition between QGP and hadronic matter. That implies the existence of the critical end-point in the diagram. The search for this point is being carried out at RHIC Beam-Energy Scan (BES) program as well as in SPS. So far no conclusive evidence for the existence of the critical point has been found.

As can be seen from Fig. 2 NICA will provide competitive experimental conditions for the exploration of the phase diagram area of interest. The $\sqrt{s_{NN}}$ range of 4-11 GeV for heavy-ion collisions allows to explore the area of maximum baryon density. The expected luminosity is at least an order of magnitude larger than in other facilities at similar $\sqrt{s_{NN}}$. Moreover, the MPD has collider geometry, with large and uniform acceptance which changes little with collision energy, allowing for measurements of several key observables with reduced systematic uncertainty. In general the discovery potential at NICA is unique and complementary to similar measurements at lower energies at FAIR as well as at higher energies at SPS and RHIC BES.

Initial measurements at MPD will need precise determination of several key characteristics in data. The track multiplicity in the TPC as well as energy deposit in FHCal will enable determination of the event centrality. To associate the experimentally determined quantities to the geometrical parameters, such as impact parameter and number of participants/spectators, a Modified Wounded Nucleon model will be used, also known as MC-Glauber [10]. Multiplicity density per unit of pseudorapidity is related to the achieved energy density in the collision [11]. Its measurement will allow to test, whether densities needed for the creation of QGP are reached in the collisions at NICA. When combined with PID it will also determine the transverse energy density, which is related to the internal pressure in the extremely dense matter produced in collisions of heavy-ions [12].

The ratio of production of identified hadrons with respect to their anti-particles is directly related to the charge transport mechanism in the collision. The ratio will be measured separately for π^+/π^-, K^+/K^-, and $p/\bar{p}$, as a function of collision energy and centrality. When extrapolated to full 4π acceptance, the data will provide total yields of various particle types, including the ones reconstructed via their decay topology. Transverse momentum spectra for all those particle types will be measured in wide acceptance in η and p_T. A wealth of information can be deduced from such data, including the freeze-out temperature and baryochemical potential of the produced system. This will allow precise positioning of each collision category (e.g. vs. centrality and collision energy) on the phase diagram of QCD. The reconstruction of particles with non-zero strangeness, including multi-strange baryons such as Ξ and Ω, will allow to address critical questions, such as onset of deconfinement, the possible existence of the critical point as well as the degree of thermalization in the system.

The study will be extended to reconstruction of strongly decaying resonances, such as $\rho(770)$, $K^*(892)$, $\phi(1020)$, $\Sigma(1385)$, and $\Lambda(1520)$. They will be included in the fits with the statistical model mentioned above, but they can also provide unique information on the rescattering phase. Resonances' lifetime is comparable with the duration of the phase, during which they can undergo rescattering and regeneration. By comparing the observed yield of resonances with respect to the statistical model expectation, the duration of the phase can be inferred [13, 14].

QCD matter created in heavy-ion collisions is strongly interacting, and is subject to intense density gradients, leading to large pressure and development of collective behavior of matter. It manifests itself as various types of flow. In the transverse plane the matter generally flows outwards from the collision axis, resulting in radial flow. It modifies strongly the inclusive transverse momentum spectra of particles with different mass, from which its measurement can be inferred. Just after the collision the spectator nucleons still remain close to the created system, preventing the flow of matter in their direction. This results in the so-called directed flow v_1. Its measurement in RHIC BES energies [15] reveals intriguing non-monotonic behavior, which will be investigated in MPD thanks to its excellent acceptance and PID capabilities. In the transverse plane the cross-section of colliding nuclei has an almond shape, leading to the development of elliptic flow v_2. Its measurement reveals how efficient is the transfer of spatial anisotropy to the momentum one, allowing for the determination of crucial properties of the system, such as shear viscosity to entropy ratio η/s, as well as possible changes to the speed of sound c_s near the phase transition, resulting in the softening of the EOS.

The size and dynamic evolution of the system created in heavy-ion collisions can be studied via the femtoscopy technique, relying on the measurement of two-particle correlation as a function of pair relative momentum. Analysis for pairs of identical charged pions will be performed at MPD vs. centrality and pair transverse momentum. The size of the system will be measured. The dependence of this size on the momentum is interpreted as a consequence of dynamical radial expansion. It has also been argued [16], that a specific ratio of system sizes in the transverse plane is directly sensitive to the existence of the first order phase transition.

The event-by-event fluctuations of total event multiplicity, mean transverse momentum and conserved charges, as well as multipicities of identified hadrons should show a non-monotonic behavior in the vicinity of a critical point. MPD, as a collider experiment, offers large acceptance (at least $|\eta| < 1.2$ and $0.2 < p_T < 3.0$ for identified particles), approximately constant with respect to collision energy. Such conditions allow to significantly reduce systematic uncertainties in fluctuation measurements making MPD ideally suited to perform this search.

The EMCal will provide data on photon production at NICA. Of special interest is the measurement of low to intermediate p_T direct photons, which carry information about the initial temperature of the system [17, 18, 19], which can be compared to the critical temperature of the phase transition to the deconfined phase of approximately 160 MeV. Photon measurement will also allow for identification of π^0 and η mesons as a function of centrality and p_T, which will serve as a cross-check for charged meson data. EMCal, when combined with tracking and charged particle rejection from TPC and TOF will also provide a clean sample of electrons.

This will allow a study of light vector meson, as well as extensive investigation of the di-lepton production at a wide range of invariant mass, as well as the total transverse momentum of the pair. NICA offers a favorable conditions for such studies, as the expected background from charm production is expected to be significantly reduced, when compared to collisions at higher energies.

5. Summary

The status of the preparation of the NICA Accelerator Complex has been briefly presented. The Nuclotron, as well as two main injector chains are commissioned and operational. Booster auxiliary accelerator is in final stages of assembly. The civil construction for NICA is progressing according to schedule. First beams in NICA are expected in 2022. The civil construction for MPD Hall has been finished and the assembly of detector magnet has started. Production of subdetector components for TPC, TOF, ECal, FHCal and FFD are in progress with commissioning of the MPD setup expected to start in 2021.

The physics program of MPD has been briefly discussed with the emphasis on data from initial NICA operation, where collisions of heavy ions are expected. Measurements of collision centrality and event plane, together with full tracking are foreseen. Full characterization of the particle production in the soft regime is planned, with an important component of electromagnetic probes. Data will allow for the search for the signatures of the decnfined matter, the existence and nature of the phase transition to ordinary hadronic matter, as well as possible existence of a ctitical point in the phase diagram of the QCD matter. We are looking forward to first data-taking at NICA with MPD.

Acknowledgments

Author is supported by the Polish NCN grant 2017/27/B/ST2/01947 and the grants of the Polish Plenipotientiary for JINR.

References

[1] S. Afanasiev *et al.* [NA49], Nucl. Instrum. Meth. A **430** (1999), 210-244 doi:10.1016/S0168-9002(99)00239-9
[2] S. Afanasiev *et al.* [NA49], Phys. Rev. C **66** (2002), 054902 doi:10.1103/PhysRevC.66.054902 [arXiv:nucl-ex/0205002 [nucl-ex]].
[3] N. Abgrall *et al.* [NA61/SHINE], Phys. Rev. C **84** (2011), 034604 doi:10.1103/PhysRevC.84.034604 [arXiv:1102.0983 [hep-ex]].
[4] N. Abgrall *et al.* [NA61/SHINE], Eur. Phys. J. C **76** (2016) no.2, 84 doi:10.1140/epjc/s10052-016-3898-y [arXiv:1510.02703 [hep-ex]].
[5] J. Adams *et al.* [STAR], Nucl. Phys. A **757** (2005), 102-183 doi:10.1016/j.nuclphysa.2005.03.085 [arXiv:nucl-ex/0501009 [nucl-ex]].
[6] J. Adam *et al.* [STAR], [arXiv:1906.03732 [nucl-ex]].
[7] L. Adamczyk *et al.* [STAR], Phys. Rev. Lett. **121** (2018) no.3, 032301 doi:10.1103/PhysRevLett.121.032301 [arXiv:1707.01988 [nucl-ex]].
[8] B. P. Abbott *et al.*, Phys. Rev. Lett. **119**, no. 16, 161101 (2017) doi:10.1103/PhysRevLett.119.161101 [arXiv:1710.05832 [gr-qc]].
[9] B. P. Abbott *et al.*, Astrophys. J. **848**, no. 2, L12 (2017) doi:10.3847/2041-8213/aa91c9 [arXiv:1710.05833 [astro-ph.HE]].
[10] C. Loizides, J. Nagle and P. Steinberg, SoftwareX **1-2** (2015), 13-18 doi:10.1016/j.softx.2015.05.001 [arXiv:1408.2549 [nucl-ex]].
[11] K. Aamodt *et al.* [ALICE], Phys. Rev. Lett. **105** (2010), 252301 doi:10.1103/PhysRevLett.105.252301 [arXiv:1011.3916 [nucl-ex]].
[12] R. Sahoo, A. N. Mishra, N. K. Behera and B. K. Nandi, Adv. High Energy Phys. **2015** (2015), 612390 doi:10.1155/2015/612390 [arXiv:1408.5773 [nucl-ex]].
[13] S. Acharya *et al.* [ALICE], Phys. Rev. C **99** (2019) no.6, 064901 doi:10.1103/PhysRevC.99.064901 [arXiv:1805.04365 [nucl-ex]].
[14] B. Abelev *et al.* [STAR], Phys. Rev. Lett. **97** (2006), 132301 doi:10.1103/PhysRevLett.97.132301 [arXiv:nucl-ex/0604019 [nucl-ex]].

Journal of Physics: Conference Series **1602** (2020) 012021 doi:10.1088/1742-6596/1602/1/012021

[15] L. Adamczyk *et al.* [STAR], Phys. Rev. Lett. **120** (2018) no.6, 062301 doi:10.1103/PhysRevLett.120.062301 [arXiv:1708.07132 [hep-ex]].
[16] P. Batyuk, I. Karpenko, R. Lednicky, L. Malinina, K. Mikhaylov, O. Rogachevsky and D. Wielanek, Phys. Rev. C **96** (2017) no.2, 024911 doi:10.1103/PhysRevC.96.024911 [arXiv:1703.09628 [nucl-th]].
[17] A. Adare *et al.* [PHENIX], Phys. Rev. Lett. **104** (2010), 132301 doi:10.1103/PhysRevLett.104.132301 [arXiv:0804.4168 [nucl-ex]].
[18] A. Adare *et al.* [PHENIX], Phys. Rev. C **91** (2015) no.6, 064904 doi:10.1103/PhysRevC.91.064904 [arXiv:1405.3940 [nucl-ex]].
[19] J. Adam *et al.* [ALICE], Phys. Lett. B **754** (2016), 235-248 doi:10.1016/j.physletb.2016.01.020 [arXiv:1509.07324 [nucl-ex]].

Recent results on light (anti-)nuclei production with ALICE at the LHC

Esther Bartsch for the ALICE Collaboration
Goethe University Frankfurt, Max-von-Laue-Straße 1, 60438 Frankfurt, Germany
E-mail: esther.bartsch@cern.ch

Abstract. The high-statistics and high-quality data samples in pp, p–Pb and Pb–Pb collisions at various collision energies collected by the ALICE Collaboration, and the excellent tracking and particle identification capabilities of the Inner Tracking System, the Time-Projection Chamber and the Time-of-Flight detector allow for the detection of rarely produced light (anti-)(hyper-)nuclei. The first (anti-)triton p_{T} spectra in Pb–Pb at the LHC as well as the first (anti-)^{4}He p_{T} spectra are presented. Moreover, the p_{T}-integrated production yield ratios of deuteron over proton as well as ^{3}He and triton over proton as a function of charged-particle multiplicity are shown. In addition coalescence parameters B_2 and B_3 as a function of p_{T}/A and charge particle multiplicity for several collision systems and energies are discussed. Furthermore, light-nuclei radial and elliptic flow are presented and compared to the flow of lighter particles. These various results are compared with statistical hadronization and coalescence model predictions to provide insight in the production mechanism of light (anti-)nuclei in heavy-ion collisions.

1. Introduction

In ultrarelativistic heavy-ion collisions at the Large Hadron Collider (LHC) a state of deconfined strongly interacting matter consisting of quarks and gluons, called the quark-gluon plasma (QGP), is created. Afterwards, the QGP expands and cools down. When the chemical freeze-out temperature is reached, hadronization takes place and from then on the hadron yields are fixed, but there can still be elastic interactions between the particles. After the kinetic freeze-out the momentum spectra of the particles also do not change anymore. The abundances of different particle species after hadronization provide information about their production mechanism. Among these particles, light (anti-)(hyper-)nuclei are of special interest since they are loosely bound objects. Their binding energies are very small compared to the chemical and kinetic freeze-out temperatures and their production mechanism is still not completely understood. There are two classes of models available to describe nuclei production: the statistical hadronization model and the coalescence model (see for instance [1] and [2], respectively).

In the statistical hadronization or thermal model, the production of nuclei happens at the chemical freeze-out in statistical equilibrium with all other hadrons and scales with the particle mass. In heavy-ion collisions, the system can be described by a grand-canonical ensemble where the free parameters are the average particle number $\langle N \rangle$, the volume V and the temperature T at chemical freeze-out. As the system exchanges particles, the baryochemical potential μ_{B} has to be introduced to ensure the average conservation of particle numbers. For a certain collision energy, a fit to the measured particle yields can be performed to determine V, T and μ_B. However at LHC energies μ_B is zero. For small systems, i.e. pp and p-Pb, a canonical approach is used, where the quantum numbers are locally conserved rather than on average.

In the coalescence model nuclei are produced at the kinetic freeze-out and the production depends on the wave functions of the nucleus' species. A phenomenological parameter of the coalescence model is the coalescence parameter B_A, which is related to the probability to form a nucleus with mass number A via coalescence. It is calculated by the ratio of the invariant yield of a given nucleus to the nucleon invariant yield to the power of A, where the nucleon yield is measured at the corresponding fraction of the nucleus momentum.

Light (anti-)(hyper-)nuclei are produced at the LHC in pp, p–Pb and, in particular, Pb–Pb collisions. The large high-quality data samples in Pb–Pb collisions at $\sqrt{s_{NN}} = 5.02$ TeV as well as in pp and p–Pb collisions at several collision energies collected by the ALICE Collaboration provide a unique opportunity to study the production mechanism of these loosely bound objects. This is shown in these proceedings exemplarily for the transverse momentum spectra of (anti-)triton (t,$\bar{t}$) and (anti-)^{4}He, and the elliptic-flow coefficient of (anti-)^{3}He.

Furthermore, charged-particle multiplicity dependences of p_T-integrated production yields over the proton yield and coalescence parameters B_A for A=2 and A=3 nuclei for different collision systems and energies are presented.

2. Light nuclei spectra and production yields

The first (anti-)triton transverse-momentum (p_T) spectra in Pb–Pb collisions at the LHC were obtained from the high-statistics data set from 2018 at $\sqrt{s_{NN}} = 5.02$ TeV. The nuclei were identified using the energy-loss measurement in the ALICE TPC, combined with the time-of-flight information provided by the TOF detector. The (anti-)triton spectra were measured in four centrality intervals (see Fig. 1), where an ordering of the yields can be observed. As there are many knocked-out tritons at low p_T, emerging from spallation processes in the beam pipe or the detector material, the triton spectra were only extracted above 2.4 GeV/c (or 2 GeV/c in the most peripheral centrality interval). In the overlap region, the measured t and $\bar{t}$ yields are compatible. The difference in the last two p_T bins of the 0-10% centrality interval is less than 2σ. The spectra exhibit an increase of the average p_T with increasing centrality.

In the same data set, the first (anti-)^{4}He p_T spectra were measured (see Fig. 2). These were obtained in the 0-10% most central collisions in four p_T bins from 2 to 6 GeV/c for $^4\overline{\text{He}}$ and in three p_T bins from 3 to 6 GeV/c for ^{4}He. The ^{4}He spectrum is starting at higher p_T for the same reason as for the tritons. The difference between $^4\overline{\text{He}}$ and ^{4}He for $4 < p_T < 5$ GeV/c is about 2σ.

In Fig. 3 the p_T spectra of π [4], K [4], p [4], d, ^{3}He, $\bar{t}$ and $^4\overline{\text{He}}$ in the 0-10% most central Pb–Pb collisions at $\sqrt{s_{NN}} = 5.02$ TeV are shown and simultaneously fitted with a Blast-Wave function [3]. The fit describes all particles quite well, the spectra show clear radial flow with a common flow velocity $\langle\beta\rangle$ and kinetic freeze-out temperature T_{kin}. One can also see that the ^{3}He and $\bar{t}$ spectra are compatible. Although these two different particle species are treated separately, the fitted curves lie on top of each other. At higher p_T, the spectra are not expected to be described by the Blast-Wave fit as the shape is dominated by high-p_T processes, like jets.

From individual Blast-Wave fits the p_T-integrated production yield dN/dy was extracted for several particles [4] including light (anti-)(hyper-)nuclei and compared to different variants of the thermal model (see Fig. 4) [5, 6, 7, 8, 9, 10]. Although the newly measured $\bar{t}$ and $^4\overline{\text{He}}$ are not yet included in the fit, the yield is in good agreement with the model predictions from the fit to the other light-flavoured particles.

In addition, the ratio of the p_T-integrated yield dN/dy for deuterons (see Fig. 5) as well as ^{3}He and tritons (see Fig. 6) relative to the proton yield has been studied as a function of charged particle multiplicity (dN_{ch}/dη) for different collision systems and center of mass energies. A

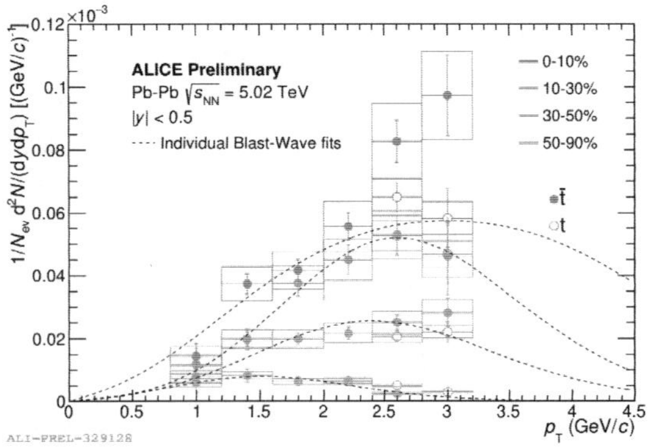

Figure 1. p_T spectra of (anti-)triton in four centrality intervals together with individual Blast-Wave fits. The $\bar{t}$ data points are presented by the full circles, whereas the open circles represent the t data points.

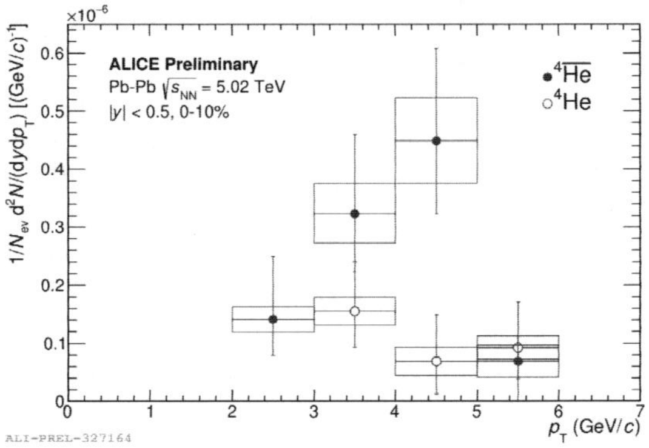

Figure 2. p_T spectra of (anti-)^{4}He in the 0-10% centrality interval. The $^4\overline{He}$ data points are presented by the full circles, whereas the open circles represent the ^{4}He data points.

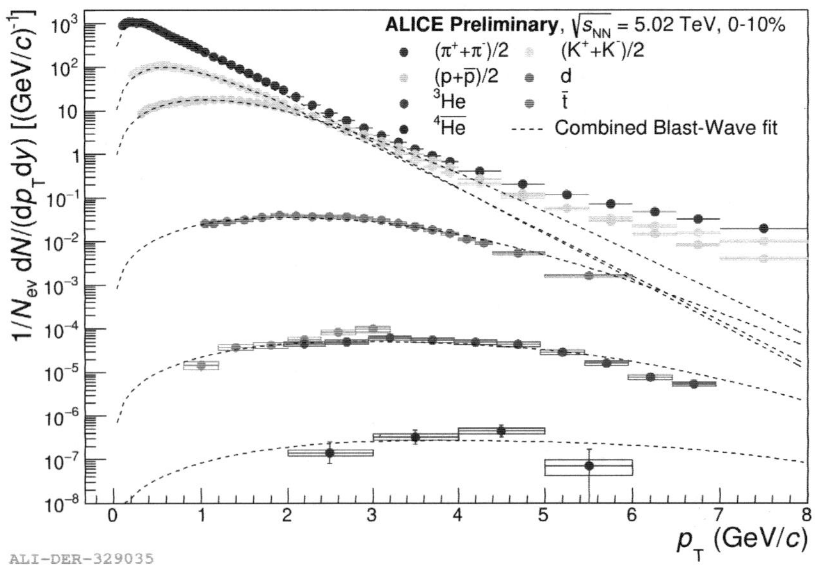

Figure 3. Combined Blast-Wave fit of several particle species, including light nuclei, in the 0-10% centrality interval.

universal trend can be observed, showing an increase of the ratio with increasing multiplicity from pp to p–Pb and a saturation in Pb–Pb collisions. This trend is rather well described by the canonical statistical [11] as well as the coalescence [12] models. The canonical statistical model is shown for two different volumes V_c (dotted and solid black lines). In Fig. 6, two-body and three-body coalescence is shown (magenta and orange lines, respectively). In two-body coalescence, the coalescence happens between a deuteron and a proton or neutron forming a ^{3}He or t, respectively. In the case of three-body coalescence, the coalescence takes place between two protons and one neutron or two neutrons and one proton.

3. Coalescence parameters

As already mentioned, the probability to form a nucleus via coalescence can be quantified by the coalescence parameter B_A, where A is the mass number of the respecting nucleus. It can be

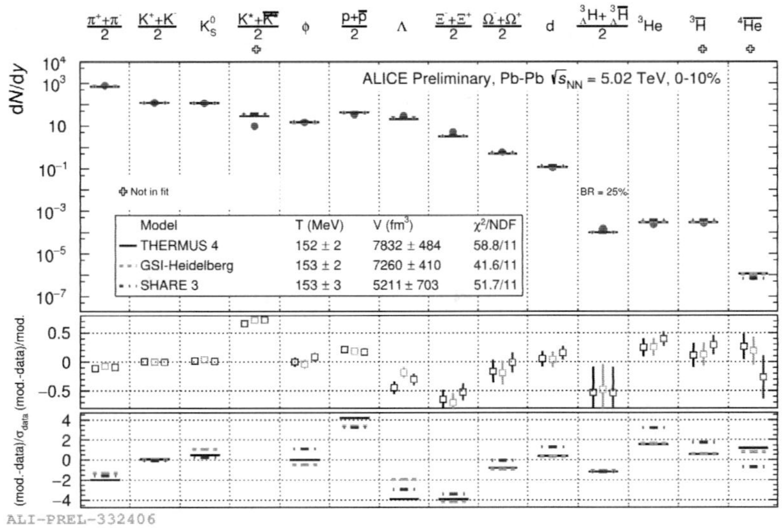

Figure 4. p_{T}-integrated production yield $\mathrm{d}N/\mathrm{d}y$ fitted with three thermal model implementations in the 0-10% centrality interval.

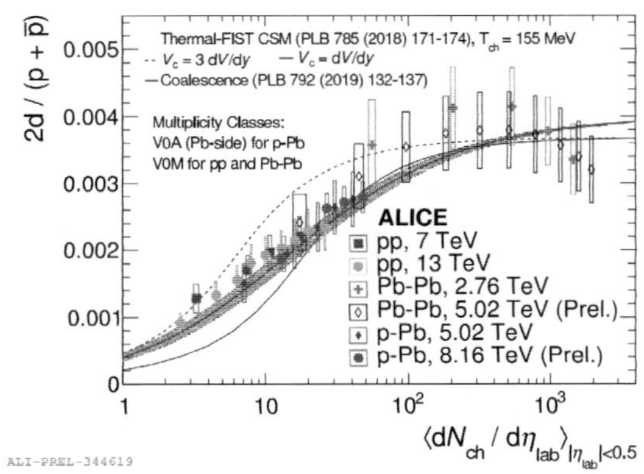

Figure 5. Ratio of the p_{T}-integrated production yield of deuterons over the proton production yield versus multiplicity compared to theoretical model predictions.

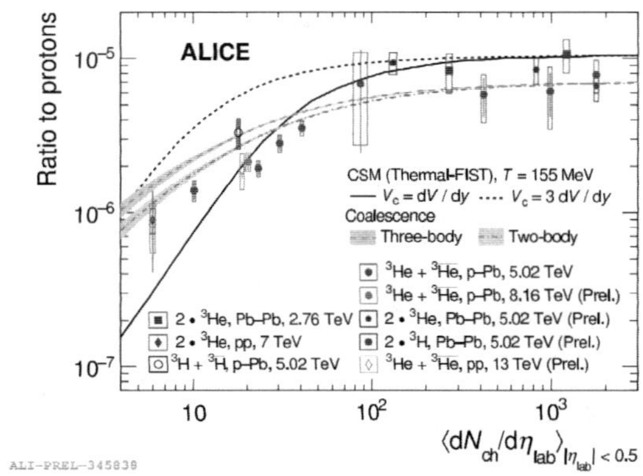

Figure 6. Ratio of the p_{T}-integrated production yield of A = 3 nuclei over the proton production yield versus multiplicity compared to theoretical model predictions.

calculated according to the following expression:

$$B_A = \frac{\frac{1}{2\pi p_{\mathrm{T}}^{A}} \frac{\mathrm{d}^2 N_A}{\mathrm{d}y \mathrm{d}p_{\mathrm{T}}^{A}}}{\left(\frac{1}{2\pi p_{\mathrm{T}}^{\mathrm{p}}} \frac{\mathrm{d}^2 N_{\mathrm{p}}}{\mathrm{d}y \mathrm{d}p_{\mathrm{T}}^{\mathrm{p}}}\right)^A} \qquad \text{with } p_{\mathrm{p}} = \frac{p_A}{A} \tag{1}$$

The invariant yield of a given nucleus (in the following d and ^{3}He) is divided by the invariant proton yield to the power of A. It is assumed that neutron and proton yields are the same. The proton yield is measured at the corresponding fraction ($\frac{1}{2}$ for d, $\frac{1}{3}$ for ^{3}He) of the nucleus momentum.

B_2 and B_3 is measured versus p_{T}/A for various collision systems and energies. Exemplarily the B_2 of deuterons in Pb–Pb collisions and the B_3 of (anti-)^{3}He in p–Pb collisions, both at $\sqrt{s_{\mathrm{NN}}}$ = 5.02 TeV, are shown in Fig. 7 and 8, respectively. The B_2 is measured in several centrality intervals. It shows a rise with p_T that becomes milder going from central to more peripheral collisions. The B_2 in inelastic pp collisions at $\sqrt{s} = 13$ TeV is also shown in Fig. 7 (black data points). Like for peripheral Pb–Pb collisions, a mild rise is visible with p_T. The B_3 (see Fig. 8) was measured in different possible multiplicity intervals and also in p–Pb collisions a rise with p_T can be observed.

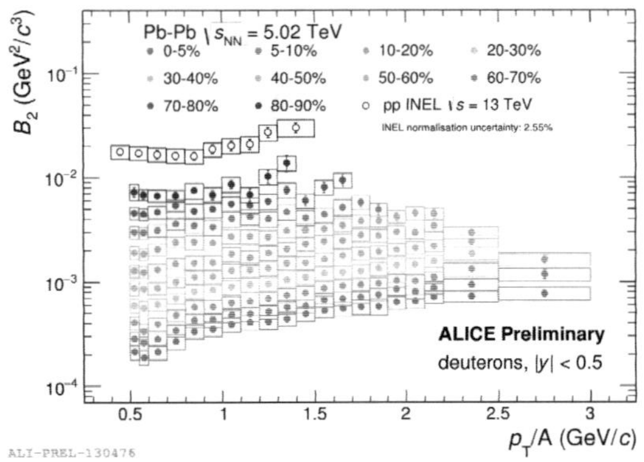

Figure 7. B_2 of deuterons in Pb–Pb collisions at 5.02 TeV versus p_T/A for several centrality intervals compared to pp collisions at 13 TeV.

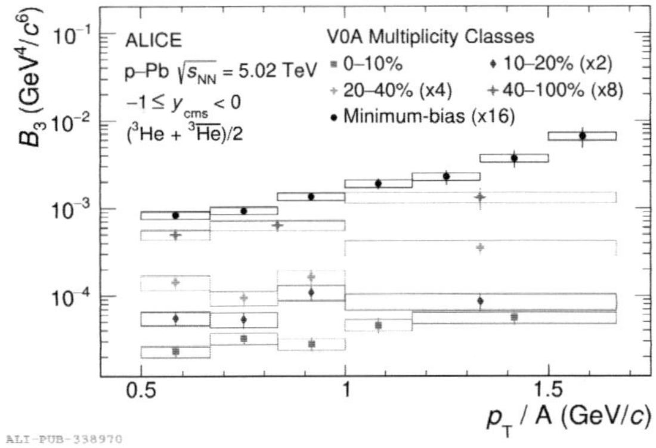

Figure 8. B_3 of (anti-)^{3}He in p–Pb collisions at 5.02 TeV versus p_T/A for several centrality intervals.

To further investigate the production mechanism, it is for instance possible to study, for a certain value of p_T/A, the evolution of B_2 and B_3 versus charged-particle multiplicity $\langle dN_{ch}/d\eta \rangle$ as various collision systems and centre-of-mass energies have been studied. This is shown for B_2 at p_T/A=0.75 GeV/c and for B_3 at p_T/A=0.9 GeV/c in Fig. 9 and 10, respectively. Both plots show a smooth evolution with multiplicity across different collision systems. Therefore the production mechanism seems to depend only on the system size, which can be expressed in terms of the charged-particle multiplicity. At lower multiplicities, where the system size is smaller than the nucleus, a flat behaviour is observed, slightly decreasing going from pp to p–Pb collisions, when the system size becomes larger. At higher multiplicities, where the system size is larger than the nucleus, a decreasing trend is observed. The overall trend is described by the coalescence model [13], where two different parametrizations of HBT (Hanbury-Brown, Twiss) radii [14] have been used to define the source dimension (dotted and solid lines). The B_3 (Fig. 10) is also compared to predictions of the canonical and grand-canonical statistical hadronization model (blue and red dashed lines, respectively). It appears that no model can describe the values of B_2 and B_3 in the full range of multiplicities studied.

4. Light-nuclei elliptic flow

The observation of radial flow for light nuclei motivates the study of elliptic flow (v_2). Therefore, the (anti-)^{3}He elliptic flow has been measured in Pb–Pb collisions at $\sqrt{s_{NN}} = 5.02$ TeV [15]. It was extracted in three centrality intervals (0-20%, 20-40% and 40-60%). The v_2 shows to be larger in semi-peripheral collisions, where the eccentricity of the system is higher, and features an increase with pT.

The v_2 of (anti-)^{3}He was compared in all three centrality intervals to the v_2 of π, K and p at the same center-of-mass energy (see Fig. 11). A clear mass ordering is observed, i.e. the maximum of the distribution is shifted to higher p_T with increasing mass, as expected from hydrodynamics.

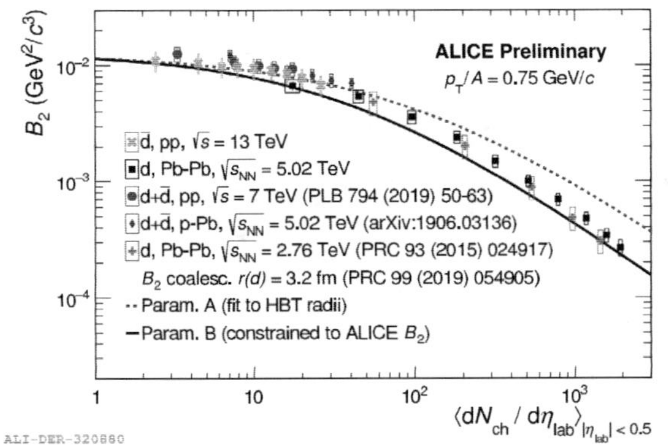

Figure 9. Multiplicity dependence of B_2 for different collision systems and energies, compared to two coalescence model variants.

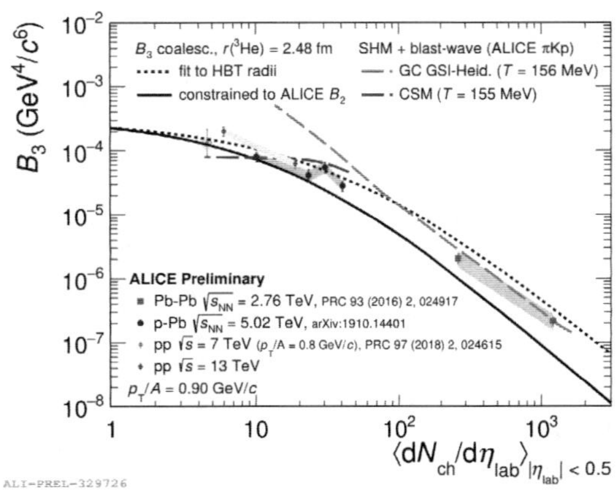

Figure 10. Multiplicity dependence of B_3 for different collision systems and energies, compared to two coalescence model variants and canonical and grand-canonical thermal model curves.

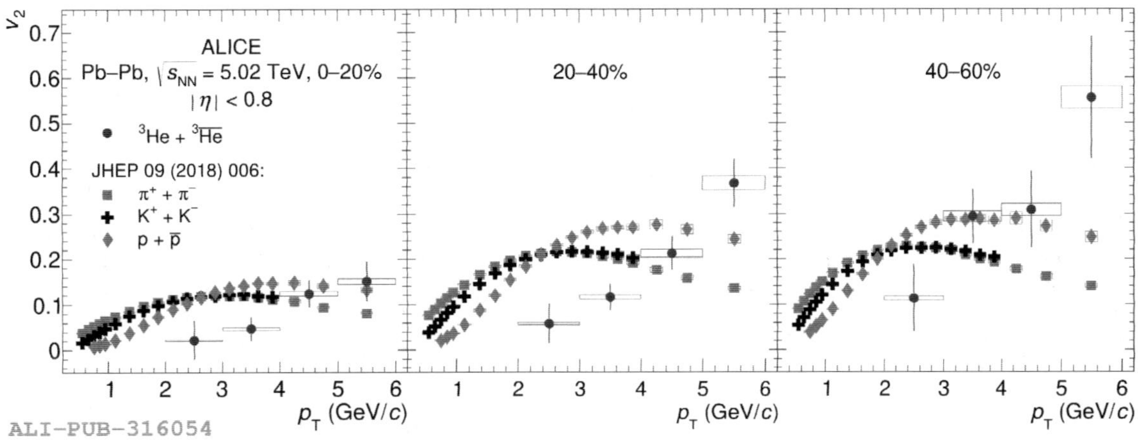

Figure 11. v_2 of (anti-)^{3}He compared to the v_2 of π, K, and p at the same center-of-mass energy for three centrality intervals.

Furthermore, the data in the 0-20% and 20-40% centrality intervals were compared to a more sophisticated coalescence model (iEBE-VISHNU hybrid model with AMPT initial conditions [16]), where the phase-space distribution of protons and neutrons is taken into account. The hybrid model includes viscous hydrodynamics and the nuclei are formed by final-state coalescence. As seen in Fig. 12 the data and the model are in good agreement in both centrality intervals.

5. Conclusion

Among other collision systems and energies, new results from the high-quality Pb–Pb data sets at a center-of-mass energy of $\sqrt{s_{\rm NN}} = 5.02$ TeV, collected in 2015 and 2018, have been presented. The first $p_{\rm T}$ spectra of $\bar{\rm t}$ and t as well as of $^4\overline{\rm He}$ and ^{4}He in Pb–Pb at the LHC were shown. The light nuclei exhibit a significant radial and elliptic flow, consistent with the flow of the

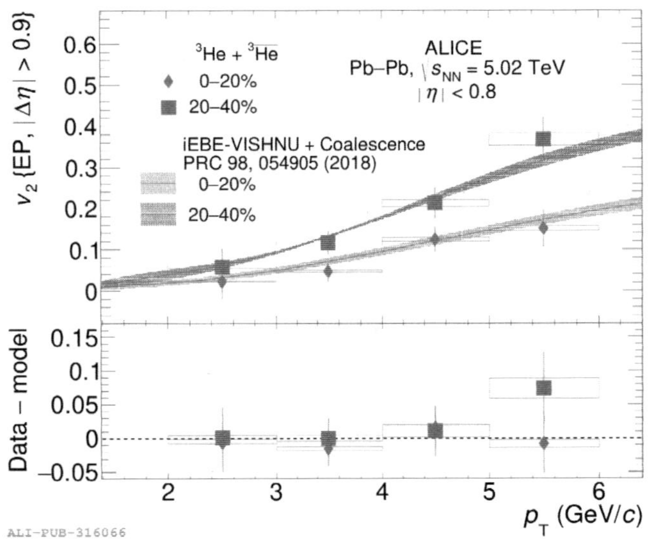

Figure 12. v_2 of (anti-)^{3}He in the 0-20% and 20-40% centrality intervals compared to iEBE-VISHNU hybrid model with AMPT initial conditions [16].

lighter particles. The production yields of light nuclei as well as the lighter particles are well described by the thermal model. The production yield ratios of d as well as ^{3}He and t over proton versus multiplicity show an increasing trend going from pp to p–Pb collisions with a saturation in Pb–Pb collisions. This trend is rather well described by the canonical statistical as well as the coalescence models. The coalescence parameters B_2 and B_3 in Pb–Pb and p–Pb collisions exhibit a rise versus p_T/A. In Pb–Pb collisions, this rise becomes milder going from central to more peripheral collisions. Looking at B_2 and B_3 versus multiplicity, which is related to the system size, a clear trend can be observed. At low multiplicities a more flat and slightly decreasing trend going from pp to p–Pb collisions, and at higher multiplicities a more pronounced decreasing trend is seen. The overall behaviour is described by the coalescence model. Future data taking periods in 2021-2024 and 2027-2030 will increase the statistics significantly and will hopefully help to solve the current ambiguity between the discussed production models.

References

[1] A. Andronic, P. Braun-Munzinger, J. Stachel, H. Stöcker, Phys. Lett. B697 (2011) 203–207. arXiv:1010.2995, doi:10.1016/j.physletb.2011.01.053.
[2] J. I. Kapusta, Phys. Rev. C21 (1980) 1301–1310. doi:10.1103/PhysRevC.21.1301.
[3] E. Schnedermann, J. Sollfrank, U. W. Heinz, Phys. Rev. C48 (1993) 2462–2475. arXiv:nucl-th/9307020, doi:10.1103/PhysRevC.48.2462.
[4] S. Acharya, et al. (ALICE Collaboration), Physical Review C 101 (4). arXiv:1910.07678, doi:10.1103/physrevc.101.044907.
[5] S. Wheaton, J. Cleymans, Comput. Phys. Commun. 180 (2009) 84–106. arXiv:hep-ph/0407174, doi:10.1016/j.cpc.2008.08.001.
[6] A. Andronic, P. Braun-Munzinger, J. Stachel, Phys. Lett. B673 (2009) 142–145, [Erratum: Phys. Lett.B678,516(2009)]. arXiv:0812.1186, doi:10.1016/j.physletb.2009.02.014,10.1016/j.physletb.2009.06.021.
[7] A. Andronic, P. Braun-Munzinger, K. Redlich, J. Stachel, Nature 561 (7723) (2018) 321–330. arXiv:1710.09425, doi:10.1038/s41586-018-0491-6.
[8] G. Torrieri, S. Steinke, W. Broniowski, W. Florkowski, J. Letessier, J. Rafelski, Comput. Phys. Commun. 167 (2005) 229–251. arXiv:nucl-th/0404083, doi:10.1016/j.cpc.2005.01.004.
[9] G. Torrieri, S. Jeon, J. Letessier, J. Rafelski, Comput. Phys. Commun. 175 (2006) 635–649. arXiv:nucl-th/0603026, doi:10.1016/j.cpc.2006.07.010.
[10] M. Petran, J. Letessier, J. Rafelski, G. Torrieri, Comput. Phys. Commun. 185 (2014) 2056–2079. arXiv:1310.5108, doi:10.1016/j.cpc.2014.02.026.

[11] V. Vovchenko, B. Dönigus, H. Stöcker, Phys. Lett. B785 (2018) 171–174. arXiv:1808.05245, doi:10.1016/j.physletb.2018.08.041.
[12] K.-J. Sun, C. M. Ko, B. Dönigus, Phys. Lett. B792 (2019) 132–137. arXiv:1812.05175, doi:10.1016/j.physletb.2019.03.033.
[13] F. Bellini, A. P. Kalweit, Phys. Rev. C 99 (2019) 054905. doi:10.1103/PhysRevC.99.054905.
[14] R. Scheibl, U. Heinz, Physical Review C 59 (3) (1999) 15851602. doi:10.1103/physrevc.59.1585.
[15] S. Acharya, et al. (ALICE Collaboration), Phys. Lett. B805 (2020) 135414. arXiv:1910.09718.
[16] W. Zhao, L. Zhu, H. Zheng, C. M. Ko, H. Song, Phys. Rev. C 98 (2018) 054905. doi:10.1103/PhysRevC.98.054905.

Journal of Physics: Conference Series **1602** (2020) 012023 doi:10.1088/1742-6596/1602/1/012023

Sea Quark Chemistry and Correlations in the Initial Conditions of Heavy Ion Collisions

Matthew D. Sievert
University of Illinois at Urbana-Champaign, Urbana, IL 61801, USA
E-mail: msievert@illinois.edu

Abstract. QCD dynamics are directly studied in two different but intersecting contexts: high-temperature QCD in heavy-ion collisions and zero-temperature hadronic structure in electron-hadron collisions. The approaches to high-energy nuclear physics in these two contexts are very different, however, and bringing the two fields into dialogue can reveal new opportunities. In this paper I highlight one interesting intersection between the two: leveraging the sea quark chemistry of the proton to study charge transport in top-energy heavy-ion collisions.

1. ICCING: Initial Conserved Charges in Nuclear Geometry

The equations of motion for hydrodynamics solve the gross conservation laws for the energy-momentum tensor $T^{\mu\nu}$ and the conserved currents N_i^μ where $i = B, S, Q$ denotes the conserved charges baryon number, strangeness, and electric charge, respectively:

$$\partial_\mu T^{\mu\nu} = 0 \tag{1}$$

$$\partial_\mu N_i^\mu = 0 \tag{2}$$

with initial conditions $T_0^{\mu\nu}$, $N_{i\,0}^\mu$ specified at at a fixed-proper-time hypersurface τ_0. A range of initial condition models exist for $T_0^{\mu\nu}$ with varying levels of theoretical input, from models with only nucleonic degrees of freedom [1] to those which incorporate calculations or models of nucleonic substructure [2, 3]. All models must specify the energy density T^{00}, while some models also incorporate the initial momentum flux T^{0i} or the initial shear tensor T^{ij}. At top collider energies, baryon stopping is suppressed and the QGP carries no net charge: $\mu_B = \mu_S = \mu_Q = 0$; for this reason, the initial charge currents $N_{i\,0}^\mu$ are usually set to zero.

However, while the *total* charge of the QGP at top collider energies may be zero, that does not mean that the charge is homogeneously zero, *locally*. Consider the initial state from the point of view of perturbative QCD. At small x (corresponding to high energy), the dominant degrees of freedom are gluons, as seen in the parton distribution functions extracted from HERA (Fig. 1). The steep decrease in the valence quark contributions xu_v, xd_v at small x is equivalent to the suppression of baryon stopping in this regime. Since gluons carry zero baryon number, strangeness, and electric charge, an initial state energy profile $T_0^{\mu\nu}$ with no charge currents $N_i^\mu = 0$ can be naturally interpreted as composed of gluons. This is indeed the correct leading-order picture of the initial state.

But as can already be seen at the level of the PDFs, there are substantial corrections to the initial state from sea quark fluctuations, which *do* carry the conserved charges B, S, Q. These $q\bar{q}$

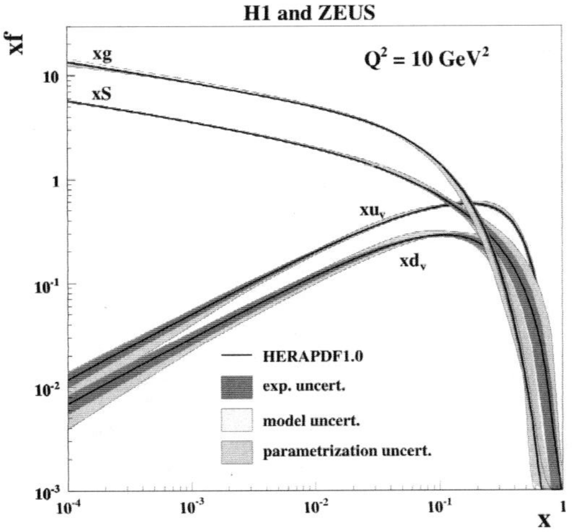

Figure 1. Parton distribution functions (PDFs) extracted from HERA showing the comparable growth rates of gluons and sea quarks at small x. This figure reproduced following Creative Commons Attribution License guidelines from Fig. 19(b) of Ref. [4] .

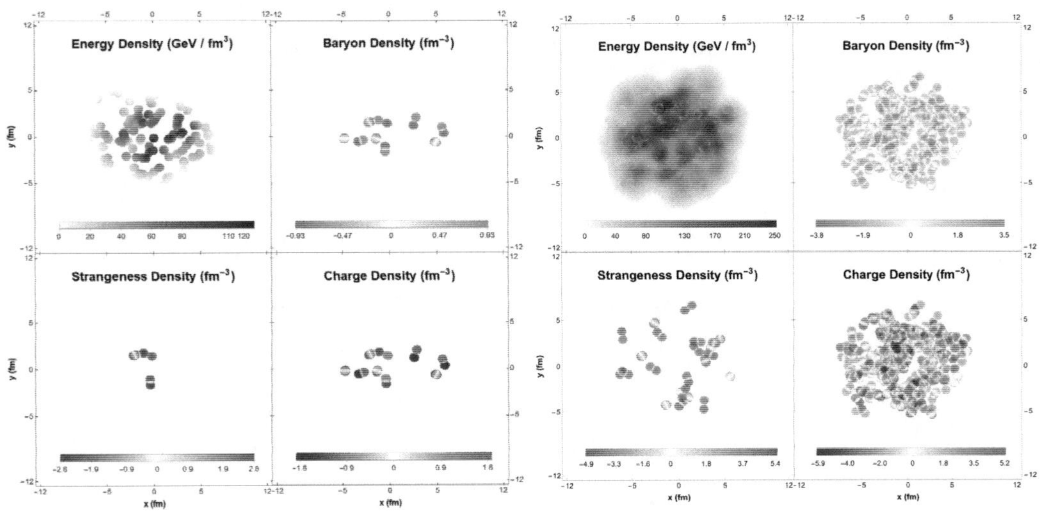

Figure 2. Left: the ICCING algorithm as it begins to resample an event, selecting blobs of energy and giving them the chance to split into $q\bar{q}$ pairs. Right: a different event after being fully resampled by the ICCING algorithm, resulting in a reconstructed energy density as well as new distributions of the three conserved charges. Note that some artifacts of the energy redistribution can be seen in the modified energy density.

pairs carry no net charges, but because the quark and antiquark are displaced from one another, they create regions of local positive and negative charge. So even at top collider energies where $\mu_B = \mu_S = \mu_Q = 0$, there can still be nontrivial local fluctuations which constitute a nonzero $N^{\mu}_{i\,0}$.

This idea is the key concept behind the Monte Carlo event generator ICCING [5, 6] (Initial Conserved Charges in Nuclear Geometry), as illustrated in Fig. 2. Intended as an additional layer of simulation on top of an existing initial condition for the energy density $\epsilon_0 = T_0^{00}$, ICCING uses input on the differential probability for gluons to split into quark-antiquark pairs to resample the initial condition. By redistributing charge and energy in space, ICCING produces an initial condition for the conserved charge densities $\rho_i = N_i^0$ out of the initial energy density ϵ_0.

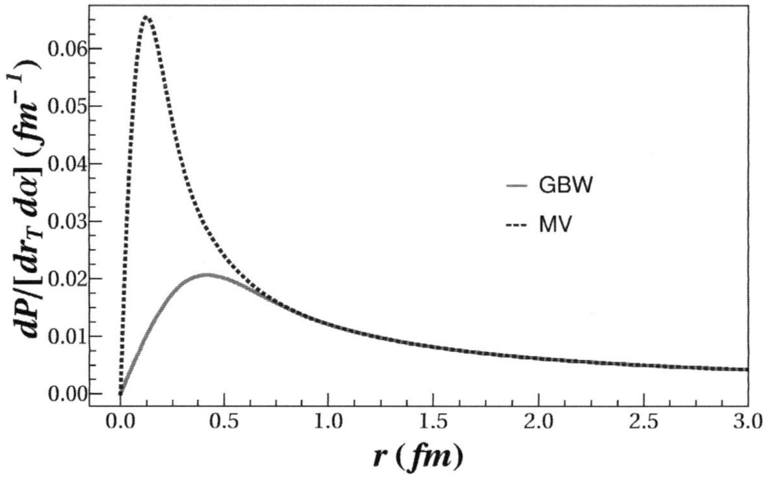

Figure 3. Differential $q\bar{q}$ splitting probability as a function of the distance r_T between them. In this plot we use the down quark mass $m = 4.8$ MeV, representative values of the splitting fraction $\alpha = 0.3$ and saturation scale $Q_s = 1.5$ GeV, and we compare the model shown in (3) (GBW) with another alternative (MV). See Ref. [6] for details.

As input, ICCING takes information on the differential probability distribution to produce a quark-antiquark pair with transverse separation r_T and fractional energy α carried by the quark. In a simple dilute-dense calculation [7] in the Color Glass Condensate (CGC) framework (see e.g. Ref. [8] and references therein), one obtains the distribution

$$\frac{dP}{dr_T d\alpha} = \frac{\alpha_s}{4\pi} m^2 r_T \left[1 - \exp\left(-\tfrac{1}{4}[\alpha^2 + (1-\alpha)^2]\, r_T^2 Q_s^2(\vec{u}_\perp)\right)\right] \times \left[(\alpha^2 + (1-\alpha)^2) K_1^2(m r_T) + K_0^2(m r_T)\right] \tag{3}$$

shown in Fig. 3. We note that the r_T distribution is controlled by the Bessel functions in the last line; these terms correspond precisely to the vacuum DGLAP splitting kernels [9, 10, 11] (light-front wave functions) in coordinate space. In this calculation, all the CGC effects are contained in the prefactor $[1 - \exp(\cdots)]$, which leads to a slight dependence of the flavor chemistry (the total probability to produce $q\bar{q}$ pairs of a given flavor) on the saturation scale Q_s (see Fig. 4).

The energy/charge redistribution algorithm is illustrated in Figs. 5 and 6. A random point from the input energy grid is chosen as the seed for a "gluon," which in this model is a circular blob of energy. First the total probabilities to split into the various quark flavors are sampled to determine the outcome of the splitting. If the outcome is that the gluon *does not* split into a $q\bar{q}$ pair, then its energy is pasted into the output grid as shown in Fig. 5. If instead the gluon splits into quarks of a given flavor, then a second sampling of the differential probability distribution (see Fig. 3) is performed to determine the distance r_T and energy fraction α of the splitting. Then "quark" and "antiquark" blobs of energy and charge are deposited at the

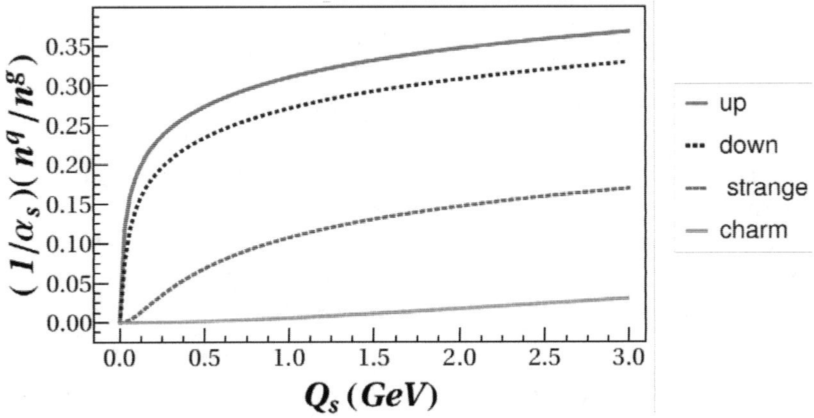

Figure 4. The flavor chemistry of the initial state in the calculation (3). Plotted are the ratios of quark to gluon multiplicities (normalized by the strong coupling α_s) as a function of the saturation scale Q_s.

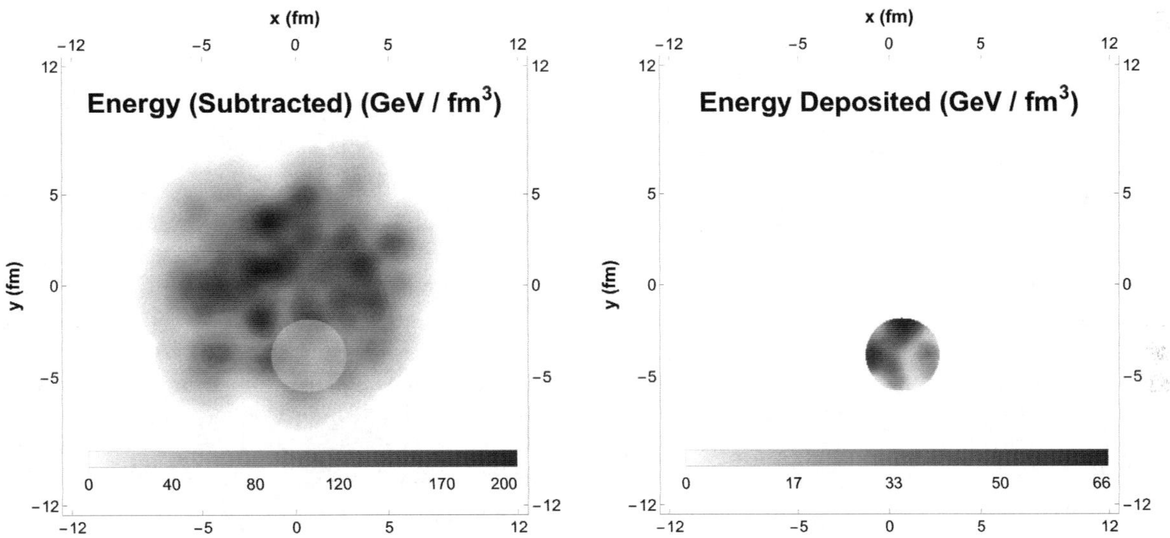

Figure 5. Illustration of how the ICCING algorithm transfers energy when the gluon *does not* split into a $q\bar{q}$ pair. The energy is deducted from the input grid (left) and deposited in the output grid (right) as shown. The energy transfer is done point by point and proportionately to the total enclosed energy. As a result, the transferred energy retains the underlying geometric structure of the original energy density, as seen in both the input energy grid after subtraction and the output energy grid after deposition. The gluon radius here has been greatly increased to clearly show these details.

displaced positions. In this way, the algorithm redistributes the energy and separates a region of zero charge density into two displaced regions of separate positive and negative charge.

The final output of the ICCING algorithm is a fully resampled event as shown in the right

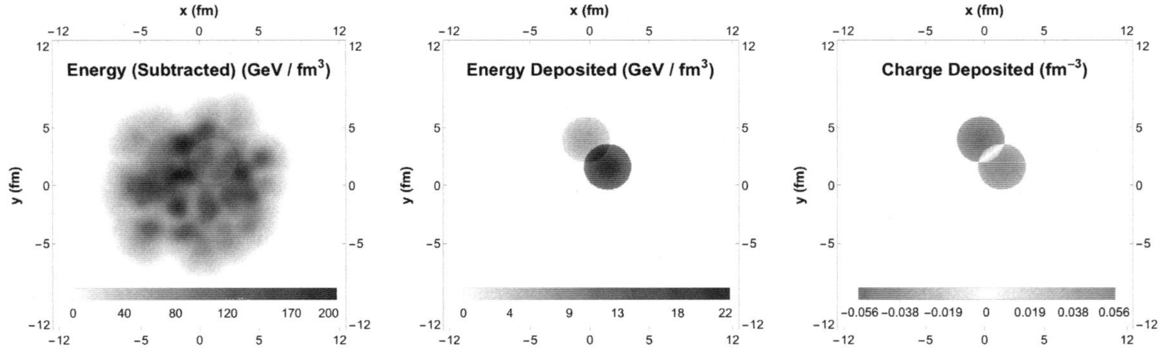

Figure 6. Illustration of how the ICCING algorithm transfers energy when the gluon *does* split into a $q\bar{q}$ pair. The energy is deducted from the input grid (left plot) proportionately, preserving the underlying geometry in the input grid. But it is deposited in Gaussian blobs for the $q\bar{q}$ pair, which are displaced relative to the original gluon position. This modifies the energy density distribution (center plot) and also leads to a net displacement of positive and negative charge (here baryon density, right plot). Note that the energy is in general shared unequally between the quark and antiquark; in this case, the quark carried about 75% of the original gluon energy, as visible in the center plot. Here both the radii and overall $q\bar{q}$ displacement have been greatly increased to clearly show these details.

panel of Fig. 2. In addition to a somewhat modified energy density profile, the algorithm has generated nontrivial spatial profiles of all three conserved charge densities B, S, and Q. Some general features are immediately visible. The spatial distributions of ρ_B and ρ_Q visibly reflect the spatial profile of the bulk energy density ϵ, but the strangeness density ρ_S is visibly different. There are much fewer strange quarks produced than up and down quarks (which dominate the ρ_B, ρ_Q distributions) because only gluons with energies exceeding the strange quark mass threshold $2m_s \approx 200$ MeV can produce $s\bar{s}$ pairs. For the same reason, the strange quark pairs are produced with a qualitatively different spatial distribution than the other quarks: the $s\bar{s}$ pairs are only produced at hot spots capable of exceeding their mass threshold. As a result, the spatial distribution of ρ_S can be used to access different information about the collision geometry than can be learned from the bulk energy density.

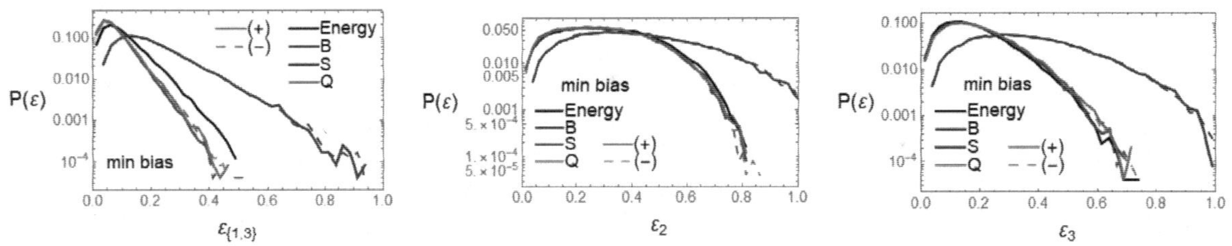

Figure 7. Histograms of the eccentricities for positive and negative baryon number, strangeness, and electric charge. For comparison, the distribution of the energy density eccentricities is shown.

These features are seen quantitatively in the eccentricity histograms shown in Fig. 7. The eccentricities $\varepsilon_{n,m}$ are measures which quantify the shape of the initial state distributions of some quantity f:

$$\varepsilon_{n,m} = \left| \frac{\int d^2r\, r^m\, e^{in\phi}\, f(\vec{r}_\perp)}{\int d^2r\, r^m\, f(\vec{r}_\perp)|} \right| . \tag{4}$$

with $\varepsilon_2 \equiv \varepsilon_{2,2}$ and $\varepsilon_3 \equiv \varepsilon_{3,3}$ for brevity. In Fig. 7 we plot the eccentricities for the energy density ϵ as well as for the conserved charges B, S, Q. For the charge densities we separately calculate the eccentricities of the positive regions $\rho_i > 0$ and negative regions $\rho_i < 0$ of charge. As can be immediately seen from Fig. 7, the strange quark distribution is far more eccentric in all measures than the bulk geometry. Additionally, one readily confirms the picture from Fig. 2 that the ρ_B, ρ_Q geometries mirror the energy density ϵ.

A fully-developed framework to study the charge transport properties of the QGP at top collider energies would truly open the door to "multi-messenger femtoscopy": the use of multiple, independent degrees of freedom to reveal different slices of the initial state geometry and subsequent transport.

2. PDF Chemistry in ICCING

While the particular probabilities used in (3) are specific to the model chosen, it is clear that the mass threshold effect responsible for the different behavior of the strangeness distribution is much more general. In fact, the spatial distribution in (3) is dominated by the vacuum-like DGLAP kernels which are fundamental ingredients in QCD. For this reason, it seems natural to relax the CGC input to ICCING and consider other ways to specify the quark flavor chemistry and the quark-antiquark spatial correlations.

This seems like a natural opportunity to introduce the quark PDFs to fix the partonic flavor chemistry. The simplest thing to do is to would be to use an extraction of the various quark and gluon PDFs to set the flavor chemistry for the appropriate kinematics of the collision. For the spatial correlations, one can simply use the vacuum DGLAP kernels directly, rather than ones modified by CGC effects as in (3). The subtlety is in what scales one should reasonably evaluate the PDFs at to be relevant for a heavy-ion collision.

While the longitudinal momentum fraction x is fairly straightforward to specify based on the energy of the collision, the resolution scale Q^2 is far more subtle. On one hand, the Q^2 evolution of the PDFs is very slow (logarithmic) so the chemistry dependence of the PDFs on Q^2 is weak. But the inability to set the hard resolution scale Q^2 reflects a deeper mismatch between a perturbative point of view and the reality of heavy ion collisions. The absence of a clear scale for Q^2 is an indication that there is no external scale in the initial condition which justifies a perturbative treatment. Put differently, a justification for thinking of the initial state in terms of perturbative "gluons" is hard to come by. By this measure, the CGC framework for gluon saturation stands alone. Because the high gluon density in a boosted nucleus gives rise to a semi-hard saturation scale Q_s, it provides a rare justification for the use of perturbative degrees of freedom in the initial conditions of heavy-ion collisions.

Acknowledgments

The author gratefully acknowledges support from the US-DOE Nuclear Science Grant No. DE-SC0019175, the Alfred P. Sloan Foundation, and the Illinois Campus Cluster, a computing resource that is operated by the Illinois Campus Cluster Program (ICCP) in conjunction with the National Center for Supercomputing Applications (NCSA) and which is supported by funds from the University of Illinois at Urbana-Champaign.

References

[1] J. S. Moreland, J. E. Bernhard and S. A. Bass, *Alternative ansatz to wounded nucleon and binary collision scaling in high-energy nuclear collisions*, *Phys. Rev.* **C92** (2015) 011901, [1412.4708].

[2] B. Schenke, P. Tribedy and R. Venugopalan, *Fluctuating Glasma initial conditions and flow in heavy ion collisions*, *Phys. Rev. Lett.* **108** (2012) 252301, [1202.6646].

[3] J. Moreland, J. E. Bernhard, W. Ke and S. A. Bass, *Flow in small and large quark-gluon plasma droplets: the role of nucleon substructure*, *Nucl. Phys. A* **967** (2017) 361–364, [1704.04486].

[4] H1 and ZEUS Collaboration collaboration, F. Aaron et al., *Combined Measurement and QCD Analysis of the Inclusive e+- p Scattering Cross Sections at HERA*, *JHEP* **1001** (2010) 109, [0911.0884].

[5] M. Martinez, M. D. Sievert, D. E. Wertepny and J. Noronha-Hostler, *Initial state fluctuations of QCD conserved charges in heavy-ion collisions*, 1911.10272.

[6] M. Martinez, M. D. Sievert, D. E. Wertepny and J. Noronha-Hostler, *Toward Initial Conditions of Conserved Charges Part II: The ICCING Monte Carlo Algorithm*, 1911.12454.

[7] M. Martinez, M. D. Sievert and D. E. Wertepny, *Toward Initial Conditions of Conserved Charges Part I: Spatial Correlations of Quarks and Antiquarks*, *JHEP* **07** (2018) 003, [1801.08986].

[8] Y. V. Kovchegov and E. Levin, *Quantum Chromodynamics at High Energy*. Cambridge University Press, 2012.

[9] V. N. Gribov and L. N. Lipatov, *Deep inelastic e p scattering in perturbation theory*, *Sov. J. Nucl. Phys.* **15** (1972) 438–450.

[10] G. Altarelli and G. Parisi, *Asymptotic Freedom in Parton Language*, *Nucl. Phys.* **B126** (1977) 298.

[11] Y. L. Dokshitzer, *Calculation of the Structure Functions for Deep Inelastic Scattering and e^+e^- Annihilation by Perturbation Theory in Quantum Chromodynamics*, *Sov. Phys. JETP* **46** (1977) 641–653.

From small to large x: toward a unified description of high energy collisions

Jamal Jalilian-Marian

Department of Natural Sciences, Baruch College, 17 Lexington Ave., New York, NY 10010, USA

E-mail: jamal.jalilian-marian@baruch.cuny.edu

Abstract. Inclusive particle production at high p_t (equivalently intermediate to large x) in high energy hadronic collisions is successfully described by perturbative QCD, using the collinear-factorization formalism. On the other hand at very high energies and not so high p_t the Color Glass Condensate (CGC) formalism has been quite successful in describing particle production at small Bjorken x. Here we propose a way to unify the two approaches which enables one to apply it to particle production at all x (all p_t).

1. Introduction

Perturbative QCD has been very successful in describing high p_t particle production in high energy collisions. Its use depends on the concept of collinear-factorization where the particle production cross section can be written as a convolution of several independent quantities; parton distribution functions describing probabilities f_1, f_2 for finding partons of energy fractions x_1 and x_2 in the two incoming hadrons, the hard scattering of the two partons and a subsequent hadronization of the partons described by the parton-hadron fragmentation function $D_{p/h}$. This can be symbolically written as

$$d\sigma \sim f_1(x_1, Q^2) \otimes f_2(x_2, Q^2) \otimes d\sigma_{parton} \otimes D_{p/h} \tag{1}$$

where the parton distribution and fragmentation functions satisfy the DGLAP evolution equation [1] which describes further radiation of a parton. This formalism is ideal at high p_t (equivalently at small distances) where parton density, number of partons per unit area, is small which justifies ignoring the fact that there are other partons in the colliding hadrons. While theoretically well-defined and on solid grounds at high p_t, collinear-factorization is expected to break down at small x due to large parton densities.

It is an experimental fact that the number of partons, specially gluons, increases very fast as $x \to 0$. Due to this large number of partons occupying the same transverse area of a hadron the parton density becomes large (referred to as gluon saturation) which makes it very likely that more than one parton from each hadron will participate in the scattering which breaks the underlying approximations made in the collinear-factorization approach. This is specially significant for nuclei due to longitudinal coherence length of small x partons which can be of the order of or larger than the nuclear size. If so this parton will see the effective parton density of the nucleus which is enhanced by $A^{1/3}$ compared to a proton. This indicates that parton density

Journal of Physics: Conference Series **1602** (2020) 012024 doi:10.1088/1742-6596/1602/1/012024

effects can be large for large nuclei and that one needs to go beyond collinear-factorization for processes dominated by small x kinematics.

Due to large parton densities at small x the concept of a quasi-free parton is not very useful anymore and it may be more appropriate to treat the system of partons collectively, i.e. as fields. In this approach, known as the Color Glass Condensate (CGC) formalism [2] the large x degrees of freedom in the hadron act as sources of color charge which radiate a classical gluon field representing the collective small x (gluon) degrees of freedom. One then includes quantum loop effects by re-summing the large logs of x which become $\mathcal{O}(1/\alpha_s)$ at small x such that $\alpha_s \log 1/x \sim 1$. Due to the large parton densities at small x multiple scattering (as compared to single scattering in pQCD) must be taken into account. This is efficiently done via the use of Wilson lines which re-sum multiple soft scatterings of a projectile from a dense target, represented by the strong classical color field. In the hybrid formulation [3] of particle production in CGC formalism the cross section is given in terms of correlations of products of Wilson lines [4] which satisfy the JIMWLK evolution equation [5]. For example, single inclusive hadron production in proton-nucleus collisions can be symbolically written as

$$d\sigma \sim f_1(x_1, Q^2) \otimes \sigma_{dipole} \otimes D_{p/h} \tag{2}$$

where σ_{dipole} is the quark anti-quark diople scattering cross section on the target which satisfies the JIMWLK equation. In this formalism the dipole cross section contains all the dependence on the target. When expanded in powers of density its first term contains the convolution of Leading Log gluon distribution function with the parton-gluon scattering cross section. In this sense, and very loosely speaking the dipole cross section contains the target parton distribution function and the parton-parton scattering cross section.

While each of the two formalisms are valid in their respective kinematic regions there is no single formalism which contains both in the appropriate limits. In addition to being theoretically unsatisfactory it also leads to practical problems which prevents one from unequivocally establishing the presence of gluon saturation at small x. For instance it is still not clear what region of x contributes to single inclusive hadron production in the forward rapidity region of RHIC as the two competing formalisms are able to fit the data, and yet yield vastly different kinematics [6]. While some of these issues can be resolved experimentally in the proposed Electron-Ion Collider [7] it is essential to have a more general formalism that is valid in both kinematic regions.

2. Scattering from both large and small x gluons in the target

In [8] we proposed a more general formalism which includes scattering of a projectile from both large and small x degrees of freedom. As the parton density is large at small x we include multiple scatterings of the projectile from the small x gluons of the target. On the other hand the large x parton density of the target is not large so that it is enough to consider a single scattering from the large x partons of the target. In addition to the standard eikonal scattering employed in CGC formalism, there are three other classes of diagrams where 1) a projectile will multiply scatter from the soft gluons of the target both before and after a hard scattering from the large x gluons of the target; 2) The large x gluon itself will also interact with the soft gluon background, and finally, 3) both the scattered projectile and the large x gluon will scatter from the soft background field. The amplitude is most easily evaluated using the spinor helicity formalism [9]. To show how spin asymmetries arise in our formalism, we just show our results for the Dirac numerators of the amplitude for scattering of a quark projectile on a target hadron or nucleus (ignoring the Wilson lines which encode the target information),

$$\mathcal{N}_{1,a}^{+} = p^{+}\sqrt{\frac{q^{+}}{p^{+}}}\left\{2\,A_a^{-}(x) - A_a^{i}(x)\left[\frac{k_{1i} - i\epsilon_{ij}\,k_1^{j}}{q^{+}} + \frac{k_i + i\epsilon_{ij}\,k^{j}}{p^{+}}\right]\right\}$$

$$
\begin{aligned}
\mathcal{N}^{+}_{2,a} &= \frac{p^+}{q_\perp^2}\sqrt{\frac{q^+}{p^+}}\Big\{\Big(1+\frac{q^+}{p^+}\Big)\, q_\perp\cdot A^b_\perp(x) + i\Big(1-\frac{q^+}{p^+}\Big)\,\epsilon^{ij}\, q_i\, A^a_j(x)\Big\}\\
\mathcal{N}^{+}_{3,a} &= \mathcal{N}^{+}_{2,a}(q_i\to p_{1i})\\
&= \frac{p^+}{p_{1\perp}^2}\sqrt{\frac{q^+}{p^+}}\Big\{\Big(1+\frac{q^+}{p^+}\Big)\, p_{1\perp}\cdot A^b_\perp(x) + i\Big(1-\frac{q^+}{p^+}\Big)\,\epsilon^{ij}\, p_{1i}\, A^a_j(x)\Big\} \qquad (3)
\end{aligned}
$$

and

$$\mathcal{N}^{-a}_{1,2,3} = \Big[\mathcal{N}^{+a}_{1,2,3}\Big]^{\star} \qquad (4)$$

where p, q are the momenta of the incoming and outgoing quark respectively, k and k_1 are the momenta of the internal quark line before and after the hard scattering which is integrated over and a is a color index. To get the cross sections for polarized scattering one needs to square the above amplitudes. Here we show one of the terms, specifically, the term $|\mathcal{N}_2^+|^2$ which is the simplest,

$$
\begin{aligned}
|\mathcal{N}_2^{+,\,b\,c}|^2 &= \frac{q^+}{p^+}\frac{1}{q_\perp^4}\Big\{\Big[(4p^+\,q^+)\, q_\perp\cdot A^b_\perp(x)\, q_\perp\cdot A^c_\perp(y) + (p^+-q^+)^2\, q_\perp^2\, A^b_\perp(x)\cdot A^c_\perp(y)\Big]\\
&+\ i\,\epsilon^{ij}\,[(p^+)^2-(q^+)^2]\Big[q_i\, A^b_j(x)\, q_\perp\cdot A^c_\perp(y) - q_i\, A^c_j(y)\, q_\perp\cdot A^b_\perp(x)\Big]\Big\} \qquad (5)
\end{aligned}
$$

Using the fact that $\mathcal{N}_2^{-b} = \Big[\mathcal{N}_2^{+b}\Big]^\star$ we see that there is asymmetry in the polarized cross section, for example in the double longitudinal spin asymmetry A_{LL}

$$A_{LL} \equiv \frac{d\sigma^{++} - d\sigma^{--}}{d\sigma^{++} + d\sigma^{--}} \qquad (6)$$

due to the presence of terms proportional to $i\epsilon^{ij}$. We note that effects like this are already present [10] at small x, the novel thing here is their appearance in our unified treatment in the intermediate x.

Inclusion of scattering from the large x gluons of the target also leads to angular asymmetries in unpolarized scattering addition to the spin asymmetry. This is already seen in previous studies of sub-eikonal corrections to eikonal scattering [11, 12] at small x. The difference between our results here and those in earlier studies is in the kinematic region where these corrections appear; in our formalism these corrections appear in the intermediate-large x region of the target which corresponds to intermediate to high p_t region of particle production.

Furthermore, this formalism leads to a rapidity shift of the projectile parton (commonly referred to as beam rapidity loss) which is absent in the eikonal approximation made in the standard CGC approach [13]. This is due to exchange of longitudinal momenta between the projectile and target which is not included in the standard CGC approach.

It will be interesting to apply this formalism to study cold matter energy loss. While the fully coherent energy loss is contained in parton radiation in the CGC formalism as part of higher order corrections (in α_s) to inclusive production, the standard CGC approach does not contain partially coherent energy loss as in the small x kinematics a radiated parton can not resolve the longitudinal structure of the target nucleus. In our formalism this will not be the case, however, due to the fact that there is partially coherent radiation before and after hard scattering. To see this explicitly one will need to construct the gluon propagator [14] the same way as done for the projectile quark.

It will be easier to start with radiation of a real or virtual photon [15] first, as unlike gluons, photons will not interact with the target color fields. This should already exhibit the QED like radiative energy loss pattern. Furthermore it will allow us to investigate two-particle correlations in this simpler process [16] before considering the more involved di-hadron or di-jet correlations.

Finally, and in the long term, it will be very fruitful to try to generalize this formalism by using an effective action approach [17]. This will allow one to use this formalism to study the early dynamics of high energy heavy ion collisions in a more robust way. Currently there is no formalism that can treat production of both the soft Quark-Gluon Plasma and the hard particles that probe its properties concurrently. Due to inclusion of large x partons in the wave-function of a hadron or nucleus one will not only have the soft gluons forming the medium but also the high p_t probes corresponding to the produced large x partons.

In summary, we have introduced a new formalism which generalizes the Color Glass Condensate approach to particle production in high energy collisions by including large x partons of a hadron or nucleus in the scattering process. We have shown that this leads to spin asymmetries in polarized scattering, as well as rapidity loss and angular asymmetries. It will be interesting to develop the formalism further and to apply it to phenomenology of both high and low p_t particle production in high energy collisions such as those at RHIC and the LHC, as well as the proposed Electron-Ion Collider.

2.1. Acknowledgments

We acknowledge support by the DOE Office of Nuclear Physics through Grant No. DE-FG02-09ER41620 and by PSC-CUNY through grant No. 62185-0050. We would like to thank Centre de Physique Théorique of École Polytechnique for their kind hospitality during the completion of this work and the CNRS for support.

References

[1] G. Altarelli and G. Parisi, Nucl. Phys. B **126**, 298-318 (1977) doi:10.1016/0550-3213(77)90384-4 V. Gribov and L. Lipatov, Sov. J. Nucl. Phys. **15**, 438-450 (1972) IPTI-381-71; Y. L. Dokshitzer, Sov. Phys. JETP **46**, 641-653 (1977)

[2] E. Iancu and R. Venugopalan, doi:10.1142/9789812795533$_0$005 [arXiv:hep-ph/0303204 [hep-ph]]. F. Gelis, E. Iancu, J. Jalilian-Marian and R. Venugopalan, Ann. Rev. Nucl. Part. Sci. **60**, 463-489 (2010) doi:10.1146/annurev.nucl.010909.083629 [arXiv:1002.0333 [hep-ph]]. J. Jalilian-Marian and Y. V. Kovchegov, Prog. Part. Nucl. Phys. **56**, 104-231 (2006) doi:10.1016/j.ppnp.2005.07.002 [arXiv:hep-ph/0505052 [hep-ph]].

[3] A. Dumitru and J. Jalilian-Marian, Phys. Rev. Lett. **89**, 022301 (2002) doi:10.1103/PhysRevLett.89.022301 [arXiv:hep-ph/0204028 [hep-ph]], Phys. Lett. B **547**, 15-20 (2002) doi:10.1016/S0370-2693(02)02709-0 [arXiv:hep-ph/0111357 [hep-ph]]; Phys. Rev. D **70**, 114017 (2004) doi:10.1103/PhysRevD.71.079901 [arXiv:hep-ph/0405266 [hep-ph]]; C. Marquet, Nucl. Phys. A **796**, 41-60 (2007) doi:10.1016/j.nuclphysa.2007.09.001 [arXiv:0708.0231 [hep-ph]].

[4] A. Dumitru, J. Jalilian-Marian, T. Lappi, B. Schenke and R. Venugopalan, Phys. Lett. B **706**, 219-224 (2011) doi:10.1016/j.physletb.2011.11.002 [arXiv:1108.4764 [hep-ph]]; A. Dumitru and J. Jalilian-Marian, Phys. Rev. D **81**, 094015 (2010) doi:10.1103/PhysRevD.81.094015 [arXiv:1001.4820 [hep-ph]]; A. Dumitru, J. Jalilian-Marian and E. Petreska, Phys. Rev. D **84**, 014018 (2011) doi:10.1103/PhysRevD.84.014018 [arXiv:1105.4155 [hep-ph]].

[5] J. Jalilian-Marian, A. Kovner, L. D. McLerran and H. Weigert, Phys. Rev. D **55**, 5414-5428 (1997) doi:10.1103/PhysRevD.55.5414 [arXiv:hep-ph/9606337 [hep-ph]]; J. Jalilian-Marian, A. Kovner, A. Leonidov and H. Weigert, Nucl. Phys. B **504**, 415-431 (1997) doi:10.1016/S0550-3213(97)00440-9 [arXiv:hep-ph/9701284 [hep-ph]], Phys. Rev. D **59**, 014014 (1998) doi:10.1103/PhysRevD.59.014014 [arXiv:hep-ph/9706377 [hep-ph]], Phys. Rev. D **59**, 034007 (1999) doi:10.1103/PhysRevD.59.034007 [arXiv:hep-ph/9807462 [hep-ph]]; J. Jalilian-Marian, A. Kovner and H. Weigert, Phys. Rev. D **59**, 014015 (1998) doi:10.1103/PhysRevD.59.014015 [arXiv:hep-ph/9709432 [hep-ph]], Phys. Rev. D **59**, 034007 (1999) doi:10.1103/PhysRevD.59.034007 [arXiv:hep-ph/9807462 [hep-ph]]; A. Kovner, J. Milhano and H. Weigert, Phys. Rev. D **62**, 114005 (2000) doi:10.1103/PhysRevD.62.114005 [arXiv:hep-ph/0004014 [hep-ph]]; E. Ferreiro, E. Iancu, A. Leonidov and L. McLerran, Nucl. Phys. A **703**, 489-538 (2002) doi:10.1016/S0375-9474(01)01329-X [arXiv:hep-ph/0109115 [hep-ph]].

[6] A. Dumitru, A. Hayashigaki and J. Jalilian-Marian, Nucl. Phys. A **765**, 464-482 (2006) doi:10.1016/j.nuclphysa.2005.11.014 [arXiv:hep-ph/0506308 [hep-ph]]. V. Guzey, M. Strikman and W. Vo-

gelsang, Phys. Lett. B **603**, 173-183 (2004) doi:10.1016/j.physletb.2004.10.033 [arXiv:hep-ph/0407201 [hep-ph]].

[7] E. C. Aschenauer, C. Aidala, A. Bazilevsky, M. Diehl, R. Fatemi, C. Gagliardi, Z. Kang, Y. V. Kovchegov, J. Jalilian-Marian, J. Lajoie, D. V. Perepelitsa, R. Seidl, R. Sassot, E. Sichter-mann, M. Stratmann, S. Trentalange, W. Vogelsang, A. Vossen and P. Zurita, [arXiv:1602.03922 [nucl-ex]].

[8] J. Jalilian-Marian, [arXiv:1912.08878 [hep-ph]], Phys. Rev. D **99**, no.1, 014043 (2019) doi:10.1103/PhysRevD.99.014043 [arXiv:1809.04625 [hep-ph]], Phys. Rev. D **96**, no.7, 074020 (2017) doi:10.1103/PhysRevD.96.074020 [arXiv:1708.07533 [hep-ph]].

[9] L. J. Dixon, [arXiv:hep-ph/9601359 [hep-ph]]; A. Ayala, M. Hentschinski, J. Jalilian-Marian and M. E. Tejeda-Yeomans, Nucl. Phys. B **920**, 232-255 (2017) doi:10.1016/j.nuclphysb.2017.03.028 [arXiv:1701.07143 [hep-ph]], Phys. Lett. B **761**, 229-233 (2016) doi:10.1016/j.physletb.2016.08.035 [arXiv:1604.08526 [hep-ph]].

[10] Y. V. Kovchegov, D. Pitonyak and M. D. Sievert, JHEP **10**, 198 (2017) doi:10.1007/JHEP10(2017)198 [arXiv:1706.04236 [nucl-th]].

[11] G. A. Chirilli, JHEP **01**, 118 (2019) doi:10.1007/JHEP01(2019)118 [arXiv:1807.11435 [hep-ph]].

[12] P. Agostini, T. Altinoluk and N. Armesto, Eur. Phys. J. C **79**, no.9, 790 (2019) doi:10.1140/epjc/s10052-019-7315-1 [arXiv:1907.03668 [hep-ph]], Eur. Phys. J. C **79**, no.7, 600 (2019) doi:10.1140/epjc/s10052-019-7097-5 [arXiv:1902.04483 [hep-ph]].

[13] J. Jalilian-Marian, Phys. Rev. C **70**, 027902 (2004) doi:10.1103/PhysRevC.70.027902 [arXiv:nucl-th/0212018 [nucl-th]]; F. Gelis, A. M. Stasto and R. Venugopalan, Eur. Phys. J. C **48**, 489-500 (2006) doi:10.1140/epjc/s10052-006-0020-x [arXiv:hep-ph/0605087 [hep-ph]].

[14] A. Ayala, J. Jalilian-Marian, L. D. McLerran and R. Venugopalan, Phys. Rev. D **53**, 458-475 (1996) doi:10.1103/PhysRevD.53.458 [arXiv:hep-ph/9508302 [hep-ph]].

[15] E. M. Henley and J. Jalilian-Marian, Phys. Rev. D **73**, 094004 (2006) doi:10.1103/PhysRevD.73.094004 [arXiv:hep-ph/0512220 [hep-ph]]; F. Gelis and J. Jalilian-Marian, Phys. Rev. D **67**, 074019 (2003) doi:10.1103/PhysRevD.67.074019 [arXiv:hep-ph/0211363 [hep-ph]], Phys. Rev. D **66**, 094014 (2002) doi:10.1103/PhysRevD.66.094014 [arXiv:hep-ph/0208141 [hep-ph]], Phys. Rev. D **76**, 074015 (2007) doi:10.1103/PhysRevD.76.074015 [arXiv:hep-ph/0609066 [hep-ph]]; J. Jalilian-Marian, Nucl. Phys. A **753**, 307-315 (2005) doi:10.1016/j.nuclphysa.2005.02.156 [arXiv:hep-ph/0501222 [hep-ph]], Nucl. Phys. A **739**, 319-328 (2004) doi:10.1016/j.nuclphysa.2004.04.103 [arXiv:nucl-th/0402014 [nucl-th]].

[16] J. Jalilian-Marian and A. H. Rezaeian, Phys. Rev. D **86**, 034016 (2012) doi:10.1103/PhysRevD.86.034016 [arXiv:1204.1319 [hep-ph]], Phys. Rev. D **85**, 014017 (2012) doi:10.1103/PhysRevD.85.014017 [arXiv:1110.2810 [hep-ph]]; J. Jalilian-Marian, Nucl. Phys. A **770**, 210-220 (2006) doi:10.1016/j.nuclphysa.2006.02.013 [arXiv:hep-ph/0509338 [hep-ph]].

[17] J. Jalilian-Marian, S. Jeon and R. Venugopalan, Phys. Rev. D **63**, 036004 (2001) doi:10.1103/PhysRevD.63.036004 [arXiv:hep-ph/0003070 [hep-ph]]; J. Jalilian-Marian, S. Jeon, R. Venugopalan and J. Wirstam, Phys. Rev. D **62**, 045020 (2000) doi:10.1103/PhysRevD.62.045020 [arXiv:hep-ph/9910299 [hep-ph]].

Journal of Physics: Conference Series **1602** (2020) 012025 doi:10.1088/1742-6596/1602/1/012025

Quark spin – thermal vorticity alignment and the Λ, $\bar{\Lambda}$ polarization in heavy-ion collisions

Alejandro Ayala[1,2], D. de la Cruz[1,3], L. A. Hernández[1,2,4], S. Hernández-Ortíz[5] and J. Salinas[1]

[1] Instituto de Ciencias Nucleares, Universidad Nacional Autónoma de México, Apartado Postal 70-543, México, Distrito Federal 04510, Mexico
[2] Centre for Theoretical and Mathematical Physics, and Department of Physics, University of Cape Town, Rondebosch 7700, South Africa
[3] Departamento de Física, Escuela Superior de Física y Matemáticas del Instituto Politécnico Nacional, Unidad Adolfo López Mateos, Edificio 9, 07738 Ciudad de México, México
[4] Facultad de Ciencias de la Educación, Universidad Autónoma de Tlaxcala, Tlaxcala, 90000, Mexico
[5] Institute of Nuclear Theory, University of Washington, Physics-Astronomy Building, Seattle 98195-1550, United States of America

E-mail: ayala@nucleares.unam.mx

Abstract. It has been proposed that the Λ and $\bar{\Lambda}$ polarizations observed in heavy-ion collisions are due to the interaction between quark spin and thermal vorticity. In this work we report on a computation of the relaxation time required for this alignment to occur at finite temperature and baryon chemical potential, considering quarks with a finite mass. The calculation is performed after modelling the interaction by means of an effective vertex which couples the thermal gluons and quarks within the vortical medium. We show that the effect of the quark mass is to reduce the relaxation time as compared to the massless quark case. An intrinsic global polarization of quarks/antiquarks emerges which is shown to be linked with the $\Lambda/\bar{\Lambda}$ polarization.

1. Introduction

Experiments where heavy-ions collide at relativistic energies provide a unique opportunity to investigate the properties of hadronic matter under extreme conditions of temperature and density, conditions that are analogous to those found soon after the Big Bang. At experimental facilities such as RHIC and the LHC –and in coming years at J-PARC, FAIR, and NICA– two atomic nuclei collide at relativistic energies giving rise to sequential stages of strongly interacting matter dynamics, among these, the quark-gluon plasma (QGP) phase. It has been suggested that during the QGP stage a kinematic vorticity of the order of $\omega \sim 10^{22}$ s^{-1} can be achieved in collisions at RHIC

Journal of Physics: Conference Series **1602** (2020) 012025 doi:10.1088/1742-6596/1602/1/012025

and LHC energies, thus establishing QGP as the most vortical fluid known so far [1]. The vorticity is capable to transfer angular momentum to the spin degrees of freedom of primary quarks, originating the so-called global baryon polarization once polarized quarks hadronize [2]. Since QGP can be very well described using hydrodynamical models, a relativistic hydro description of the classical vorticity led to the concept of thermal vorticity [3, 4] defined as

$$\varpi_{\mu\nu} = \frac{1}{2}\left(\partial_\nu \beta_\mu - \partial_\mu \beta_\nu\right), \tag{1}$$

where $\beta_\mu = u_\mu(x)/T(x)$, $u_\mu(x)$ is the local fluid four-velocity and $T(x)$ is the local temperature. Thermal vorticity describes a rotating fluid that is able to generate a global particle polarization. An outstanding question is what is the time for the spin degrees of freedom to equilibrate with the thermal vorticity. The question is similar to that of finding the time for the momentum degrees of freedom to equilibrate and achieve a common temperature.

2. Quark Spin - Thermal Vorticity Alignment

Current hydrodynamical descriptions indicate that the QCD plasma reaches a state of local thermal equilibrium within $\tau_{\text{hydro}} \sim 1$ fm [5]. After this time, the system can be approximately identified as having a temperature T and quark chemical potential μ_q. The interaction rate Γ of a quark with four-momentum $P = (p_0, \vec{p})$ can be expressed in terms of the quark self energy Σ as

$$\Gamma(p_0) = \widetilde{f}(p_0 - \mu_q)\text{Tr}\left\{\gamma^0 \text{ Im } \Sigma\right\}, \tag{2}$$

where $\widetilde{f}(p_0 - \mu_q)$ is the Fermi-Dirac distribution. The interaction between the thermal vorticity and the quark spin is modeled by means of a phenomenological effective vertex

$$\lambda_a^\mu = g\frac{\sigma^{\alpha\beta}}{2}\varpi_{\alpha\beta}\gamma^\mu t_a, \tag{3}$$

where $\sigma^{\alpha\beta} = \frac{i}{2}[\gamma^\alpha, \gamma^\beta]$ is the quark spin operator and t_a are the color matrices in the fundamental representation. The one-loop contribution to Σ, depicted in Fig. 1, is given explicitly by

$$\Sigma = T\sum_n \int \frac{d^3k}{(2\pi)^3}\lambda_a^\mu S(\not{P} - \not{K})\lambda_b^\nu {}^*G_{\mu\nu}^{ab}(K), \tag{4}$$

where S and ${}^*G$ are the quark and effective gluon propagators, respectively. In a covariant gauge, the Hard Thermal Loop (HTL) approximation to the effective gluon propagator is given by

$${}^*G_{\mu\nu}(K) = {}^*\Delta_L(K)P_{L\,\mu\nu} + {}^*\Delta_T(K)P_{T\,\mu\nu}, \tag{5}$$

where $P_{L,T\,\mu\nu}$ are the polarization tensors for three dimensional longitudinal and transverse gluons. The gluon propagator functions for longitudinal and transverse

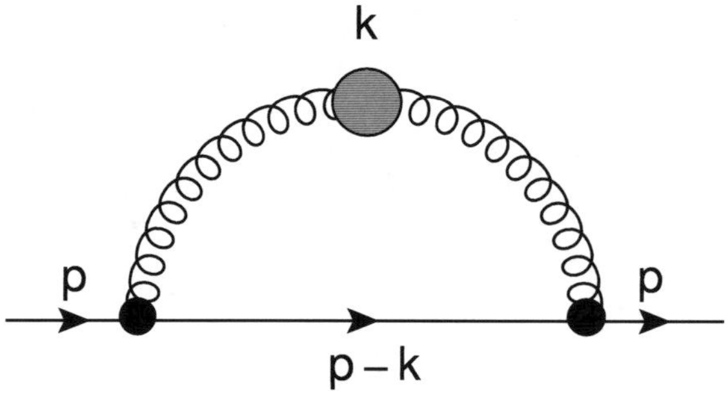

Figure 1. One-loop Feynman diagram representing the quark self-energy. The gluon line with a blob represents the effective gluon propagator at finite density and temperature. The blobs on the quark-gluon vertices represent the effective coupling between the quark spin and the vorticity.

modes, ${}^*\Delta_{L,T}(K)$, are given by

$$
\begin{aligned}
{}^*\Delta_L(K)^{-1} &= K^2 + 2m^2\frac{K^2}{k^2}\left[1-\left(\frac{i\omega_n}{k}\right)Q_0\left(\frac{i\omega_n}{k}\right)\right],\\
{}^*\Delta_T(K)^{-1} &= -K^2 - m^2\left(\frac{i\omega_n}{k}\right)\left\{\left[1-\left(\frac{i\omega_n}{k}\right)^2\right]Q_0\left(\frac{i\omega_n}{k}\right)+\left(\frac{i\omega_n}{k}\right)\right\},
\end{aligned}
\tag{6}
$$

where

$$
Q_0(x) = \frac{1}{2}\ln\frac{x+1}{x-1}, \tag{7}
$$

and m is the gluon thermal mass given by

$$
m^2 = \frac{1}{6}g^2C_AT^2 + \frac{1}{12}g^2C_F\left(T^2+\frac{3}{\pi^2}\mu^2\right), \tag{8}
$$

where $C_A = 3$ and $C_F = 4/3$ are the Casimir factors for the adjoint and the fundamental representations of $SU(3)$, respectively. The sum over Matsubara frequencies involves products of the propagator functions for longitudinal and transverse gluons ${}^*\Delta_{L,T}$ and the Matsubara propagator for the bare quark $\tilde{\Delta}_F$. Then, the sum over Matsubara frequencies can be expressed as

$$
S_{L,T} = T\sum_n {}^*\Delta_{L,T}(i\omega_n)\tilde{\Delta}_F(i(\omega_m-\omega_n)), \tag{9}
$$

which is evaluated introducing the spectral densities $\rho_{L,T}$ and $\widetilde{\rho}$ for the gluon and fermion propagators, respectively. The imaginary part of S_i (with $i = L, T$) can thus be written as

$$
\text{Im } S_i = \pi\left(e^{(p_0-\mu_q)/T}+1\right)\int_{-\infty}^{\infty}\frac{dk_0}{2\pi}\int_{-\infty}^{\infty}\frac{dp_0'}{2\pi}f(k_0)\widetilde{f}(p_0'-\mu)\delta(p_0-k_0-p_0')\,\rho_i(k_0)\,\widetilde{\rho}(p_0'), \tag{10}
$$

Journal of Physics: Conference Series **1602** (2020) 012025 doi:10.1088/1742-6596/1602/1/012025

where $f(k_0)$ is the Bose-Einstein distribution. The spectral densities $\rho_{L,T}(k_0, k)$ are obtained from the imaginary part of $^*\Delta_{L,T}(i\omega_n, k)$ after the analytic continuation $i\omega_n \to k_0 + i\epsilon$ and contain the discontinuities of the gluon propagator across the real k_0-axis. Their support depends on the ratio $x = k_0/k$. For $|x| > 1$, $\rho_{L,T}$ have support on the (time-like) quasiparticle poles. For $|x| < 1$ their support coincides with the branch cut of $Q_0(x)$. On the other hand, the spectral density corresponding to a bare quark is given by

$$\widetilde{\rho}(p_0') = 2\pi\epsilon(p_0')\delta(p_0'^2 - E_p^2), \tag{11}$$

where $E_p^2 = (p-k)^2 + m_q^2$ with m_q the quark mass. The kinematical restriction that Eq. (11) imposes on Eq. (10) limits the integration over gluon energies to the space-like region, namely, $|x| < 1$. Therefore, the parts of the gluon spectral densities that contribute to the interaction rate are given by

$$\begin{aligned}
\rho_L(k_0, k) &= \frac{x}{1-x^2}\frac{2\pi m^2\theta(k^2-k_0^2)}{\left[k^2 + 2m^2\left(1 - \frac{x}{2}\ln\left|\frac{1+x}{1-x}\right|\right)\right]^2 + \left[\pi m^2 x\right]^2},\\
\rho_T(k_0, k) &= \frac{\pi m^2 x(1-x^2)\theta(k^2-k_0^2)}{\left\{k^2(1-x^2) + m^2\left[x^2 + \frac{x}{2}(1-x^2)\ln\left|\frac{1+x}{1-x}\right|\right]\right\}^2 + \left[\frac{\pi m^2 x(1-x^2)}{2}\right]^2}.
\end{aligned} \tag{12}$$

Collecting all the ingredients, the interaction rate for a massive quark with energy p_0 to align its spin with the thermal vorticity is given by

$$\begin{aligned}
\Gamma(p_0) &= \frac{\alpha_s}{4\pi}\left(\frac{\omega}{T}\right)^2 \frac{C_F}{\sqrt{p_0^2 - m_q^2}}\int_0^\infty dk\, k\\
&\times \int_{\mathcal{R}} dk_0\, \left[1 + f(k_0)\right] \tilde{f}(p_0 + k_0 - \mu_q) \sum_{i=L,T} C_i(p_0, k_0, k)\rho_i(k_0, k),
\end{aligned} \tag{13}$$

where $\mathcal{R}$ represents the region

$$\sqrt{\left(\sqrt{p_0^2 - m_q^2} - k\right)^2 + m_q^2} - p_0 \le k_0 \le \sqrt{\left(\sqrt{p_0^2 - m_q^2} + k\right)^2 + m_q^2} - p_0. \tag{14}$$

The polarization coefficients $C_{L,T}$ come from the contraction of the polarization tensors $P_{L,T\,\mu\nu}$ with the trace of the factors involving Dirac gamma matrices from the self-energy. After implementing the kinematical restrictions for the allowed values of the angle between the quark and gluon momenta, these functions are found to be

$$C_T(p_0, k_0, k) = 8(p_0 + k_0)\left(\frac{k^2 - 2k_0 p_0 - k_0^2}{2k\sqrt{p_0^2 - m_q^2}}\right)^2, \tag{15}$$

$$C_L(p_0, k_0, k) = -8(p_0 + k_0)\left[\left(\frac{k^2 - 2k_0 p_0 - k_0^2}{2k\sqrt{p_0^2 - m_q^2}}\right)^2 - \frac{1}{2}\right] - 8\frac{p_0\, k^2}{k_0^2 - k^2}\left(\frac{k^2 - 2k_0 p_0 - k_0^2}{2k\sqrt{p_0^2 - m_q^2}}\right)^2.$$

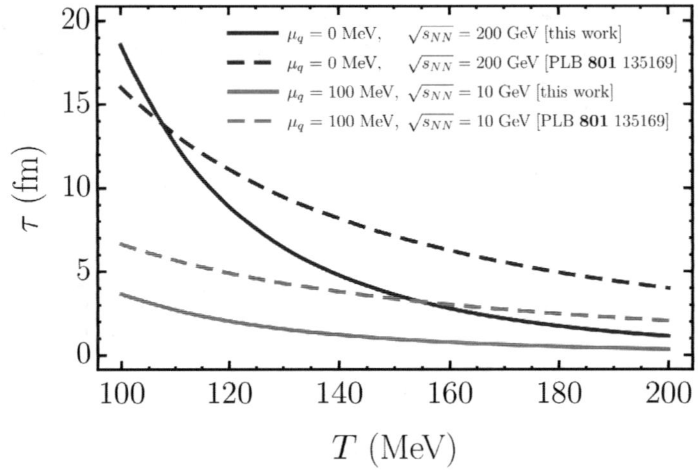

Figure 2. Relaxation time τ for quarks as a function of temperature T for semicentral collisions at an impact parameter $b = 10$ fm. In dashed lines, massless quarks [6] for $\sqrt{s_{NN}} = 10, 200$ GeV with $\omega \simeq 0.12, 0.10$ fm^{-1}, respectively. In solid lines, massive quarks for $\sqrt{s_{NN}} = 10, 200$ GeV with $\omega \simeq 0.072, 0.051$ fm^{-1}, respectively, using the findings of Ref. [7].

This result should be contrasted with Eqs. (14) of Ref. [6], *i.e.*,

$$C_T(p_0, k_0, k) = 8k_0 \left(\frac{k^2 - 2k_0p_0 - k_0^2}{2kp_0}\right)^2,$$
$$C_L(p_0, k_0, k) = -8k_0 \left[\left(\frac{k^2 - 2k_0p_0 - k_0^2}{2kp_0}\right)^2 - \frac{1}{2}\right], \tag{16}$$

which were computed for the $m_q \to 0$ and small quark momentum limit. The total interaction rate is obtained by integrating Eq. (13) over the available phase space

$$\Gamma = V \int \frac{d^3p}{(2\pi)^3}\Gamma(p_0), \tag{17}$$

where V is the volume of the overlap region in the collision. Notice that, although the available phase space for a massive quark is reduced as compared to the massless quark case, the contribution of the new terms in Eqs. (15) enhance the overall interaction rate. Recall that for the collision of symmetric systems of nuclei with radii R and a given impact parameter b, V is given by

$$V = \frac{\pi}{3}(4R + b)(R - b/2)^2. \tag{18}$$

From the expression for Γ in Eq. (17), we study the parametric dependence of the relaxation time for spin and vorticity alignment, defined as

$$\tau \equiv 1/\Gamma. \tag{19}$$

Journal of Physics: Conference Series **1602** (2020) 012025 doi:10.1088/1742-6596/1602/1/012025

3. Results

To estimate the angular velocity ω produced in semicentral collisions, we follow the findings of Ref. [7] that provide these values for given energies and impact parameters. Although ω evolves with time, we work out the computation using its initial value at full nuclei overlap. For an impact parameter $b = 10$ fm, the angular velocity was found to be $\omega \simeq 0.06, 0.04$ fm^{-1} for collision energies $\sqrt{s_{NN}} = 10, 200$ GeV, respectively.

The relaxation times for quarks of mass $m_q = 100$ MeV and quark chemical potential μ_q as a function of temperature are shown in Fig. 2 for two different energies. Notice that $\tau \lesssim 5$ fm for the temperature range 150 MeV $< T <$ 200 MeV, where the phase transition is expected. In this temperature range, the relaxation times are smaller than the ones found in Ref. [6] for most of the energy range considered. Notice that a finite quark mass produces a smaller relaxation time compared to the findings of Ref. [6].

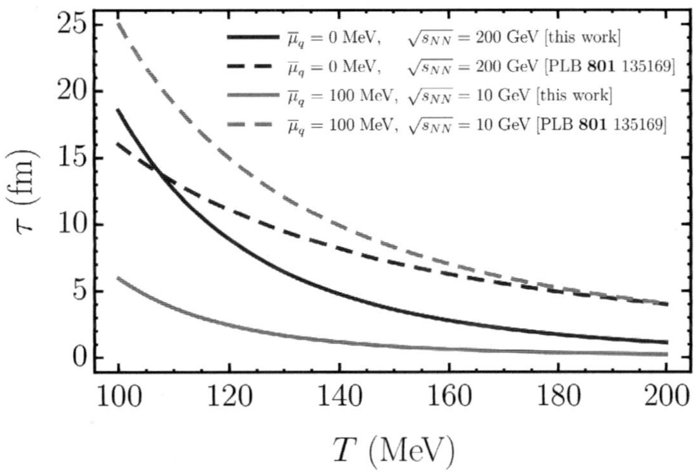

Figure 3. Relaxation time $\bar{\tau}$ for antiquarks as a function of temperature T for semicentral collisions at an impact parameter $b = 10$ fm. In dashed lines, massless quarks [6] for $\sqrt{s_{NN}} = 10, 200$ GeV with $\omega \simeq 0.12, 0.10$ fm^{-1}, respectively. In solid lines, massive quarks for $\sqrt{s_{NN}} = 10, 200$ GeV with $\omega \simeq 0.072, 0.051$ fm^{-1}, respectively, using the findings of Ref. [7].

For antiquarks with chemical potential $\overline{\mu}_q = -\mu_q$, the resulting relaxation times for $\sqrt{s_{NN}} = 10, 200$ GeV and an impact parameter of $b = 10$ fm are shown in Fig. 3. In order to obtain the quark/antiquark relaxation times as a function of collision energy, we use the freeze-out parametrization of Ref. [8]

$$\begin{aligned} T(\mu_B) &= 166 - 139\mu_B^2 - 53\mu_B^4, \\ \mu_B(\sqrt{s_{NN}}) &= \frac{1308}{1000 + 0.273\sqrt{s_{NN}}}, \end{aligned} \tag{20}$$

where the freeze-out baryon chemical potential μ_B and temperature T are given in MeV. These relaxation times are shown in Fig. 4 Notice that the relaxation times for quarks show a monotonic growth as a function of the collision energy. In contrast, the corresponding relaxation times for antiquarks have a minimum for collision energies in the range 40 GeV $\lesssim \sqrt{s_{NN}} \lesssim$ 70 GeV.

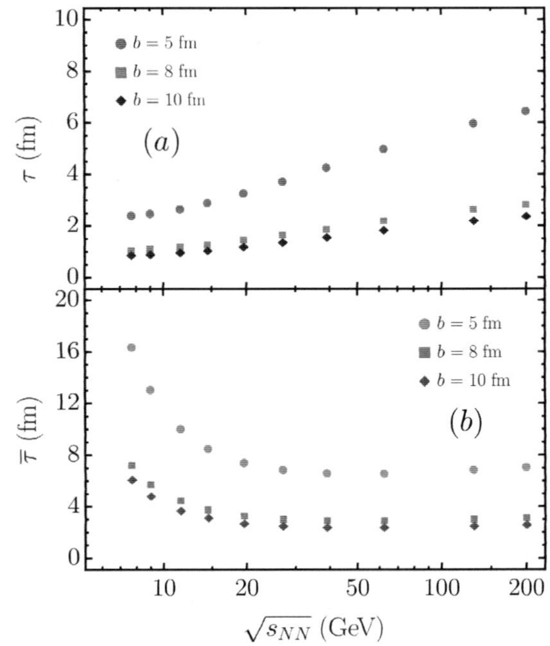

Figure 4. (a) Relaxation time τ for quarks as a function of $\sqrt{s_{NN}}$ for semicentral collisions at impact parameters $b = 5, 8, 10$ fm. (b) Relaxation time $\bar{\tau}$ for antiquarks as a function of $\sqrt{s_{NN}}$ for semicentral collisions at impact parameters $b = 5, 8, 10$ fm.

Finally, we compute the fraction of globally polarized particles as a function of time, which is given as

$$
\begin{aligned}
z &\equiv \frac{N}{N_0} = 1 - e^{-t/\tau}, \\
\bar{z} &\equiv \frac{\bar{N}}{\bar{N}_0} = 1 - e^{-t/\bar{\tau}},
\end{aligned} \tag{21}
$$

where z and $\bar{z}$ are properly referred to as the *intrinsic* global polarization for quarks and antiquarks, respectively. Notice that $\bar{z} < z$ when the impact parameter and the collision energy are the same for both quarks and antiquarks, as can be seen from Fig. 5. Both intrinsic polarizations tend to 1 for $t \simeq 10$ fm. However, a finite intrinsic global polarization for quarks and antiquarks can still be expected when the QGP phase lasts for less than 10 fm.

In conclusion, we have performed a microscopic study to estimate relaxation times for the alignment between strange quark spin and thermal vorticity at finite temperature

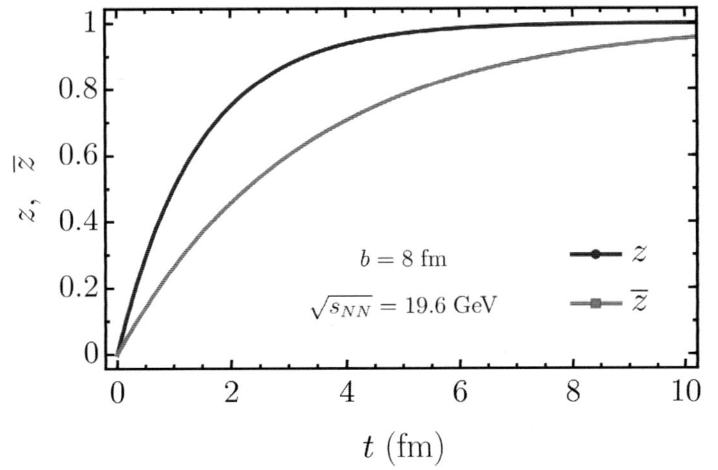

Figure 5. Intrinsic global polarization for quarks (z) and antiquarks ($\bar{z}$) as functions of time t for semicentral collisions at an impact parameter $b = 8$ fm for $\sqrt{s_{NN}} = 19.6$ GeV.

and baryon chemical potential, considering the effects of the quark mass in the alignment. When T and μ_B are increased and the initial angular velocity is fixed, relaxation times lie well within the expected life-time of the system, although this does not change the fact that relaxation times for antiquarks are larger than the corresponding relaxation times for quarks. It is interesting to note that during hadronization, further effects can enhance $\overline{\Lambda}$ polarization so as to obtain $\mathcal{P}_{\overline{\Lambda}} > \mathcal{P}_{\Lambda}$. For example, as discussed in this same proceedings [9] and in more detail in Ref. [10], this can happen when the reaction zone is modelled as composed of a high-density core and a less dense corona. The idea was put forward some time ago in Ref. [11]. Although both regions are subject to the vortical motion, Λs and $\overline{\Lambda}$s coming from one or the other regions could show different polarization properties. This can happen since their origins are different: in the core these hyperons come mainly from QGP induced processes. In the corona they come from nucleon-nucleon interaction processes. When this is considered together with a larger abundance of Λs as compared to $\overline{\Lambda}$s in the corona and a smaller number of Λs coming from the core as compared to those coming from the corona, an amplification effect for the $\overline{\Lambda}$ polarization can occur. This is more prominent for semi-central to peripheral collisions and small collision energies. Further details are provided in Refs. [9, 10].

Acknowledgments

Support for this work has been received in part by UNAM-DGAPA-PAPIIT grant number IG100219 and by Consejo Nacional de Ciencia y Tecnología grant numbers A1-S-7655 and A1-S-16215.

References

[1] Adamczyk L *et al.* (STAR Collaboration) 2017 *Nature* **548** 62.

Journal of Physics: Conference Series **1602** (2020) 012025 doi:10.1088/1742-6596/1602/1/012025

[2] Liang Z-T and Wang X-N 2005 *Phys. Lett.* B **629** 20.
[3] Csernai L P, Magas V K and Wang D J 2013 *Phys. Rev.* C **87** 034906.
[4] Becattini F, Chandra V, Del Zanna L and Grossi E 2013 *Ann. Phys.* **338** 32.
[5] Liu J, Shen C and Heinz U 2015 *Phys. Rev.* C **91** 064906.
[6] Ayala A, de la Cruz D, Hernández-Ortíz S, Hernández L A and Salinas J 2020 *Phys. Lett.* B **801** 135169.
[7] Deng X-G, Huang X-G, Ma Y-G and Zhang S 2020 Vorticity in low-energy heavy-ion collisions *Preprint* nucl-th/2001.01371.
[8] Cleymans J, Oeschler H, Redlich K and Wheaton S 2006 *Phys. Rev.* C **73** 034905.
[9] Maldonado I 2020 Two-component source to explain Λ and $\overline{\Lambda}$ global polarization in non-central heavy-ion collisions, 36th Winter Workshop on Nuclear Dynamics, Puerto Vallarta, Mexico.
[10] Ayala A *et al.* 2020 Core meets corona: a two-component source to explain Λ and $\bar{\Lambda}$ global polarization in semi-central heavy-ion collisions *Preprint* hep-ph/2003.13757.
[11] Ayala A, Cuautle E, Herrera G and Montaño L M 2002 *Phys. Rev.* C **85** 024902.

Journal of Physics: Conference Series **1602** (2020) 012026 doi:10.1088/1742-6596/1602/1/012026

Calculation of E_T from single particle spectra

Christine Nattrass, Biswas Sharma, Soren Sorensen, Ben Smith, Tanner Mengel, Charles Hughes, Nathan Webb
University of Tennessee, Knoxville, TN, USA-37996.

Abstract. In high energy heavy ion collisions, measurements of transverse energy (E_T) can constrain the initial energy density and therefore provide insight into the formation of the Quark Gluon Plasma (QGP). This is particularly interesting in regions where it is unclear if the QGP is formed. The E_T in a collision can be measured either from a calorimeter or calculated from the single particle momentum spectra. We use spectra measured by the STAR collaboration to calculate the transverse energy in heavy ion collisions in Au+Au collisions at $\sqrt{s_{\mathrm{NN}}} = 7.7$ –200 GeV. These calculations are compared to PHENIX measurements using a calorimeter.

1. Introduction

The mean transverse energy per unit pseudorapidity $\langle \mathrm{d}E_{\mathrm{T}}/\mathrm{d}\eta \rangle$ conveys information about how much of the initial longitudinal energy carried by the incoming nuclei is converted into energy carried transverse to the beam axis. The $\langle \mathrm{d}E_{\mathrm{T}}/\mathrm{d}\eta \rangle$ can provide information on the energy densities using simple geometric assumptions [1]. This is particularly important for the RHIC Beam Energy Scan (BES), as it can provide insight into when conditions are conducive to the formation of the Quark Gluon Plasma.

Only the PHENIX experiment has measured $\langle \mathrm{d}E_{\mathrm{T}}/\mathrm{d}\eta \rangle$ at all RHIC energies [2–5] using a calorimeter. The single particle spectra, $d^2N/dydp_T$, carry equivalent information, enabling a cross check between experiments. We use a method similar to the calculation of E_T from spectra in [6] to calculate E_T from previously published STAR spectra [7–13]. We define E_T to be what would be measured in a calorimeter

$$E_T = \sum_{i=1}^{N} E_i^{cal} \sin(\theta_i) \tag{1}$$

where the sum over i is over all N particles and

$$E_i^{cal} = \begin{cases} E - mc^2 & \text{for baryons} \\ E + mc^2 & \text{for anti-baryons} \\ E & \text{for all other particles} \end{cases} . \tag{2}$$

where m is the particle's mass, c is the speed of light, and E is the relativistic energy,

$$E = \sqrt{(pc)^2 + (mc^2)^2}. \tag{3}$$

These calculations provide a consistency check between the experiments but will also allow the distribution of energy among different particles to be studied.

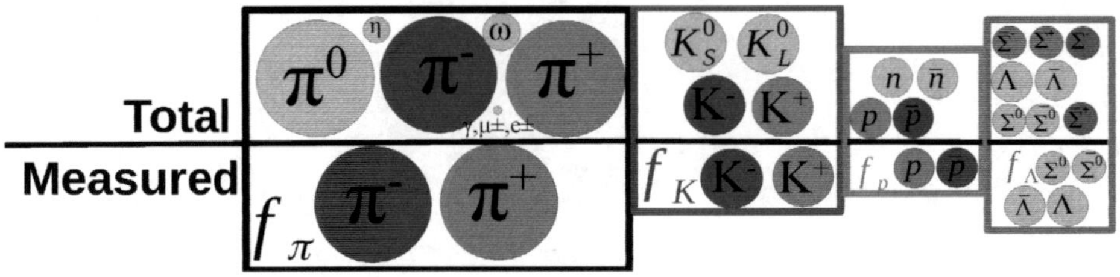

Figure 1. Schematic diagram showing which particles are included in the correction factors. The area of the circles is approximately proportional to the E_T carried by each particle.

2. Method

Measurements of the single particle double differential spectra, $d^2N/dydp_T$, are available at RHIC energies for charged pions, kaons, and protons and for lambda baryons. The publications with these measurements are summarized in table 1. The E_T can be calculated from these spectra by first calculating the average E_T per event carried by each particle type and summing over the different particle types

$$E_T = \sum_i^M E_T^i \tag{4}$$

where the sum over i is over each M different type of particle. Since the spectra are not generally available for all particle types, the E_T carried by measured particles has to be scaled up to get the total E_T using

$$E_T = f_\pi(E_T^{\pi^+} + E_T^{\pi^-}) + f_K(E_T^{K^+} + E_T^{K^-}) + f_p(E_T^{p} + E_T^{\bar{p}}) + f_\Lambda(E_T^{\Lambda} + E_T^{\bar{\Lambda}}) \tag{5}$$

where $f_{\pi^\pm}$, $f_{K^\pm}$, $f_{p,\bar{p}}$, and $f_{\Lambda,\bar{\Lambda}}$ are constants to scale up the measured E_T. Each of the scale factors used in this preliminary analysis and which particles are included are listed in table 2. This is illustrated schematically in figure 1 with the particles included in each correction indicated on the figure. Each of the correction factors are discussed in greater detail below.

Table 1. The collision energy, collaboration, particle species, reference, and maximum rapidity for midrapidity measurements of Λ, $\pi^\pm$, $K^\pm$, p and $\bar{p}$ in Au+Au.

$\sqrt{s_{NN}}$ (GeV)	Particle	Ref.	y_{max}
7.7, 11.5, 19.6, 27, 39	$\pi^\pm$, $K^\pm$, p, $\bar{p}$	[7]	< 0.1
	Λ, $\bar{\Lambda}$	[8]	< 0.5
62.4	$\pi^\pm$, $K^\pm$, p, $\bar{p}$	[9]	< 0.5
	Λ, $\bar{\Lambda}$	[10]	< 1.0
130	$\pi^\pm$, $K^\pm$, p, $\bar{p}$	[9]	< 0.5
	Λ, $\bar{\Lambda}$	[11]	< 0.5
200	$\pi^\pm$, $K^\pm$, p, $\bar{p}$	[9, 12]	< 0.5
	Λ, $\bar{\Lambda}$	[13]	< 1.0

2.1. Fits to the data

The E_T carried by each particle, $dE_T^i/d\eta$, can be calculated from the double differential spectra $d^2N^i/dydp_T$ using

$$dE_T^i/d\eta = \int_0^\infty \frac{dy}{d\eta}\frac{d^2N^i}{dydp_T}E_T^i(p_T)dp_T \tag{6}$$

Table 2. Correction factors

Correction	Particles included
$f_\pi = 1.56 \pm 0.02$	$\pi^\pm, \pi^0, \eta, \omega, e^\pm, \mu^\pm, \gamma$
$f_K = 1.80 \pm 0.20$	$K^\pm, K^0_S, K^0_L$
$f_p = 2 - 3$, see table 4	$p, \bar{p}, n, \bar{n}$
$f_{\Lambda,\bar{\Lambda}} = 1.08 \pm 0.51$	$\Lambda, \bar{\Lambda}, \Sigma^\pm, \bar{\Sigma^\pm}, \Sigma^0, \bar{\Sigma^0}$

where $\frac{d^2N}{dp_T dy}$ is the measured spectrum of particle i, $dy/d\eta$ is the Jacobian for the transformation from y to η, and $E_T^i(p_T)$ is the transverse energy for a particle of type i with mass m and transverse momentum p_T. At midrapidity, $\eta \approx 0$ so $\frac{\partial y}{\partial \eta} \approx \frac{p_T}{m_T}$ and E_T is as defined in (2). The momentum in each bin is estimated to be $p_T \approx p_T^{bc} - \frac{1}{3}w$ where p_T^{bc} is the center of the bin and $w = 50$ MeV$/c$ is the width of the bin. The data cover a finite range in momentum so a fit is used to extrapolate to both $pT = 0$ and $p_T = \infty$.

The published spectra are fit to both a Boltzmann-Gibbs Blast Wave distribution [14,15] and a variation of a modified Hagedorn [16] distribution. The Blast Wave is given by

$$\frac{d^2N}{dp_T dy} = N p_T \int_0^1 r' dr' \left(\sqrt{m^2 + p_T^2}\right) \cdot I_0\left(\frac{p_T \sinh\left(\tanh^{-1}\left(\beta_s r'^n\right)\right)}{T_{kin.}}\right) \cdot K_1\left(\frac{\sqrt{m^2 + p_T^2}\cosh\left(\tanh^{-1}\left(\beta_s r'^n\right)\right)}{T_{kin.}}\right) \tag{7}$$

where p_T is the transverse momentum, y is the rapidity, N is the normalization, m is the mass of the particle, β_s is the surface velocity, n is an exponent describing the evolution of the velocity profile, and $T_{kin.}$ is the kinetic freeze out temperature. The I_0 and K_1 are modified Bessel functions. The reduced radius, r', is integrated over from 0 to 1. The variation on a modified Hagedorn distribution is given by

$$\frac{d^2N}{dp_T dy} = \frac{A}{\sqrt{m^2 + p_T^2}}\left(1 + \frac{\sqrt{m^2 + p_T^2}}{nT}\right)^{-n} \tag{8}$$

where A, n, and T are parameters and m is the mass. The difference between these two functional forms is used as a systematic uncertainty on these extrapolations.

The Λ and $\bar{\Lambda}$ spectra are available for fewer centrality bins than $\pi^\pm$, K$^\pm$, p and $\bar{p}$. In order to allow measurements at more centralities, we interpolate $\langle dE_T/d\eta\rangle/\langle N_{part}/2\rangle$ between centrality bins by using a polynomial between neighboring centralities.

2.2. *Pion scaling factor f_p*

There are slight deviations between the yields of different pion species, dominantly due to contributions from the decays of η and ω mesons, both of which are short-lived and whose daughters are operationally primary particles. Decays from other short-lived resonances also contribute. The η and ω decays which have a branching ratio of more than 1% and their branching ratios are given in table 3. We estimate the contribution from the η and ω mesons using transverse mass scaling to estimate the shape of the spectra and matching the yields to the

Table 3. Decays of the η and ω mesons and their branching ratios for decays with a branching ratio greater than 1% [17].

decay	BR (%)
$\eta \to \gamma\gamma$	39.4
$\eta \to \pi^0\pi^0\pi^0$	32.7
$\eta \to \pi^+\pi^-\pi^0$	22.9
$\eta \to \pi^+\pi^-\gamma$	4.2
$\omega \to \pi^+\pi^-\pi^0$	89.2
$\omega \to \pi^0\gamma$	8.3
$\omega \to \pi^+\pi^-$	1.5

observed η/π^0 yield ratios [18]. This approach was also taken in [6]. We estimate $f_\pi = 1.56 \pm 0.02$. We anticipate more detailed studies of these contributions and contributions from $e^\pm$, $\mu^\pm$, and γ before publication.

2.3. Kaon scaling factor f_K

At low energies, charged and neutral kaons can have different yields (and hence carry different E_T) due to differences in kaon production and in feeddown from short-lived resonances. We calculate the ratio of $2K^0_S/(K^+ + K^-)$ using preliminary kaon yields for Au+Au collisions at $\sqrt{s_{\rm NN}} = 7.7$, 11.5, 19.6, 27, and 39 GeV. The lowest value for $2K^0_S/(K^+ + K^-)$ within the uncertainties is 0.6, leading to a lower bound on f_K of 1.6. At high energies, this ratio approaches 1.0, which we use as our upper bound leading to $f_K = 1.8 \pm 0.2$. We will update these estimates with the final data [8] before publication.

2.4. Proton scaling factor f_p

For an upper bound on the contribution from neutrons, we consider a simple picture where protons (N_p) and neutrons (N_n) at midrapidity are either generated in the collision or are primordial nucleons from the incoming nuclei. Antiprotons ($N_{\bar{p}}$) and antineutrons ($N_{\bar{n}}$) are assumed to consist entirely of generated nucleons and the number of generated particles is the same for all generated baryons and anti-baryons. In that case, the ratio of the yields can be written as

$$f_p = \frac{N_n + N_{\bar{n}} + N_p + N_{\bar{p}}}{N_p + N_{\bar{p}}} = \frac{1 + \beta + (3 - \beta)N_{\bar{p}}/N_p}{1 + N_{\bar{p}}/N_p} \tag{9}$$

where $\beta = N/Z$ where N is the number of primordial neutrons and Z is the number of primordial protons. We use this to estimate the upper bound of f_p, listed in table 2. At high energies, all (anti)protons and (anti)neutrons are generated in the collision, $\frac{N_{\bar{p}}}{N_p}$ approaches one, and f_p would approach 2.0. We use this combined with measurements of $\frac{N_{\bar{p}}}{N_p}$ to calculate the lower bound of f_p. We anticipate updating (9) to take into account the different definition of E_T for baryons and anti-baryons.

2.5. Lambda scaling factor f_Λ

There are two major corrections to the Λ and $\bar{\Lambda}$ baryons, a feeddown correction to avoid double counting of the energy from proton daughters from the decay $\Lambda \to p\pi^-$ and a scale factor to take into account Σ baryons. The dominant decays for Σ baryons are listed in table 5. The Λ and the Σ baryons have approximately the same mass and therefore are expected to carry the same amount of energy, but the Σ baryons' E_T is divided among three different isospin partners. The

$\sqrt{s_{NN}}$	$\frac{N_{\bar{p}}}{N_p}$	Ref.	f_p
7.7	$0.0073 \pm 0.0002 \pm 0.0006$	[7]	2.49 ± 0.49
11.5	$0.0331 \pm 0.0002 \pm 0.0028$	[7]	2.46 ± 0.46
14.5	$0.0641 \pm 0.0005 \pm 0.0109$	[19]	2.43 ± 0.43
19.6	$0.1216 \pm 0.0003 \pm 0.0104$	[7]	2.39 ± 0.39
27	$0.1892 \pm 0.0003 \pm 0.0162$	[7]	2.34 ± 0.34
39	$0.3204 \pm 0.0003 \pm 0.0274$	[7]	2.25 ± 0.25
62.4	0.469 ± 0.026	[9]	2.18 ± 0.18
130	0.708 ± 0.036	[9]	2.08 ± 0.08
200	0.769 ± 0.055	[9]	2.06 ± 0.06

Table 4. Antiproton to proton ratios and the correction factors calculated from them for various energies.

Table 5. Weakly decaying particle decays relevant to the calculation of E_T from strange baryons.

decay (branching ratio)	$c\tau$ (cm)
$\Lambda \to p\pi^-$ (64%) $\Lambda \to n\pi^0$ (36%)	7.9
$\Sigma^0 \to \Lambda\gamma$ (100%)	2×10^{-9}
$\Sigma^+ \to p\pi^0$ (52%) $\Sigma^+ \to n\pi^-$ (48%)	2.4
$\Sigma^- \to n\pi^-$ (100%)	4.4

Σ^0 and its antiparticle decay to a Λ and a γ too rapidly for the Λ, which carries the majority of the energy, to be distinguished from primordial Λ baryons. This means that the E_T of the Σ^0 is included in the E_T from the Λ. To get a lower estimate, we assume that the Σ and Λ baryons carry the same amount of energy, but that it is divided equally among all three Σ baryons, leading to $f_\Lambda = 1.5$. For an upper limit, we use the ratio in PYTHIA and HIJING at $\sqrt{s_{NN}}$ =2.76 [6], $f_\Lambda = 1.67$. This leads to a correction factor to include the Σ baryons of 1.585 ± 0.085. Before the final publication, we will include studies at RHIC energies.

The STAR (anti)proton spectra do not include corrections for feeddown from the (anti)lambda decay $\Lambda \to p\pi^-$. To avoid double counting, we estimate feeddown using the proton spectra with and without feeddown corrections reported in [12] for Au+Au collisions at $\sqrt{s_{NN}} = 200$ GeV. We calculate the fraction of E_T carried by protons from Λ decays as

$$\frac{E_T^{fd}}{E_T^{\Lambda}} = \frac{E_T^{p,all} - E_T^{p,fd}}{E_T^{\Lambda}} \tag{10}$$

where E_T^{Λ} is the E_T carried by Λ baryons, $E_T^{p,all}$ is the E_T calculated from the proton spectra without feeddown corrections, $E_T^{p,fd}$ is the E_T calculated from the proton spectra with feeddown corrections, and E_T^{fd} is the amount of E_T carried by protons from the decay of Λ baryons. We find $0.48 < \frac{E_T^{fd}}{E_T^{\Lambda}} < 1.02$. However, this fraction is also limited by the 64% branching ratio for this decay, so $\frac{E_T^{fd}}{E_T^{\Lambda}} < 0.68$. To be conservative, we use $\frac{E_T^{fd}}{E_T^{\Lambda}} = 0.34 \pm 0.34$ and $f_\Lambda^{fd} = 1 - \frac{E_T^{fd}}{E_T^{\Lambda}} = 0.66 \pm 0.34$. Since the Λ baryons from the Σ^0 would also contribute to the feeddown correction, this factor multiplies the correction to take the Σ^0 into account. Combining

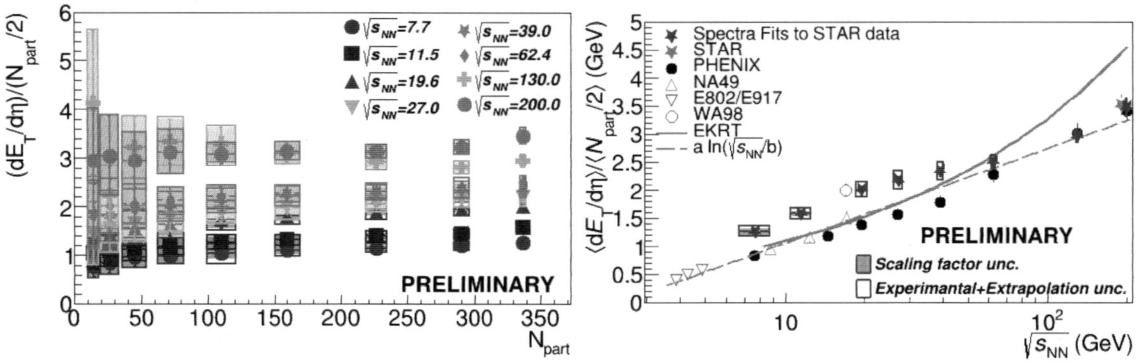

Figure 2. Left: $\langle dE_T/d\eta\rangle/\langle N_{part}/2\rangle$ vs N_{part} for Au+Au collisions at $\sqrt{s_{\rm NN}} = 7.7$, 11.5, 19.6, 27, 39, 62.4, 130 and 200 GeV. Right: Comparison of $\langle {\rm d}E_{\rm T}/{\rm d}\eta\rangle/\langle N_{\rm part}/2\rangle$ at midrapidity versus $\sqrt{s_{\rm NN}}$ in central A+A collisions at other energies [2, 20] at midrapidity. The dashed black line is a $\log(\sqrt{s_{\rm NN}})$ trend which had been seen [2] to fit the RHIC and fixed-target data, while the solid red curve is a prediction from [21, 22].

this with information from the contribution of Σ^+, Σ^- and Σ^0, we get an overall

$$f_{\Lambda,\bar{\Lambda}} = 1.08 \pm 0.51. \quad (11)$$

3. Results

Figure 2 shows $\langle dE_T/d\eta\rangle/\langle N_{part}/2\rangle$ versus N_{part} and $\sqrt{s_{\rm NN}}$. The N_{part} dependence shows similar trends to those observed by PHENIX [2–5], with a similar shape for all collision energies. The $\sqrt{s_{\rm NN}}$ dependence looks similar to that seen in previous measurements, but there is a clear discrepancy between our calculations from STAR spectra and the published PHENIX E_T measurements, particularly in the BES range.

This is unlikely to be due to the correction factors. While the contributions from neutrons and Σ^+, Σ^- and Σ^0 are not constrained by data, the uncertainties on f_p and f_Λ are large and the overall contributions from baryons are small. The contributions from kaons are larger, but measurements of the K^0_S constrain the reasonable range for this correction. The correction to pions is similarly constrained by extensive measurements of the π^0. The overall contribution from uncertainties in the scaling factors is indicated in figure 2 and is much less than the uncertainty on individual points reported by STAR.

There are several possible causes for this discrepancy. The PHENIX E_T measurements use a Jacobian for the transformation from y to η which is calculated in simulations while these calculations use the observed momenta and particle ratios. The PHENIX measurements also use simulations to determine a single scaling factor, which may be sensitive to particle ratios, the shape of the spectra at low p_T where there are few contributions to the raw measurement, and simulation of the PHENIX magnetic field. There may also be a difference in the definition of E_T, as we use what a calorimeter would make. STAR and PHENIX also use different centrality definitions, with PHENIX determining centrality from the multiplicity at forward rapidities well-separated from the midrapidity E_T measurement while STAR determines centrality at midrapidity, in the same region as the measurements. This is not likely to cause a major discrepancy in central collisions, but may lead to some differences in peripheral collisions. There may also be issues with the spectra measurements from STAR. Further cross checks include looking at multiplicities from both measurements and calculating E_T using the defintiion $E_T = m_T$.

Journal of Physics: Conference Series **1602** (2020) 012026 doi:10.1088/1742-6596/1602/1/012026

4. Conclusions

We have calculated E_T from single particle spectra measured by STAR. These calculations indicate a discrepancy between the two experiments. Given the importance of global observables in establishing the formation of the QGP and how central this question is to the Beam Energy Scan program, this warrants further investigation. We will refine our calculations to the extent possible. We also recommend that STAR and PHENIX pursue studies which could provide further insight. PHENIX could measure single particle spectra and STAR could measure E_T independently for BES energies.

References

[1] Bjorken J D 1983 *Phys. Rev.* **D27** 140–151
[2] Adler S *et al.* (PHENIX) 2005 *Phys. Rev.* **C71** 034908 (*Preprint* nucl-ex/0409015)
[3] Adcox K *et al.* (PHENIX) 2001 *Phys. Rev. Lett.* **87** 052301 (*Preprint* nucl-ex/0104015)
[4] Adler S *et al.* (PHENIX) 2014 *Phys. Rev.* **C89** 044905 (*Preprint* 1312.6676)
[5] Adare A *et al.* (PHENIX) 2015 (*Preprint* 1509.06727)
[6] Adam J *et al.* (ALICE) 2016 *Phys. Rev.* **C94** 034903 (*Preprint* 1603.04775)
[7] Adamczyk L *et al.* (STAR) 2017 *Phys. Rev.* **C96** 044904 (*Preprint* 1701.07065)
[8] Adam J *et al.* (STAR) 2019 (*Preprint* 1906.03732)
[9] Abelev B I *et al.* (STAR) 2009 *Phys. Rev.* **C79** 034909 (*Preprint* 0808.2041)
[10] Aggarwal M M *et al.* (STAR) 2011 *Phys. Rev.* **C83** 024901 (*Preprint* 1010.0142)
[11] Adler C *et al.* (STAR) 2002 *Phys. Rev. Lett.* **89** 092301 (*Preprint* nucl-ex/0203016)
[12] Abelev B I *et al.* (STAR) 2006 *Phys. Rev. Lett.* **97** 152301 (*Preprint* nucl-ex/0606003)
[13] Adams J *et al.* (STAR) 2007 *Phys. Rev. Lett.* **98** 062301 (*Preprint* nucl-ex/0606014)
[14] Ristea O, Jipa A, Ristea C, Esanu T, Calin M, Barzu A, Scurtu A and Abu-Quoad I 2013 *J. Phys. Conf. Ser.* **420** 012041
[15] Schnedermann E, Sollfrank J and Heinz U W 1993 *Phys. Rev. C* **48** 2462–2475 (*Preprint* nucl-th/9307020)
[16] Adare A *et al.* (PHENIX) 2010 *Phys. Rev. C* **81** 034911 (*Preprint* 0912.0244)
[17] Beringer J *et al.* (Particle Data Group) 2012 *Phys. Rev.* **D86** 010001
[18] Adare A *et al.* (PHENIX) 2011 *Phys. Rev.* **C84** 044902 (*Preprint* 1105.3467)
[19] Adam J *et al.* (STAR) 2019 (*Preprint* 1908.03585)
[20] Adams J *et al.* (STAR) 2004 *Phys. Rev.* **C70** 054907 (*Preprint* nucl-ex/0407003)
[21] Eskola K, Kajantie K, Ruuskanen P and Tuominen K 2000 *Nucl. Phys.* **B570** 379–389 (*Preprint* hep-ph/9909456)
[22] Renk T, Holopainen H, Paatelainen R and Eskola K J 2011 *Phys. Rev.* **C84** 014906 (*Preprint* 1103.5308)

Journal of Physics: Conference Series **1602** (2020) 012027 doi:10.1088/1742-6596/1602/1/012027

Magnetic effects of QCD parameters from finite energy sum rules

Cristián Villavicencio

Centro de Ciencias Exactas and Departamento de Ciencias Básicas, Universidad del Bío-Bío, Avda. Andrés Bello 720, Casilla 447, 3800708, Chillán, Chile.

E-mail: `cvillavicencio@ubiobio.cl`

C. A. Dominguez

Centre for Theoretical & Mathematical Physics, and Department of Physics, University of Cape Town, Rondebosch 7700, South Africa.

M. Loewe

Instituto de Fisica, Pontificia Universidad Catolica de Chile, Casilla 306, Santiago, Chile; Centro Científico Tecnológico de Valparaíso-CCTVAL, Universidad Técnica Federico Santa María, Casilla 110-V, Valparaíso, Chile; Centre for Theoretical & Mathematical Physics, and Department of Physics, University of Cape Town, Rondebosch 7700, South Africa.

Abstract. One of the advantages of the finite energy sum rules is the fact that every operator in the operator product expansion series can be selected individually by the use of an appropriate kernel function which removes other operator poles. This characteristic is maintained by QCD systems in the presence of external homogeneous magnetic field, providing interesting information about the magnetic evolution of QCD and hadronic parameters. In this work finite energy sum rules are applied on QCD in the light quark sector, combining axial and pseudoscalar channels in the presence of an external homogeneous magnetic field, obtaining the magnetic evolution of the light quark masses, pion mass, the pion decay constant, the gluon condensate and the continuum hadronic threshold.

1. Introduction

The increasing interest in the study of magnetic field effects in quantum chromodynamics (QCD), mainly in the search of the chiral magnetic effect in relativistic heavy ion collision experiments, as well as the interior of magnetars, has produced a variety of interesting and different techniques in order to obtain the modification of particle parameter due to the interaction with the medium. The main difficulty arises from trying to cover all the possible range of the magnetic field strength. A typical approximation is the use of the lowest Landau level, which is the main tool to explore QCD under highly magnetized conditions. However, it commonly exceeds the physical possible values that collider experiments or magnetars can reach. Another approximation is to expand in power series of the magnetic field, with the range of magnetic field values restricted by the scale of the model. In this work, the finite energy sum rules (FESR) will be employed in the presence of an external magnetic field in order to obtain the magnetic evolution for hadronic

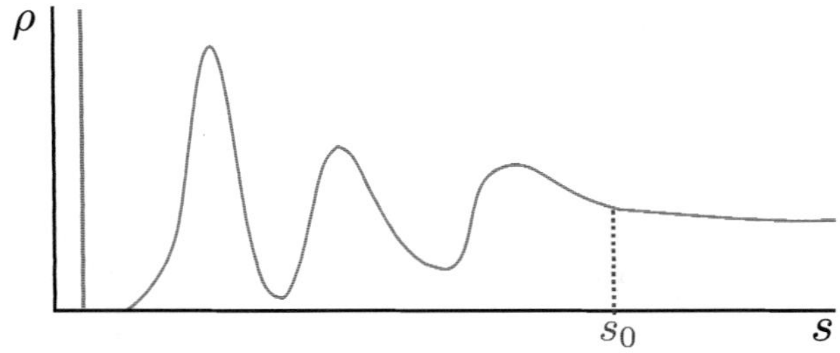

Figure 1. Schematic representation of hadronic resonances. s_0 denotes the limit where heavier and broader resonances start to overlap.

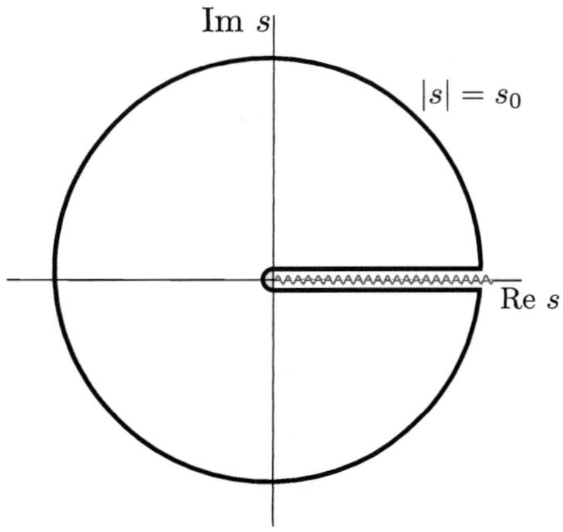

Figure 2. The *pac-man* contour. Hadronic resonances lies over the real axis, while QCD lives on the circle

and QCD parameters. More references and details an be found in [1], which is the article on which this text is mainly based.

The FESR relates the hadronic with the quarks sector based on analytic properties in the complex squared-energy s-plane. Let us consider the correlation of two currents in Fourier space

$$\Pi(p^2) = \int d^4x \, e^{ip\cdot x} \langle 0|TJ(x)J^\dagger(0)|0\rangle. \quad (1)$$

In the hadronic sector, the spectral density function can be obtained as $\rho(s) = \frac{1}{\pi}\mathrm{Im}\,\Pi(s + i\epsilon)$. Figure 1 shows a scheme of the spectral function as a function of the squared energy, where the stable particle state can be observed as a delta function and where a set of resonances appear at high energy. The parameter s_0 denotes the threshold where broader resonances start to overlap. This parameter is known as the *continuum hadronic threshold.*

The only singularities of current correlators lie on the right-hand complex plane. They are in the form of poles on the real s-axis (stable hadrons), or on the second Riemann sheet (hadronic resonances).

Next, Cauchy's theorem is invoked in the s-plane, considering the contour shown on the right side of Fig. 2, leading to finite energy QCD sum rules (FESR).

$$\frac{1}{\pi}\int_0^{s_0} ds \, K(s) \, \mathrm{Im}\,\Pi(s)|_{\mathrm{Had}} = \frac{-1}{2\pi i}\oint_{\mathrm{C}(s_0)} ds \, K(s) \, \Pi(s)|_{\mathrm{QCD}}, \quad (2)$$

where $K(s)$ is an analytic kernel. We use here $K(s) = s^{N-1}$, with N an integer number, with the advantage that it cuts the operator product expansion (OPE) of current correlators in QCD. This operator series parameterize the non perturbative sector in terms of vacuum expectation values of different operators of dimension $2n$,

$$\Pi(s)|_{\mathrm{QCD}} = \sum_{n=0} C_{2n}(s, \mu^2) \frac{\langle \hat{\mathcal{O}}_{2n}(\mu^2)\rangle}{(-s)^n}, \quad (3)$$

where μ^2 is a renormalization scale and C_{2n} are the Wilson coefficients. The $n = 0$ case corresponds to perturbative QCD which is usually a logarithmic contribution. The lowest dimension operators are $\mathcal{O}_4$, which corresponds the quark condensate (multiplied by the quark mass) and the gluon condensate. From here is easy to see that, if Wilson coefficients are constant, there will be only one specific operator contribution for each kernel used.

While the Wilson coefficients in the OPE, Eq. (3) can be computed in PQCD, the values of the vacuum condensates cannot be obtained analytically from first principles. These condensates can be determined from the QCD sum rules themselves, in terms of some experimental information input, by lattice QCD (LQCD) simulations or information provided from other models.

We consider three current correlators in the presence of an external magnetic field: the light-quark axial-vector current correlator, pseudoscalar correlator and axial-pseudoscalar correlator.

2. Vacuum FESR

The axial-vector current correlator is defined as

$$\Pi^A_{\mu\nu}(p) \;=\; i\int d^4x\, e^{ip\cdot x}\,\langle 0|T[A_\mu(x)A^\dagger_\nu(0)]|0\rangle \;=\; p_\mu p_\nu\,\Pi_0(p^2) + g_{\mu\nu}\,\Pi_1(p^2) \tag{4}$$

where $A_\mu = \bar{d}\gamma_\mu\gamma_5 u$ is the (electrically charged) axial-vector current, and p_μ is the four-momentum carried by the current. In the hadronic sector, the axial current can be expressed in terms of the pion field as $A_\mu = -\sqrt{2}f_\pi\partial_\mu\pi^+$. In this work we consider two sum rules for the form factor Π_0, i.e the FESR considering two kernel values: one sum rule with a kernel $K = 1$ and other with kernel $K = s$.[1] Therefore we need the OPE series up to order s^{-2},

$$\Pi_0(s)|_{\rm QCD} = -\,\frac{1}{4\pi^2}\ln(-s/\mu^2) - 2m_q\frac{\langle\bar{q}q\rangle}{s^2} + \frac{1}{12\pi}\frac{\langle\alpha_s\, G^2\rangle}{s^2}, \tag{5}$$

where $m_q \equiv \frac{1}{2}(m_u + m_d)$ is the average light quark mass, $G^2 \equiv G^{a\,\mu\nu}G^a_{\mu\nu}$ and $\langle\bar{q}q\rangle \equiv \frac{1}{2}(\langle\bar{u}u\rangle + \langle\bar{d}d\rangle)$ the average light quark condensate. The hadronic sector gives $\Pi_0(s)|_{\rm Had} = -2f_\pi^2/(s-m_\pi^2)$.

The second current correlator to be considered is $\Pi_{5\nu}$, involving an axial-vector current and its divergence

$$\Pi_{5\nu}(p) \;=\; i\int d^4x\, e^{ipx}\,\langle 0|T\,[i\partial^\mu A_\mu(x)\,A^\dagger_\nu(0)]|0\rangle \;=\; p_\nu\,\Pi_5(p^2). \tag{6}$$

In this case, by the use of the equation of motion in QCD as well as in the hadronic sector, the divergence of the axial-vector current turn to be $\partial^\mu A_\mu = 2m_q\,\bar{d}i\gamma_5 u$ in the QCD sector, and $\partial^\mu A_\mu = \sqrt{2}f_\pi^2 m_\pi^2\pi^+$ in the hadronic sector. As expected, the divergence of the axial-vector current is proportional to the pseudoscalar current. The expressions for $\Pi_5(s)$ in the hadronic sector and in QCD up to order $1/s$ is given by

$$\Pi_5(s)|_{\rm QCD} = -\,\frac{3}{2\pi^2}\,{m_q}^2\,\ln(-s/\mu^2) + 4m_q\frac{\langle\bar{q}q\rangle}{s}, \tag{7}$$

where we can see that the leading terms are proportional to the quark mass. The hadronic sector is given by $\Pi_5(s)|_{\rm Had} = -2f_\pi^2 m_\pi^2/(s - m_\pi^2)$.

The third form factor we will consider is the one formed by the correlation of the axial-vector divergence

$$\psi_5(p^2) = i\int d^4x\, e^{ipx}\langle 0|T\,[\partial^\mu A_\mu(x)\,\partial^\nu A^\dagger_\nu(0)]|0\rangle\,. \tag{8}$$

Its QCD expression to order $1/s$ is given by

$$\psi_5(s)|_{\rm QCD} = -\frac{3}{2\pi^2}\,{m_q}^2\,s\ln(-s/\mu^2) + 4m_q^3\frac{\langle\bar{q}q\rangle}{s} - \frac{{m_q}^2}{2\pi}\frac{\langle\alpha_s\, G^2\rangle}{s}, \tag{9}$$

[1] alternatively we can use one sum rule for Π_0 and one for Π_1 with the same kernel $K = 1$.

and in the hadronic sector, $\psi_5(s)|_{\rm QCD} = -f_\pi^2 m_\pi^4/(s - m_\pi^2)$.

The three correlators are related through the Ward identities $p^\mu \Pi^A_{\mu\nu}(p) = \Pi_{5\nu}(p) + \langle \bar{u}\gamma_\nu u\rangle - \langle \bar{d}\gamma_\nu d\rangle$ and $p^\nu \Pi_{5\nu}(p) = \psi_5(p^2) + 4m_q\langle\bar{q}q\rangle$. The selected sum rules avoid the possibility of repeated information by choosing two sum rules for Π_0 with kernels $K = 1$ and $K = s$, and one sum rule for Π_5 and Ψ_5 with a kernel $K = 1$. Notice that in this pseudoscalar channel we only consider the lowest state, which is the pion. If we do not include the next state, the a_1 resonance, the sum rules give a threshold s_0 smaller so the a_1 pole is located outside the contour of Fig. 2.

Applying the sum rules from Eq. (2) we obtain

$$2f_\pi^2 = \frac{1}{4\pi^2} s_0, \tag{10}$$

$$2f_\pi^2\, m_\pi^2 = \frac{1}{8\pi^2}\, s_0^2 - 2m_q\,\langle\bar{q}q\rangle - \frac{1}{12\pi}\langle\alpha_s G^2\rangle, \tag{11}$$

$$2f_\pi^2 m_\pi^2 = -4m_q\langle\bar{q}q\rangle + \frac{3}{2\pi^2} m_q^2\, s_0, \tag{12}$$

$$2f_\pi^2 m_\pi^4 = \frac{3}{4\pi^2} m_q^2\, s_0^2 - 4m_q^2\,\langle\bar{q}q\rangle + \frac{1}{2\pi} m_q^2\,\langle\alpha_s G^2\rangle, \tag{13}$$

where higher order quark-mass corrections $({m_q}^2/s_0)$ were neglected. Notice that Eq. (12) is the Gell-Man–Oakes–Renner (GMOR) relation [2, 3], including a higher order quark-mass correction, i.e. $\mathcal{O}(m_q^2)$. We use as an input the charged pion mass and the pion decay constant, in order to obtain, as a result, all the other parameters.

3. FESR in an external magnetic field

The presence of an external magnetic field modifies current correlators in several ways. First of all, the magnetic field interacts with quarks and hadrons through the minimal coupling. We must replace then the derivatives with the corresponding covariant derivative. Since the axial-vector current carries positive electric charge e (the elementary proton charge), its derivative is replaced by the covariant derivative $D^\mu A_\mu = [\partial^\mu - ie\mathcal{A}^\mu]A_\mu$, where $\mathcal{A}$ denotes the vector potential of the external magnetic field. Hence, the new definition of the correlators in Eq. (6) and (8) changes by the replacement of D_μ instead of ∂_μ.

Notice that the presence of the magnetic field breaks locality, and therefore $\Pi(x, y) \neq \Pi(x-y)$. However, we are free to choose any particular frame if it is the same for the hadronic and the QCD sector. The Ward identities discussed in the previous section are preserved, and their expressions are the same but contracting the $\Pi^A_{\mu\nu}(p)$ and $\Pi_{5\mu}(p)$ correlators with $p^\mu + \mathcal{A}^\mu(p)$ instead of p^μ. We consider here a constant homogeneous electromagnetic external in the Schwinger-Fock symmetric gauge $\mathcal{A}^\mu(x) = -\frac{1}{2}F^{\mu\nu}x_\nu$, and therefore the electromagnetic vector potential in momentum space is then $\mathcal{A}^\mu(p) = \frac{ie}{2}F^{\mu\nu}\frac{\partial}{\partial p^\nu}$.

Now, if we consider an homogeneous external magnetic field along the z axis, the electromagnetic field tensor can then be written in the convenient form $F_{\mu\nu} = B\epsilon^\perp_{\mu\nu}$, with the perpendicular anti-symmetric tensor defined as $\epsilon^\perp_{\mu\nu} = g_{\mu 1}g_{\nu 2} - g_{\mu 2}g_{\nu 1}$. This term will appear in all tensor structures and lead to the separation of vectors into parallel and perpendicular projections. The metric is splitted into $g_{\mu\nu} = g^{\parallel}_{\mu\nu} + g^\perp_{\mu\nu}$. With this definition we have, for example $p^2 = p_\parallel^2 + p_\perp^2$, with $p_\perp^2 = -\boldsymbol{p}_\perp^2$. The magnetic field therefore introduces several modifications in the tensor structure of the current correlators. Basically it consists of combinations of p_μ $g_{\mu\nu}$ and $\epsilon^\perp_{\mu\nu}$ which produces a rich variety of new independent components, usually associated with new condensates. For instance, for $\Pi^A_{\mu\nu}(p)$ the possible structures are $g^\parallel_{\mu\nu}$, $\epsilon^\perp_{\mu\nu}$, $g^\perp_{\mu\nu}$, and pair combinations of $p^\parallel_\mu$, $p^\perp_\mu$ and $\epsilon^\perp_{\mu\nu}p^\nu$. Similarly, the possible structures for $\Pi_{5\nu}(p)$ are the three mentioned vectors.

In the hadronic sector we use the axial-vector current field description from chiral perturbation theory (χPT) in terms of charged pion fields $A_\mu = -\sqrt{2}(f_\pi^{\parallel} D_\mu^{\parallel} + f_\pi^{\perp} D_\mu^{\perp})\pi^+$, where the parallel and perpendicular components of the pion decay constant are related with the transverse velocity as $v_\perp^2 = f_\pi^\perp / f_\pi^\parallel$. In this case the covariant divergence of the axial-vector current is $D^\mu A_\mu = \sqrt{2} f_\pi^\parallel m_\pi^2 \pi^+$. This relation is obtained from the new equations of motion for the charged pion, $(D_\parallel^2 + v_\perp^2 D_\perp^2 + m_\pi^2)\pi^+ = 0$.

We will consider the frame $p_\perp = 0$, and the specific structures considered are the ones proportional to $p_\mu^\parallel p_\nu^\parallel$ from the axial-axial correlator, and $p_\mu^\parallel$ from the pseudoscalar-axial correlator.

The presence of a magnetic field is to be reflected in the charged particle propagators. These will be expressed in a power series involving the magnetic field [4]. The quark and the pion green functions become $G_q(x,y) = e^{ie_q\phi_q(x,y)} S(x-y)$ and $G_\pi(x,y) = e^{ie_\pi \phi(x,y)} D(x-y)$, respectively, where $e_i\phi$ is the Schwinger phase with e_i the corresponding particle-charge and $\phi(x,y) = -\frac{1}{2} F_{\mu\nu} x^\mu y^\nu$ in the symmetric gauge.

The local part of the Green functions in Fourier space are then expanded in powers of B. The only terms needed for magnetic corrections in QCD are the following:

$$S_q^{(0)}(k) = i\frac{\not{k} + m_q}{k^2 - m_q^2} \tag{14}$$

$$S_q^{(1)}(k) = -\gamma_1\gamma_2(e_qB)\frac{(\not{k}_\parallel + m_q)}{(k^2 - m_q^2)^2} \tag{15}$$

$$S_q^{(2)}(k) = \frac{2i\,(e_qB)^2}{(k^2 - m_q^2)^4}\left[k_\perp^2(\not{k} + m_q) - \not{k}_\perp(k_\parallel^2 - m_q^2)\right], \tag{16}$$

where the superscript (n) refers to the power in the field B^n. In the case of the pion, the only contribution is that of the propagator at zero magnetic field $D_\pi^{(0)}(p) = i/(p^2 - m_\pi^2)$. This is because the next term is $D_\pi^{(1)} = 0$, and the other terms do not survive in the FESR under consideration.

Once all the aforementioned ingredients are considered, we have to calculate the same diagrams needed in the vacuum case, extended with the magnetic Green functions and the covariant derivative. The integrals involved are usually multi-loop diagrams (one-loop in our case) which can be handled by the introduction of Feynman parameters in the usual way. There will be infrared divergences from the magnetic contributions, which are safely controlled by the magnetic quark masses. Hence, it is necessary to keep finite quark masses to leading order in expansions in terms of m_q^2/s_0. It is important to notice that even in the chiral limit there will be magnetic mass generation. Before integrating in the Feynman parameter it is more convenient first to integrate in the external momentum through the FESR contour. The magnetic contribution to the contour integral in the complex squared-energy s-plane is given by

$$\oint_{C(s_0)} \frac{ds}{2\pi i} s^{N-1}\left[\int_0^1 dx \frac{f(x)}{[s - \mathcal{M}^2]^n}\right] = c_{N,n}\theta_{N,n}\int_0^1 dx f(x)(\mathcal{M}^2)^{N-n}\theta(s_0 - \mathcal{M}^2), \tag{17}$$

where f is an arbitrary function of x, and $\mathcal{M}^2 = m_q^2/x(1-x)$ and $c_{N,n}$ is a rational number. The usual Heaviside function is denoted as $\theta(\xi)$, and here we define a discrete theta function $\theta_{N,n} = 1$ for $N \geq n$ and $\theta_{N,n} = 0$ for $N < n$. The fact that this integral vanishes for $N < n$ is one of the most important features in this approach. Higher order contributions in powers of the magnetic field enters only if we consider a FESR kernel with higher powers in s.

Journal of Physics: Conference Series **1602** (2020) 012027 doi:10.1088/1742-6596/1602/1/012027

Some of the magnetic Feynman diagrams involved are infrared divergent for massless quarks because $f(x)$ is singular. To deal with it, notice that the contour integral in Eq. (17) is non vanishing only if $\mathcal{M}^2 < s_0$. This condition leads to the modification of the Feynman parameters integration limits

$$\int_0^1 dx\, f(x)\, (\mathcal{M}^2)^{N-n} \theta\left(s_0 - \mathcal{M}^2\right) = \int_{x_-}^{x_+} dx\, f(x) (\mathcal{M}^2)^{N-n} \tag{18}$$

where the limits $x_\pm = \frac{1}{2}\left(1 \pm \sqrt{1 + 4m_q^2/s_0}\right)$ handle the IR divergences. After the expansion in quark masses, there will appear logarithmic terms. These logarithms are strictly magnetically generated and vanish for $B = 0$.

4. Results

The FESR involving magnetic field corrections are

$$2f_\pi^{\parallel 2} = \frac{s_0}{4\pi^2} \tag{19}$$

$$2f_\pi^{\parallel 2} m_\pi^2 = \frac{1}{8\pi^2}\left\{s_0^2 - \frac{2}{9}(eB)^2\left[10\ln(s_0/{m_q}^2) - 27\right]\right\} + m_q\langle\bar{q}q\rangle - \frac{1}{12\pi}\langle\alpha_s G^2\rangle \tag{20}$$

$$2f_\pi^{\parallel 2} m_\pi^2 = -\,4m_q\langle\bar{q}q\rangle + \frac{3}{2\pi^2} m_q^2 s_0 \tag{21}$$

$$2f_\pi^{\parallel 2} m_\pi^4 = \frac{3m_q^2}{4\pi^2}\left\{s_0^2 - \frac{20}{27}(eB)^2\left[\ln(s_0/{m_q}^2) - 1\right]\right\} - 4m_q^3\langle\bar{q}q\rangle + \frac{m_q^2}{2\pi}\langle\alpha_s G^2\rangle \tag{22}$$

where all the terms are functions of the magnetic field. The restriction ${m_q}^2 \ll s_0$, remains valid for all values of eB under consideration. There are six parameters to be determined, i.e. m_q, m_π, f_π, s_0, $\langle\bar{q}q\rangle$, and $\langle\alpha_s G^2\rangle$. Since there are only four independent FESR, two inputs are required. In the case of vacuum, the pion's mass and decay width are the more precise inputs. The case with finite magnetic field requires different inputs for solving the magnetic evolution of the parameters. The most studied object in this case is the quark condensate, which in our case is obtained from Nambu–Jona-Lasinio (NJL) model results [5], which agree with LQCD [6], so we will use as an input the NJL chiral condensate at finite magnetic field normalized by its value at $B = 0$.

We need another "magnetic" input. During the realization of this work there wasn't much choice in the literature for the range of values of the magnetic field we are taking into account, so we consider three possible scenarios:

(i) The magnetic evolution of the charged pion mass provided by NJL calculations [5].

(ii) m_q/m_π^2 =constant. This scenario is natural from the Nambu-Goldstone realization of chiral $SU(2) \times SU(2)$ symmetry where we have $m_\pi^2 = 2\mathcal{B}m_q$. So here we assume that $\mathcal{B}$ is independent of the magnetic field.

(iii) m_q = constant.

Figure 3 shows the magnetic quark mass evolution and Fig. 4 shows the magnetic evolution of the pion mass. The similitude between the behaviour of both masses from the condition m_q/m_π^2 = constant (dashed line) is expected. The case with the pion mass obtained from NJL model or LQCD seems to be the same situation (continuous line), however both masses evolve differently and do not follow the relation $m_q \propto m_\pi^2$. It is worth mentioning that the GMOR relation is broken here as stronger the magnetic strength. This can be seen from Eq. (21), where the presence of last term turn to be more relevant as the magnetic field increases. In Fig. 4 we also show the result using χPT from [7] to compare with. All the pion mass results increase with the magnetic field, except for the case of constant m_q. This result reinforces the importance of the magnetic field behavior of m_q.

Figure 5 shows the magnetic evolution of the charged pion decay constant (parallel component) considering the three inputs described previously, compared with the result of the charged pion decay constant from χPT [7]. Notice that the two first conditions generates almost the same result for the decay constant. Also notice from Fig. 4 that the pion mass from χPT is very similar from the one obtained in NJL model, however the pion decay constant from χPT differs from the results obtained in this work. The case of constant quark mass presents a faster growing of the magnetic evolution of the decay constant.

Because of Eq. (19), the same plot describes the magnetic evolution of $\sqrt{s_0}$. This plot allows us to explore the validity of the magnetic field strength, because the a_1 resonance must be included if $\sqrt{s_0} \gtrsim 1.2$ GeV, . Considering that $\sqrt{s_0(0)} = 0.82$ GeV we can see that $\sqrt{s_0(B)} \approx 1.2$ GeV for $eB \approx 0.5$ GeV in the case of the results obtained using the conditions $m_\pi(B)$ from NJL, or $m_q/m_\pi^2 =$ constant. In the case of the condition $m_q =$ constant, the maximum magnetic field strength $eB \approx 0.35$ GeV. The increasing hadronic threshold confirms a previous work with FESR in the chiral limit [8].

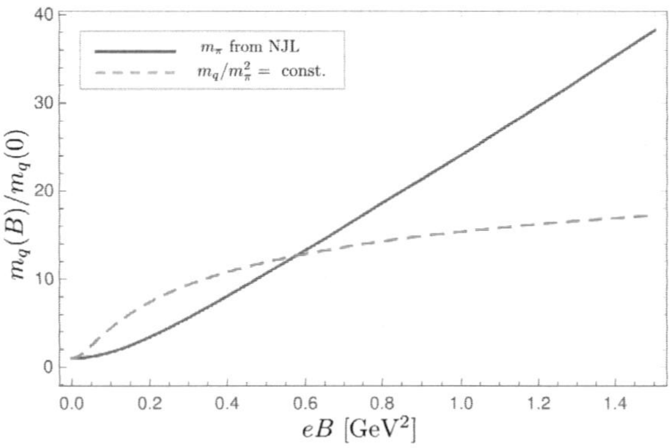

Figure 3. Magnetic evolution of the light quarks average mass. The plots show two possible inputs.

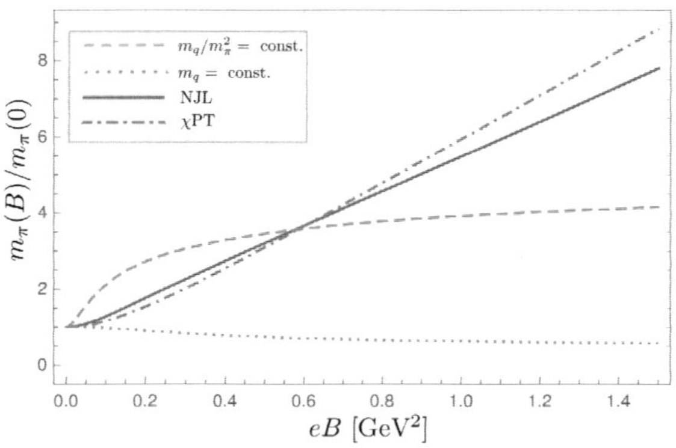

Figure 4. Magnetic evolution of the pion mass. The plot shows two possible inputs and results from NJL and χPT.

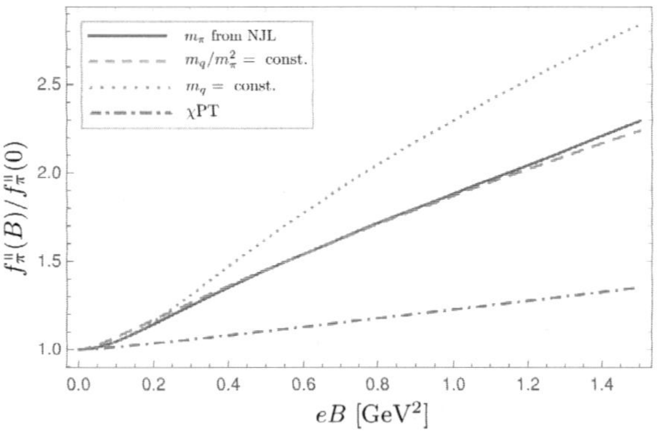

Figure 5. Magnetic evolution of $f_\pi^{\parallel}$ (or equivalently $\sqrt{s_0}$), using three possible inputs and compared with χPT results.

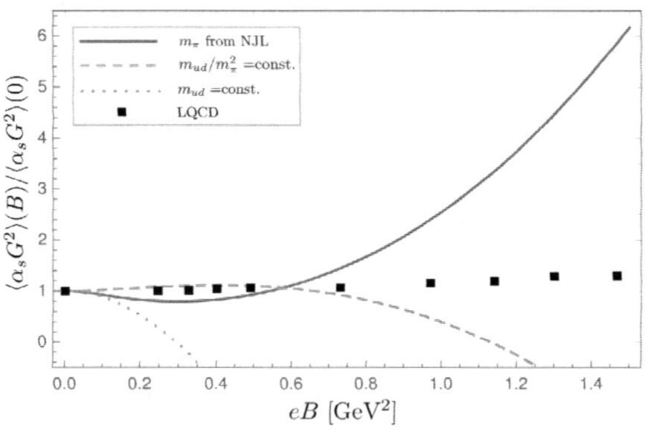

Figure 6. Magnetic evolution of gluon condensate considering three possible inputs and compared with LQCD results.

Figure 6 shows the magnetic evolution of the gluon condensate in the three conditions used as inputs and including lattice results to compare with. For a constant quark mass the gluon condensate drops dramatically. This strongly suggests that a constant quark mass is not a valid approximation, although other works has founded also a decreasing behavior of the gluon condensate [9]. Although it seems to turn negative, as we pointed previously, the maximum magnetic field strength for the case with constant quark mass is ~ 0.35 GeV.

The other values seems to present completely different behaviors. The NJL pion mass as an input gives us an initially decreasing and then increasing gluon condensate, which is in accordance with the results in [8]. With the condition of having a constant quotient m_q/m_π^2, the gluon condensate increases slightly for then decreasing gently with increasing magnetic field. Compared with Lattice results seems to deviate a for high magnetic field. However, the maximum allowed magnetic field value is ~ 0.5 GeV. The deviation in this region is less than 15%, so it do not differ so much from lattice results.

The results obtained here and the detailed procedure are explained in [1]. There are many structures and new condensates that can be explored and were skipped in this work. One of the interesting structures is the separation of the parallel and perpendicular projections of the gluon condensate, separated in chromo-electric and chromo-magnetic components. Another condensate we did not consider here is the polarization of the quark condensate $\langle \bar{q}\sigma_{12}q \rangle$. We are working on these topics and will be reported elsewhere.

aknowledgement
The authors acknowledge financial support from Fondecyt (Chile) under grants 1190192, 1170107 and 1200483.

References

[1] Dominguez C, Loewe M and Villavicencio C 2018 *Phys. Rev. D* **98** 034015 (*Preprint* arXiv:1806.10088)
[2] Gell-Mann M, Oakes R and Renner B 1968 *Phys. Rev.* **175** 2195–2199
[3] Bordes J, Dominguez C, Moodley P, Penarrocha J and Schilcher K 2010 *JHEP* **05** 064 (*Preprint* arXiv:1003.3358)
[4] Chyi T K, Hwang C W, Kao W, Lin G L, Ng K W and Tseng J J 2000 *Phys. Rev. D* **62** 105014 (*Preprint* arXiv:hep-th/9912134)
[5] Coppola M, Gómez Dumm D and Scoccola N 2018 *Phys. Lett. B* **782** 155–161 (*Preprint* arXiv:1802.08041)
[6] Bali G, Bruckmann F, Endrodi G, Fodor Z, Katz S and Schafer A 2012 *Phys. Rev. D* **86** 071502 (*Preprint* arXiv:1206.4205)
[7] Andersen J O 2012 *JHEP* **10** 005 (*Preprint* arXiv:1205.6978)
[8] Ayala A, Dominguez C, Hernandez L, Loewe M, Rojas J C and Villavicencio C 2015 *Phys. Rev. D* **92** 016006 (*Preprint* arXiv:1504.01308)
[9] Agasian N O and Shushpanov I 2000 *Phys. Lett. B* **472** 143–149 (*Preprint* hep-ph/9911254)

Journal of Physics: Conference Series **1602** (2020) 012028 doi:10.1088/1742-6596/1602/1/012028

Heavy-flavour correlations and jets with ALICE at the LHC

Antonio Carlos Oliveira da Silva, for the ALICE Collaboration
Physics Department, University of Tennessee - Knoxville, USA
E-mail: antonio.silva@cern.ch

Abstract.
This contribution summarises the results on heavy-flavour correlations and jets measured with ALICE detector. Studies of D^0-tagged jets are presented in various collision systems and energies. The measurement of the fraction of jet momentum carried by the D^0 meson in the direction of the jet axis ($z_{||}^{\mathrm{ch,jet}}$) in pp collisions at $\sqrt{s} = 5.02$ and 13 TeV as well as the nuclear modification factor in p–Pb and Pb–Pb collisions are presented. The measurements of the production of jets tagged by electrons from heavy-flavour decays in pp and p–Pb are compared to expectations from POWHEG+PYTHIA8 simulations. The p_{T}-differential cross section of b-tagged jets in p–Pb collisions is also reported and compared to expectations from POWHEG event generator. The azimuthal correlation of D mesons with charged particles in pp and p–Pb collisions are also presented.

1. Introduction

Heavy quarks are produced in hard scatterings in the first stages of hadronic collisions. Therefore, they experience the whole evolution of the quark-gluon plasma (QGP) in heavy-ion collisions. Furthermore, because of their relatively large mass, heavy-quark production is well described by perturbative QCD calculations. These features make heavy-quarks ideal probes of the hot medium. Measurements of azimuthal correlations of D mesons with charged particles and heavy-flavour jets in pp collisions are a valuable test of pQCD-based models describing heavy-quark production and also serve as reference for measurements in heavy-ion collisions that provide information on cold and hot nuclear matter effects. In this proceedings the latest results on heavy-flavour correlations and jets are presented. Jets in this contribution are reconstructed using anti-k_{t} algorithm as implemented in FASTJET [1].

2. Analysis methods and results

2.1. D^0-tagged jets

The D^0 mesons are reconstructed through their hadronic decay channel $D^0 \rightarrow K^-\pi^+$ (BR = 3.89%) exploiting the excellent particle-identification capabilities of the ALICE detector [2]. After replacing the 4-momentum vector of the D^0-decay daughter particles with the D^0 one, jets are reconstructed with resolution parameter $R = 0.3$. The feed-down contribution from bottom quark decays is removed using POWHEG+PYTHIA6 [3, 4] simulations. A Bayesian unfolding [5] is performed to correct the jet momentum for detector effects and fluctuations in the underlying-event energy density. Additional details on the method can be found in [6]. The

Journal of Physics: Conference Series **1602** (2020) 012028 doi:10.1088/1742-6596/1602/1/012028

momentum fraction of the D^0 with respect to the jet is defined as $z_{||}^{\mathrm{ch,jet}} = \frac{\vec{p}_{\mathrm{D}} \cdot \vec{p}_{\mathrm{ch,jet}}}{||\vec{p}_{\mathrm{ch,jet}}||^2}$, where $\vec{p}_D$ and $\vec{p}_{ch,jet}$ are the momenta of the D^0 and the jet, respectively.

The $z_{||}^{\mathrm{ch,jet}}$ probability density for D^0-tagged jets in pp collisions at a center-of-mass energy $\sqrt{s} = 5.02$ and 13 TeV are presented in Fig. 1 and 2.

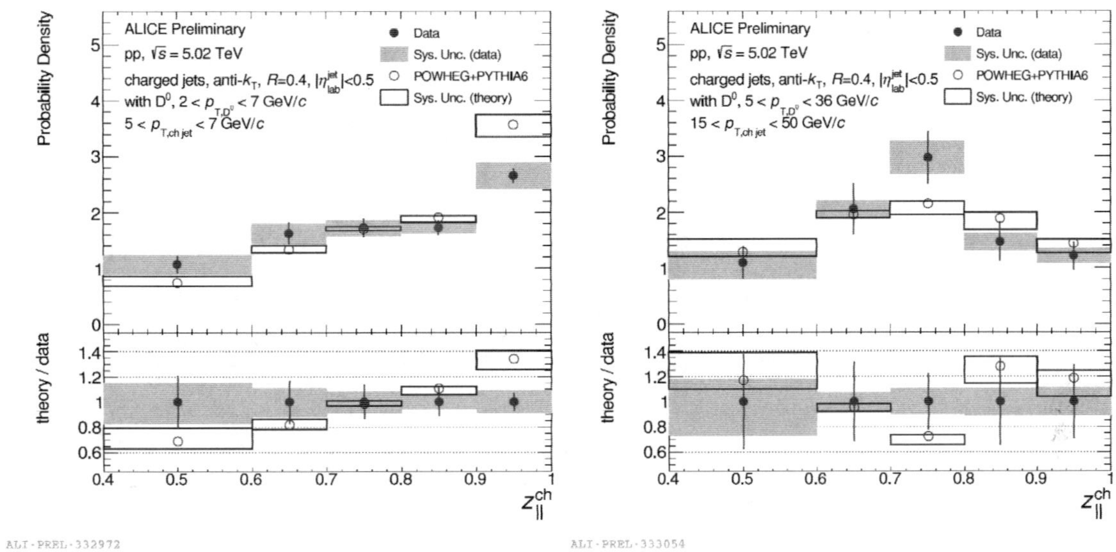

(a) Lower jet momentum region. (b) Higher jet momentum region.

Figure 1: Momentum fraction of D^0-tagged jets in pp collisions at $\sqrt{s} = 5.02$ TeV.

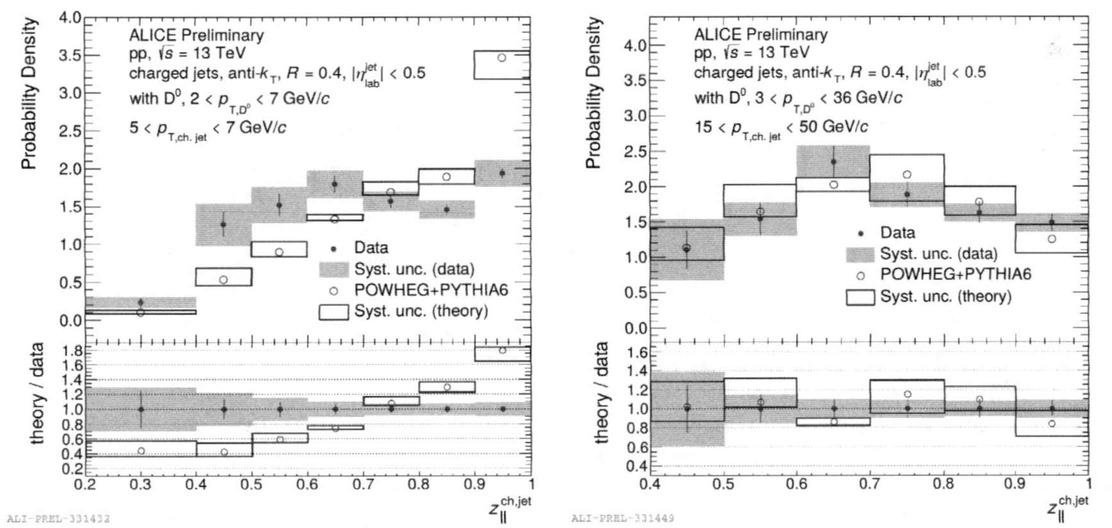

(a) Lower jet momentum region. (b) Higher jet momentum region.

Figure 2: Momentum fraction of D^0-tagged jets in pp collisions at $\sqrt{s} = 13$ TeV.

At low jet momentum, the measured probability densities favor large values of $z_{||}^{\mathrm{ch,jet}}$, indicating a hard fragmentation of the charm quark. The higher jet momentum region (Fig. 1b and 2b) presents a flatter $z_{||}^{\mathrm{ch,jet}}$ distribution in comparison to the lower momentum region. When compared with POWHEG+PYTHIA6 simulations, the data of the $5 < \mathrm{p_{T,ch.jet}}$ are above the simulation at small $z_{||}^{\mathrm{ch,jet}}$ and below for larger values of momentum fraction. The simulation provides a better description of the data for the higher jet-momentum interval.

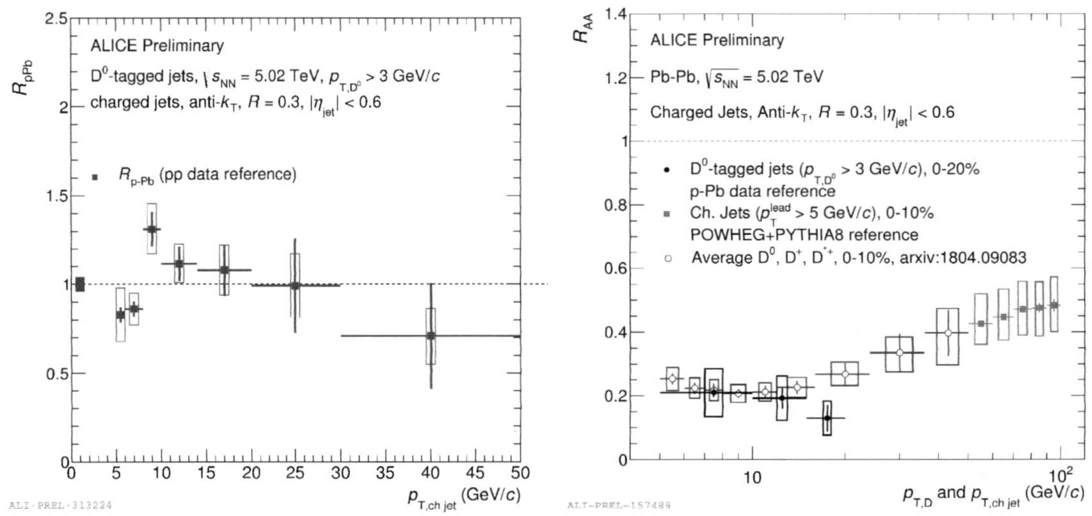

(a) R_{pPb}. (b) R_{AA}.

Figure 3: Nuclear modification factor in p–Pb (R_{pPb}) and Pb–Pb (R_{AA}) collisions at $\sqrt{s_{\mathrm{NN}}} =$ 5.02 TeV.

The nuclear modification factors (Fig. 3) were measured for $\mathrm{D^0}$-tagged jets in p–Pb and Pb–Pb collisions. This observable is defined as the ratio of the p_{T}-differential yields of jets in heavy-ion and pp collisions scaled by the Pb atomic mass, in case of p–Pb collisions, or the number of binary collisions [7], in case of Pb–Pb collisions. The R_{pPb} is shown in Fig. 3a and it is consistent within uncertainties with unity, indicating that cold nuclear matter effects are small or negligible. Figure 3b presents the R_{AA} of $\mathrm{D^0}$-tagged jets (black markers). A strong suppression of $\mathrm{D^0}$-tagged jets is observed in central (0-20%) Pb–Pb collisions. The comparison with the R_{AA} of the D mesons (blue markers) indicates similar suppression up to 20 GeV/c. At higher jet transverse momentum the R_{AA} of inclusive jets (red markers) and the average of D mesons show a similar trend.

2.2. Jets tagged by electrons from heavy-flavour decays

Jets containing an electron from heavy-flavour hadron decays (charm and bottom) were reconstructed using anti-k_{t} algorithm with resolution parameter $R = 0.3$. An invariant mass analysis of $\mathrm{e^+e^-}$ pairs is performed to remove the contribution of "photonic" electrons from gamma conversion and π^0 and η Dalitz decays. Pairs with invariant mass smaller than 0.1 GeV/c^2 are labeled as photonic electrons and removed from the sample. Ultimately, an unfolding procedure is performed to correct for detector effects. In p–Pb collisions unfolding also takes into account the fluctuations of the underlying event energy density.

The transverse momentum dependent spectra of jets tagged with electrons from heavy-flavour decays are presented in Fig. 4 in pp (Fig. 4a) and p–Pb (Fig. 4b) collisions at $\sqrt{s_{\mathrm{NN}}} = 5.02$ TeV. A good description of the data is provided by POWHEG+PYTHIA8 simulations.

Journal of Physics: Conference Series **1602** (2020) 012028 doi:10.1088/1742-6596/1602/1/012028

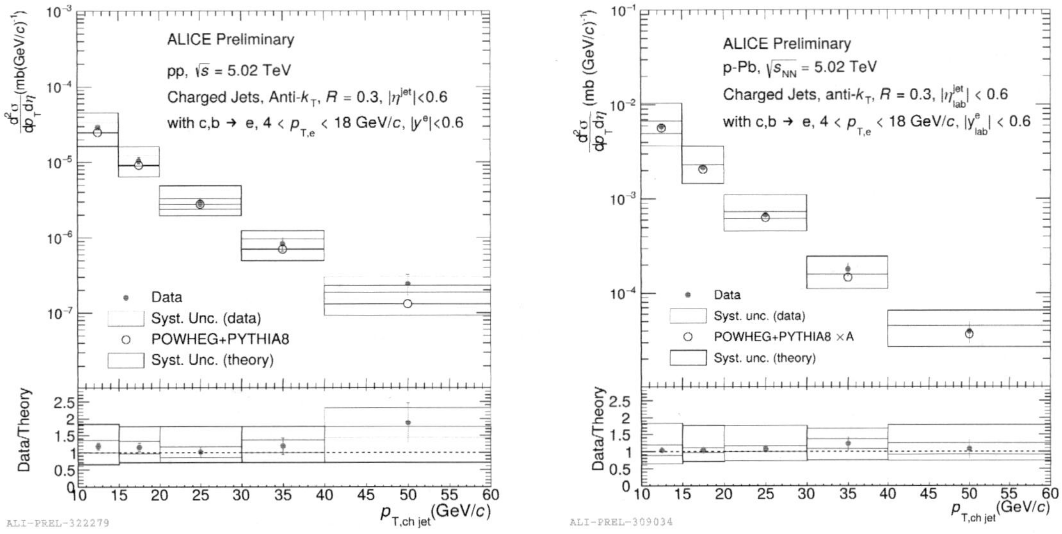

(a) pp collisions. (b) p–Pb collisions.

Figure 4: p_{T}-differential cross section of charm and bottom jets tagged with electrons from heavy-flavour decays in pp and p–Pb collisions at $\sqrt{s_{\mathrm{NN}}} = 5.02$ TeV.

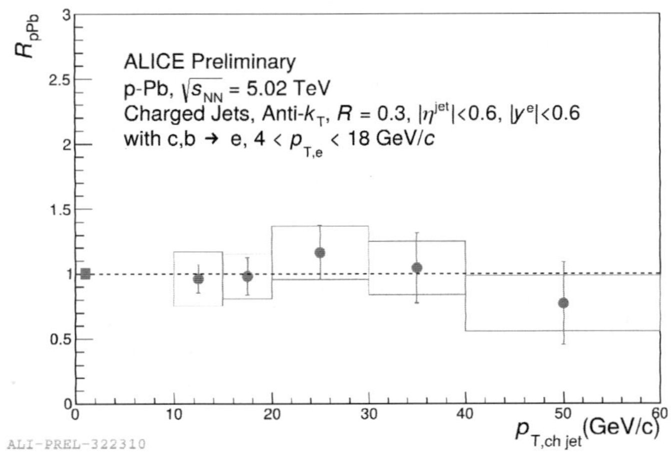

Figure 5: Nuclear modification factor (R_{pPb}) of heavy-flavour jets tagged by electrons from heavy-flavour decays in p–Pb collisions at $\sqrt{s_{\mathrm{NN}}} = 5.02$ TeV.

The measured R_{pPb} shown in Fig. 5 is compatible with unity, indicating no significant modifications from cold nuclear matter effects.

2.3. b-jets

Jets from bottom quark fragmentation are identified by the presence of a decay vertex, displaced from the collision point, reconstructed with 3 particles within the constituents of the jet. A selection exploiting the relatively long decay length of hadrons with bottom quarks is applied to enhance the discriminating power of the b-tagging. This selection is done mainly using the significance $L_{xy}/\sigma_{L_{xy}}$, where L_{xy} is the distance between the primary and secondary vertices in the xy-plane and $\sigma_{L_{xy}}$ is its uncertainty. The variable $\sigma_{SV} = \sqrt{\sum_{i=1}^{3} d_i^2}$, where d is the distance of closest approach between each of the three tracks and the position of the secondary vertex is also powerful to identify jets from bottom quarks. A resolution parameter $R = 0.4$ was used for jet finding.

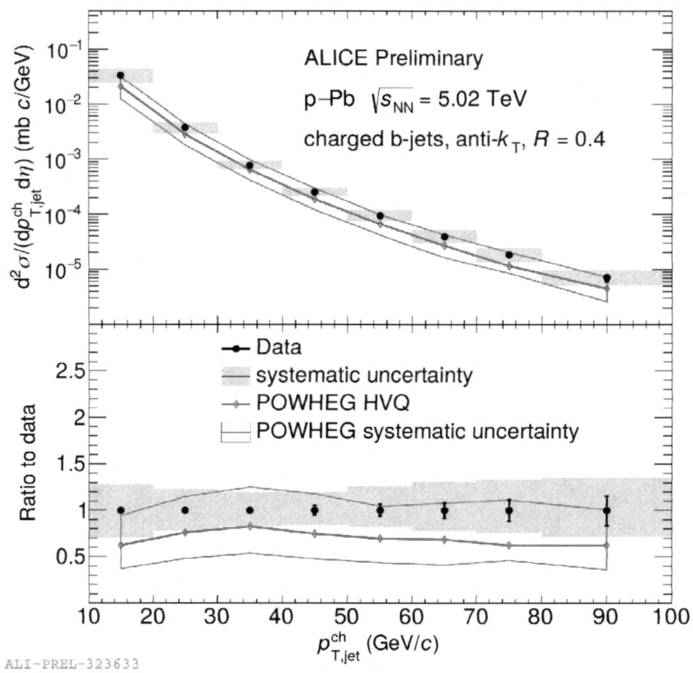

Figure 6: p_T-differential cross section of b-jets in p–Pb collisions at $\sqrt{s_{NN}} = 5.02$ TeV.

The p_T-differential cross section of jets from bottom quarks is presented in Fig. 6. The tagging efficiency was estimated with a Monte Carlo simulation in which a PYTHIA event containing a b-$\bar{b}$ pair was injected on top of a p–Pb collision simulated with EPOS [8]. The purity was calculated using POWHEG simulations and has a approximate value of 0.4, but varying as a function of jet p_T. Detector finite resolution and jet background fluctuations are corrected for with a SVD unfolding. The data are described within uncertainties by POWHEG-HVQ[9] simulations.

2.4. D-hadron correlations

The reconstruction of D mesons (D^0, D^+ and D^{*+}) is performed as in subsection 2.1 and are correlated with primary charged particles. The angular correlation distribution is corrected for the finite acceptance of the ALICE detector using an event-mixing technique. The D-meson combinatorial background is removed via an invariant mass analysis. The feed-down contribution from bottom quark decays is removed using FONLL calculations [10] and templates of the correlation function obtained from simulations performed with PYTHIA event generator. The D-hadron azimuthal correlation distribution is fitted using two gaussian functions for the near and away side peaks and a constant function for the baseline [11].

The associated yields and the standard deviation ($\sigma_{fit,NS}$) of the Gaussian function of the near side peak as a function of D-meson transverse momentum are presented in Fig. 7 for pp

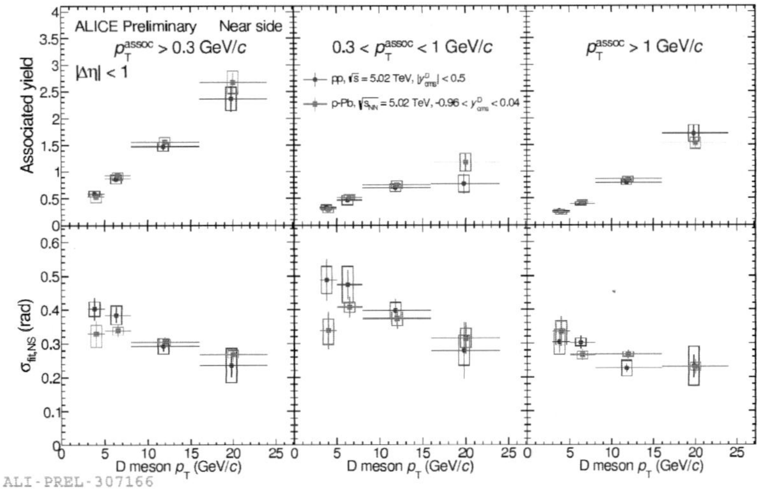

Figure 7: Associated yields and standard deviation of the near-side peak obtained from D-hadron angular correlations in pp and p–Pb collisions at $\sqrt{s_{NN}} = 5.02$ Tev.

and p–Pb collisions at $\sqrt{s_{NN}} = 5.02$ TeV. The associated yields are studied in three intervals of charged particle transverse momentum. For all intervals an enhancement of the associated yield in the near side peak is observed for increasing D-meson transverse momentum (p_T). This is an indication that the higher the D-meson p_T, the higher is the number of particles close to the D-meson in the $\eta - \phi$ phase space, which provides complementary piece of information to the change of the momentum fraction for large jet transverse momentum presented in subsection 2.1. The standard deviation of the near-side Gaussian function also shows that particles in the heavy-flavour jet are more collimated for large D-meson transverse momentum. No significant difference between pp and p–Pb measurements were observed in the associated yield or standard deviation of the near-side Gaussian function, indicating that they are not sensible to cold nuclear matter effects.

3. Summary

The probability distributions of the D^0-tagged jet momentum fraction in pp collisions at $\sqrt{s} =$ 5.02 and 13 TeV were presented. An indication of softer fragmentation of charm quarks at the lower jet momentum region is observed. The R_{pPb} of D^0-tagged jets indicate that cold nuclear matter effects are not significant. The R_{AA} indicates a strong suppression of D^0-tagged jet production in central (0–20%) Pb–Pb collisions. The R_{pPb} of jets tagged with electrons from heavy-flavour decays was measured at $\sqrt{s_{NN}} = 5.02$ TeV and no significant cold nuclear matter effects were observed. The transverse momentum distribution of b-jets in p-Pb collisions at $\sqrt{s_{NN}} = 5.02$ TeV was reported in the kinematic range $10 <$ jet $p_T < 100$ GeV/c. D-hadron correlations studies in pp and p–Pb at $\sqrt{s_{NN}} = 5.02$ TeV were presented. No significant cold nuclear matter effects were observed.

References

[1] Matteo Cacciari, Gavin P. Salam, and Gregory Soyez. Fastjet user manual. *The European Physical Journal C*, 72(3), Mar 2012.

[2] Elena Botta. Particle identification performance at ALICE. In *5th Large Hadron Collider Physics Conference*, 9 2017.

[3] Stefano Frixione, Paolo Nason, and Carlo Oleari. Matching NLO QCD computations with parton shower simulations: the POWHEG method. *Journal of High Energy Physics*, 2007(11):070–070, nov 2007.

[4] Torbjörn Sjöstrand, Stephen Mrenna, and Peter Skands. PYTHIA 6.4 physics and manual. *Journal of High Energy Physics*, 2006(05):026–026, may 2006.
[5] Harrison B. Prosper and Louis Lyons, editors. *Proceedings, PHYSTAT 2011 Workshop on Statistical Issues Related to Discovery Claims in Search Experiments and Unfolding, CERN,Geneva, Switzerland 17-20 January 2011*, CERN Yellow Reports: Conference Proceedings, Geneva, 1 2011. CERN.
[6] Shreyasi Acharya et al. Measurement of the production of charm jets tagged with D^0 mesons in pp collisions at $\sqrt{s} = 7$ TeV. *JHEP*, 08:133, 2019.
[7] Michael L. Miller, Klaus Reygers, Stephen J. Sanders, and Peter Steinberg. Glauber modeling in high-energy nuclear collisions. *Annual Review of Nuclear and Particle Science*, 57(1):205–243, 2007.
[8] T. Pierog, Iu. Karpenko, J. M. Katzy, E. Yatsenko, and K. Werner. Epos lhc: Test of collective hadronization with data measured at the cern large hadron collider. *Phys. Rev. C*, 92:034906, Sep 2015.
[9] Stefano Frixione, Paolo Nason, and Giovanni Ridolfi. The POWHEG-hvq manual version 1.0. 7 2007.
[10] Matteo Cacciari, Mario Greco, and Paolo Nason. The P(T) spectrum in heavy flavor hadroproduction. *JHEP*, 05:007, 1998.
[11] Shreyasi Acharya et al. Azimuthal correlations of prompt D mesons with charged particles in pp and p-Pb collisions at $\sqrt{s_{NN}} = 5.02$ TeV. 10 2019.

Journal of Physics: Conference Series **1602** (2020) 012029 doi:10.1088/1742-6596/1602/1/012029

PHENIX Results on J/ψ Production in p+Al, p+Au and ^{3}He+Au Collisions

Anthony D Frawley for the PHENIX collaboration
Department of Physics, Florida State University, Tallahassee, Florida 32306, USA
E-mail: afrawley@fsu.edu

Abstract.
Results are presented of measurements of the J/ψ modification in p+Au, d+Au and ^{3}He+Au collisions at forward and backward rapidity ($1.2 < |y| < 2.2$) by the PHENIX experiment. The rapidity, transverse momentum and collision centrality dependence of the modifications are presented, and compared between the three light projectile collision systems. The modification for p+Al is found to be small. For p+Au it is depends strongly on collision centrality at forward rapidity. In both cases models describe the data well. The modifications for d+Au and ^{3}He+Au are very similar to those for p+Au, although there is a hint of slightly increased suppression for ^{3}He+Au over p+Au at backward rapidity.

1. Introduction

The effects that modify the J/ψ production probability in nuclear collisions can be broadly separated into two categories, based on the time scale in which they occur. Those that occur during the expansion and hadronization of the energy produced in the collision are generally referred to as hot matter effects. Those that occur on a time scale smaller than the crossing time of the colliding nuclei are frequently referred to as cold nuclear matter (CNM) effects. The latter include modification of the nuclear-parton-distribution functions (nPDFs) in a nucleus [1, 2], initial state parton energy loss [3], coherent gluon saturation [4, 5], breakup of the forming charmonium in collisions with target nucleons [6, 7], and transverse momentum broadening [8].

Prior to the last five years, it was widely assumed that the energy produced in $p(d)$+A collisions did not cause significant modification of charmonium yields in the final state. This was called into question by two observations. One was the observation of flow-like behavior in p+Pb collisions at LHC (see for example [9]) and later in d+Au collisions at RHIC [10, 11]. This suggested that a quark-gluon plasma of small size is created in high energy collisions of light systems. The second was the observation of strong suppression of the weakly bound $\psi(2S)$ relative to the J/ψ in central d+Au collisions [12], and later in p+Pb collisions [13]. This can not be explained by known CNM effects, but explanations invoking final state effects due to energy produced in the collision have been successful [14, 15, 16]. A transport model [15] that reproduces the differential $\psi(2S)$ suppression also predicts a J/ψ suppression of $\sim 20\%$ in central d+Au collisions.

To further explore the effects of energy production on J/ψ yields in light systems, the PHENIX collaboration has recently measured the modification in three collision systems involving small projectiles, p+Al, p+Au and ^{3}He+Au [17]. The p+Au and ^{3}He+Au data, combined with the earlier PHENIX d+Au data [18, 19], provide a direct comparison of inclusive charmonium production for projectiles with one, two and three nucleons on a Au target. We will focus here mostly on the p+Au and ^{3}He+Au results.

2. Experiment

The $J/\psi \rightarrow \mu^+\mu^-$ data for the p+Au, ^{3}He+Au and reference p+p collision systems were measured in the PHENIX muon arms [20, 21, 22] during the RHIC runs in 2014 (^{3}He+Au) and 2015 (p+Au and p+p). All data were recorded at $\sqrt{s_{NN}}$ = 200 GeV. The muon spectrometers cover the range $(-2.2 < \eta < -1.2)$ and $(1.2 < \eta < 2.2)$. Collision centrality was measured using the Beam-Beam Counters (BBC) located at $3.1 < |\eta| < 3.9$. The corresponding integrated luminosity is 47 pb^{-1} for p+p, 590 nb^{-1} for p+Al, 138 nb^{-1} for p+Au, and 18 nb^{-1} for ^{3}He+Au collisions. The details of the measurement and analysis are described in [17].

3. Results

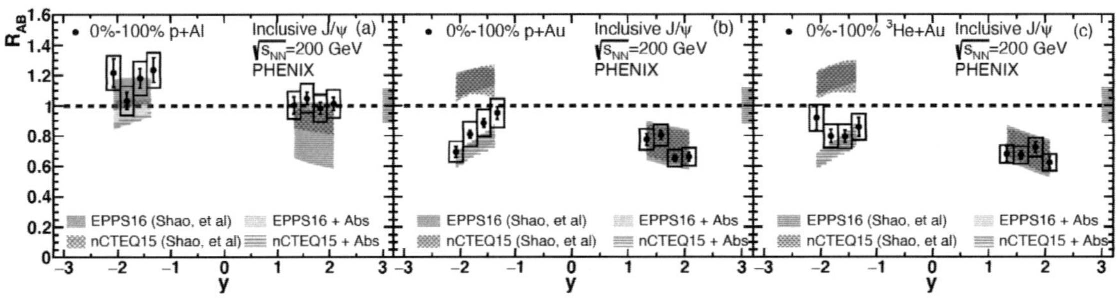

Figure 1. Nuclear modification factor of inclusive J/ψ as a function of rapidity for 0–100% p+Al (a), p+Au (b), and ^{3}He+Au (c) collisions. Bars (boxes) around data points represent point-to-point uncorrelated (correlated) uncertainties. The theory bands are discussed in the text.

There were significant changes to the PHENIX muon arm configuration and to the PHENIX simulation framework between the time that the d+Au data set was recorded in 2008 and when the p+Al, p+Au, ^{3}He+Au and their reference p+p data were recorded in 2014 and 2015. In addition, the d+Au data set used a different reference p+p measurement. Therefore the relative systematic uncertainties between the modifications measured for p+Al, p+Au and ^{3}He+Au are smaller, and we focus on those for comparisons.

Figure 1 shows the measured rapidity dependence of the modification for p+Al, p+Au and ^{3}He+Au, integrated over p_T and for 0-100% centrality. The data are compared with model calculations by Shao *et al* [23, 24, 25, 26] that include only CNM effects. The calculations use the EPPS16 [1] and nCTEQ15 [2] NLO gluon shadowing parameterizations. In the calculations, the gluon shadowing parameterizations for the Au target are modified using a Bayesian reweighting method that imposes J/ψ constraints from p+Pb data at the LHC [23]. This results in considerably reduced uncertainties of the modifications. The theoretical uncertainties (68% CL) are dominated by the uncertainty in the factorization scale.

The forward rapidity data are well described for all three systems, as are the backward rapidity p+Al data. However the backward rapidity p+Au and ^{3}He+Au data are not reproduced by the shadowing calculation alone. At RHIC energies a significant suppression is expected at backward rapidity due to breakup of the forming charmonium in collisions with nucleons that have not yet passed through the production point (commonly referred to as "nuclear absorption"). An estimate of this effect has been folded into the shadowing calculation and is included in Figure 1.

The absorption estimate is based on a model [7] that has been fitted to absorption cross sections derived from shadowing corrected data measured at a broad range of beam energies [6]. In the model, the absorption cross section depends on the size of the evolving charmonium, and thus on the proper time before it escapes the target nucleus. Including the absorption estimate brings the backward rapidity calculation into reasonable agreement with the p+Au and ^{3}He+Au data.

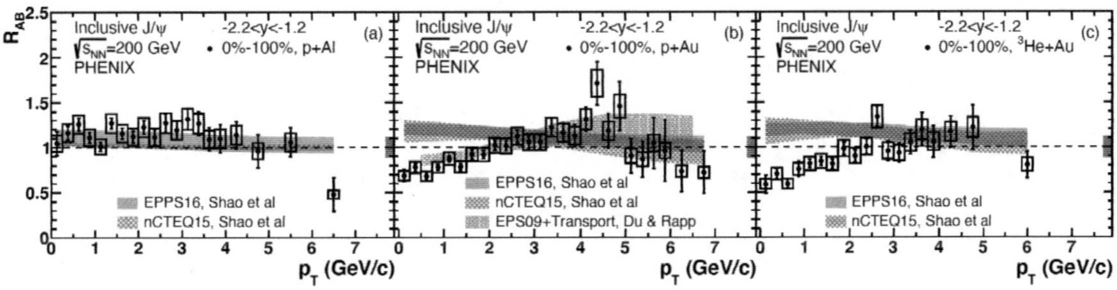

Figure 2. Nuclear modification factor of inclusive J/ψ as a function of p_T for 0–100% p+Al (a), p+Au (b), and ^{3}He+Au (c) collisions at backward rapidity. Bars (boxes) around data points represent point-to-point uncorrelated (correlated) uncertainties. The theory bands are discussed in the text.

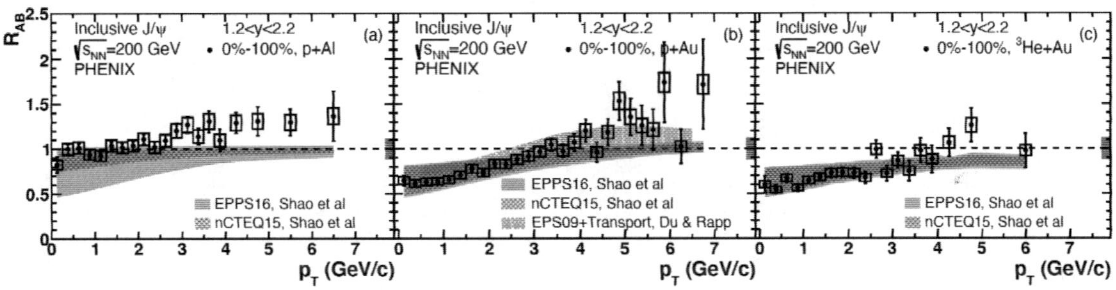

Figure 3. Nuclear modification factor of inclusive J/ψ as a function of p_T for 0–100% p+Al (a), p+Au (b), and ^{3}He+Au (c) collisions at forward rapidity. Bars (boxes) around data points represent point-to-point uncorrelated (correlated) uncertainties. The theory bands are discussed in the text.

The p_T dependence of the centrality integrated J/ψ modification is shown for p+Al, p+Au and ^{3}He+Au in Figure 2 (backward rapidity) and Figure 3 (forward rapidity). The calculations of Shao *et al* are compared with the data for the three collision systems. As was the case for the rapidity dependence in Figure 1, the calculations do a reasonable job of describing the data at forward rapidity for all three systems, and at backward rapidity for p+Al. The calculations do not include transverse momentum broadening, probably explaining why they fall below the data above about 3 GeV/c. For p+Au and ^{3}He+Au at backward rapidity and low p_T, the calculations substantially overpredict the modification. Qualitatively this is expected because the calculation contain only enhancement from anti-shadowing, but do not include suppression from nuclear absorption.

Also shown for p+Au in Figure 2 and Figure 3 are calculations from a transport model by Du and Rapp [15]. This is adapted from a model by Zhao and Rapp [27] that was developed to describe A+A collisions. The calculation includes a fireball generated by a glauber Monte-Carlo

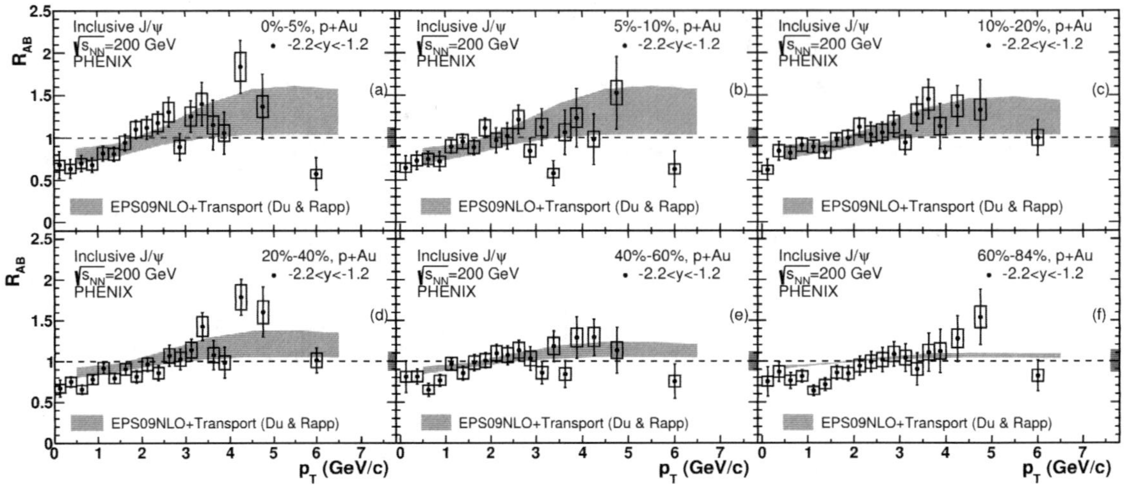

Figure 4. Nuclear modification factor of inclusive J/ψ as a function of p_T at $-2.2 < y < -1.2$ in six centrality bins for p+Au collisions. Bars (boxes) around data points represents point-to-point uncorrelated (correlated) uncertainties. The theory bands are discussed in the text. Note that the theory bands compared with the 0–5% and 5–10% centrality data are for 0–10%.

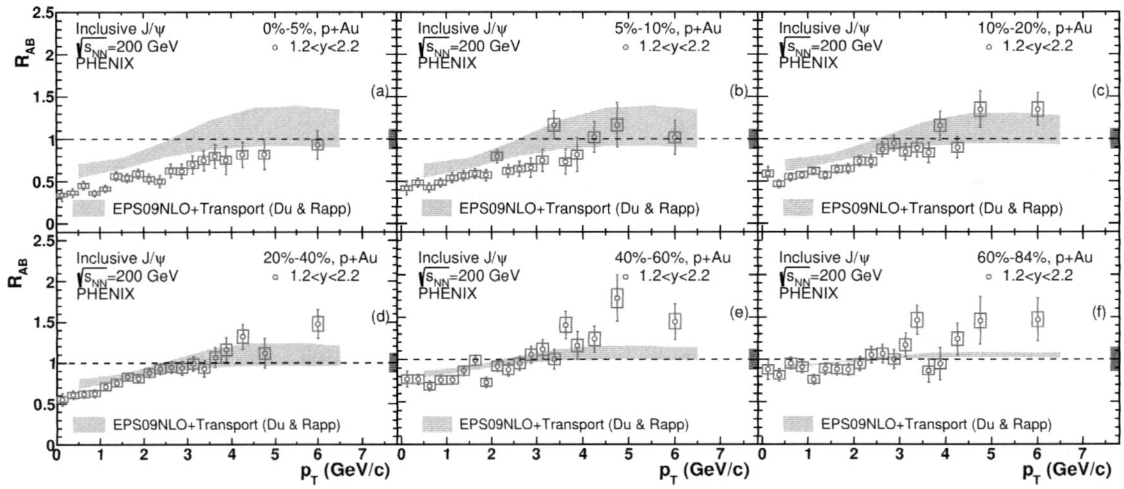

Figure 5. Nuclear modification factor of inclusive J/ψ as a function of p_T at $1.2 < y < 2.2$ in six centrality bins for p+Au collisions. The theory bands are discussed in the text. Note that the theory bands compared with the 0–5% and 5–10% centrality data are for 0–10%.

model [28], an NLO gluon shadowing parameterization from EPS09 [29], transverse momentum broadening [30], and an estimate of the J/ψ absorption cross section constrained by PHENIX d+Au data [12]. These calculations do a good job at backward rapidity, but produce too little suppression at low p_T at forward rapidity. The forward rapidity suppression is dominated by the gluon shadowing, so this indicates that the EPS09 shadowing used in the model is not strong enough at forward rapidity to describe the data. It is worth noting that at low Bjorken x, which is the regime relevant to the forward rapidity J/ψ data, the newer EPPS16 parameterization produces stronger shadowing than EPS09. Also, the Bayesian reweighting used by Shao *et al* strengthens the low x shadowing further, resulting in a satisfactory description of the forward rapidity centrality integrated data, as seen in Figure 1(b) and Figure 3(b). On the other hand, in the Bjorken x regime relevant to the backward rapidity J/ψ data, the gluon modifications are similar for EPS09 and EPPS16.

The p_T dependence, in six collision centrality bins, is presented at backward rapidity in Figure 4 and at forward rapidity in Figure 5. The 0-5% bin contains the most central collisions. At backward rapidity there is weak dependence of the modification on centrality. However at forward rapidity the data show very strong centrality dependence, with the most central bin at low p_T having a suppression of a factor of almost three.

In Figure 4 and Figure 5 calculations of the p_T dependence from the transport model by Du and Rapp are compared with the data. At backward rapidity the calculations describe the p_T dependence of the data well at all centralities. At forward rapidity and low p_T, the calculations show much less suppression than is seen in the data. As already mentioned, because the modification of J/ψ production in the transport model is small at forward rapidity, the model suppression is dominated by the EPS09 shadowing parameterization. It is important to note here that the EPS09 parameterization is integrated over all centrality, and contains no prediction for the centrality dependence of the shadowing. The centrality dependence was assumed by Du and Rapp to be linear in the nuclear thickness.

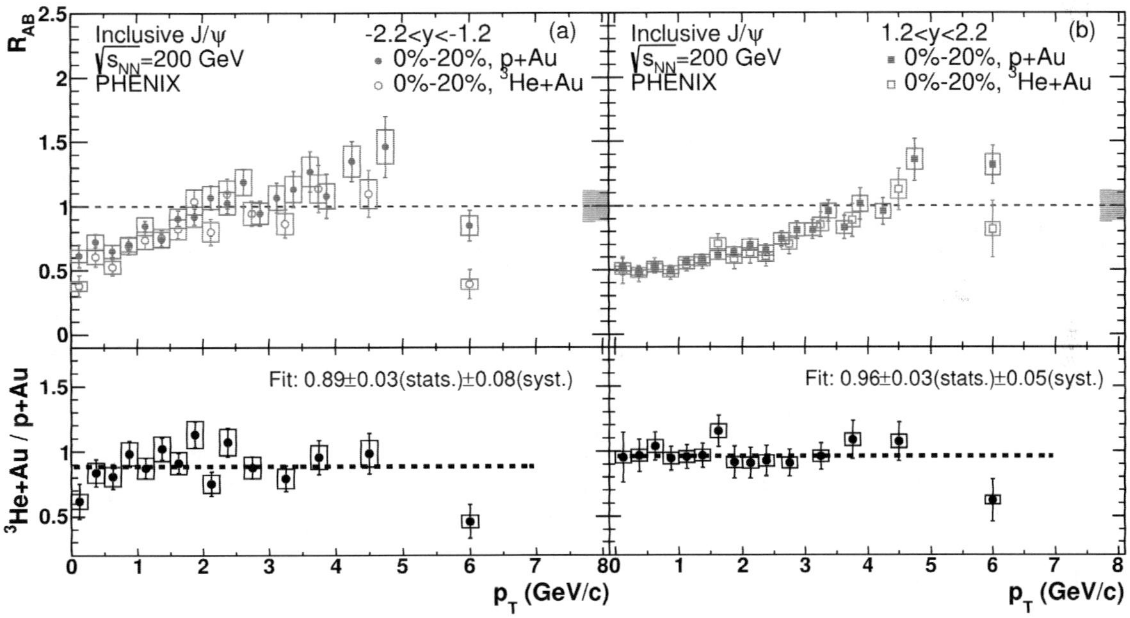

Figure 6. Comparison of nuclear modification factor of J/ψ as a function of p_T in 0–20% centrality p+Au and ^{3}He+Au collisions. Bars (boxes) around data points represent point-to-point uncorrelated (correlated) uncertainties.

At $\sqrt{s_{NN}} = 200$ GeV the charged particle multiplicity observed by PHENIX in 0-20% central ^{3}He+Au collisions is approximately twice that observed in p+Au collisions [31]. CNM effects are expected to be very similar for the two collision systems. To investigate whether the difference in charged particle multiplicity has an effect on the J/ψ modification, the measured modifications as a function of p_T in the two systems are compared directly in Figure 6. The ratio of the modifications is presented in the lower panel. At forward rapidity the average ratio is found to be

Journal of Physics: Conference Series **1602** (2020) 012029 doi:10.1088/1742-6596/1602/1/012029

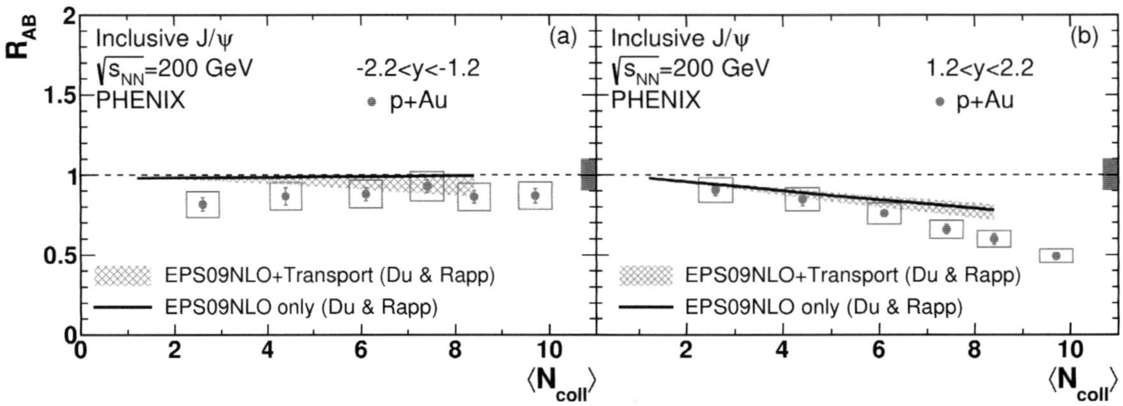

Figure 7. Nuclear modification factor of J/ψ as a function of $\langle N_{coll}\rangle$ for p+Au collisions compared with the transport model. Bars (boxes) around data points represent point-to-point uncorrelated (correlated) uncertainties.

$$\overline{R_{^3\mathrm{HeAu}}/R_{\mathrm{pAu}}} = 0.96 \pm 0.03(\mathrm{stat}) \pm 0.05(\mathrm{syst}), \quad (1)$$

which is consistent with unity. At backward rapidity it is

$$\overline{R_{^3\mathrm{HeAu}}/R_{\mathrm{pAu}}} = 0.89 \pm 0.03(\mathrm{stat}) \pm 0.08(\mathrm{syst}). \quad (2)$$

which is suggestive of additional suppression in ^{3}He+Au, but the systematic uncertainty does not allow a strong statement to be made.

Figure 7 shows the N_{coll} dependence of the J/ψ modification for p+Au, integrated over p_T, for each of the muon arms. At backward rapidity the modification is not much smaller than one, and is almost independent of centrality. At forward rapidity the modification decreases strongly with centrality, reaching 0.5 for the most central collisions. Also shown in Figure 7 is the corresponding calculated modification from Du and Rapp [15]. At backward rapidity, the calculated modification involves an almost exact cancelation between the anti-shadowing and nuclear absorption effects, with an additional suppression from transport effects that is of order 10% for the most central collisions. At forward rapidity the calculation substantially underpredicts the suppression in central collisions. As already mentioned, the suppression from EPS09 shadowing is insufficient in the model to explain the data at forward rapidity.

Finally, a comparison of the p_T integrated p+Au modification versus rapidity for the most central collisions and the most peripheral collisions illustrates two points. This is shown in Figure 8. First, at backward rapidity the measured modifications are the same within uncertainties. This can be understood as a cancelation between anti-shadowing and absorption, since those two effects have strong, but opposite, centrality dependence (see also Figure 1). Second, at forward rapidity the 0-5% most central collisions show strong suppression, which increases as the rapidity increases. At the most forward rapidity the p_T integrated modification is about 0.35.

4. Summary and Conclusions

For p+Al, p+Au and ^{3}He+Au collisions, the measured J/ψ modifications as a function of rapidity, integrated over p_T, was presented. The data are reasonably well described by Bayesian reweighted EPPS16 and nCTEQ15 calculations [23, 24, 25, 26], with estimates of nuclear absorption added by PHENIX at backward rapidity.

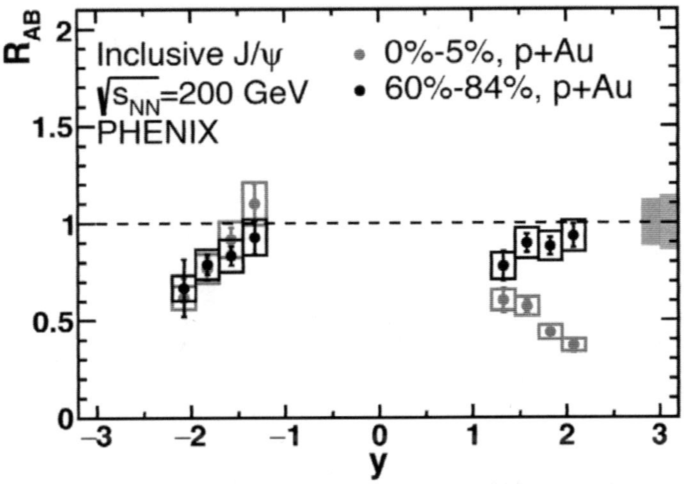

Figure 8. Comparison of the nuclear modification factor of J/ψ for p+Au collisions in two centrality classes, as a function of rapidity. Bars (boxes) around data points represent point-to-point uncorrelated (correlated) uncertainties.

The p_T dependence of the J/ψ modification integrated over centrality was presented, and compared with the same Bayesian reweighted EPPS16 and nCTEQ15 calculations. They describe the data well at low p_T, but fall below the data above 3 GeV/c, likely because they do not contain any transverse momentum broadening.

The p_T dependence of the J/ψ modification in p+Au collisions was presented in six centrality bins at forward and backward rapidity. At backward rapidity the data show little dependence on centrality, and are well described by a model [15] that combines shadowing, nuclear absorption, transverse momentum broadening and (a small) suppression due to transport effects in the fireball created in the collision. At forward rapidity the measured modification shows very strong suppression at low p_T in central collisions, reaching 0.35 at 0-5% centrality. The model substantially underpredicts the suppression at low p_T in more central collisions, indicating that the shadowing component of the model gives insufficient suppression there.

The J/ψ modifications measured for central p+Au and ^{3}He+Au collisions were compared, to look for evidence of final state effects caused by the increased charged particle multiplicity in ^{3}He+Au. The ratio of the modifications at forward rapidity is consistent with unity. At backward rapidity the ratio is suggestive of a small additional suppression in ^{3}He+Au, however the systematic uncertainty precludes a strong statement

The centrality dependence of the p_T integrated modification in the two muon arms was presented. At backward rapidity it is close to one and almost independent of centrality. At forward rapidity is has a strong centrality dependence, dropping to 0.5 for 0-5% central collisions. The transport model calculation describes the data reasonably well at backward rapidity. At forward rapidity the model produces too little suppression in central collisions, again indicating that the suppression from shadowing is insufficient.

The rapidity dependence of the modification for peripheral and central p+Au collisions was compared. It shows no change in the modification at backward rapidity, suggesting essentially complete cancelation between the enhancement due to anti-shadowing and the suppression due to nuclear absorption plus transport suppression. At forward rapidity the suppression is strong for central collisions, with the modification dropping to approximately 0.35 at the largest rapidity.

Journal of Physics: Conference Series **1602** (2020) 012029 doi:10.1088/1742-6596/1602/1/012029

References

[1] K. J. Eskola, P. Paakkinen, H. Paukkunen, and C. A. Salgado. EPPS16: Nuclear parton distributions with LHC data. *Eur. Phys. J. C*, 77(3):163, 2017.

[2] K. Kovarik et al. nCTEQ15 - Global analysis of nuclear parton distributions with uncertainties in the CTEQ framework. *Phys. Rev. D*, 93(8):085037, 2016.

[3] I. Vitev. Non-Abelian energy loss in cold nuclear matter. *Phys. Rev. C*, 75:064906, 2007.

[4] D. Kharzeev and K. Tuchin. Signatures of the color glass condensate in J/ψ production off nuclear targets. *Nucl. Phys. A*, 770:40, 2006.

[5] H. Fujii, F. Gelis, and R. Venugopalan. Quark pair production in high energy pA collisions: General features. *Nucl. Phys. A*, 780:146, 2006.

[6] D. C. McGlinchey, A. D. Frawley, and R. Vogt. Impact parameter dependence of the nuclear modification of J/ψ production in d+Au collisions at $\sqrt{s_{NN}}$ = 200 GeV. *Phys. Rev. C*, 87(5):054910, 2013.

[7] F. Arleo, P. B. Gossiaux, T. Gousset, and J. Aichelin. Charmonium suppression in p-A collisions. *Phys. Rev. C*, 61:054906, 2000.

[8] J. W. Cronin, H. J. Frisch, M. J. Shochet, J. P. Boymond, R. Mermod, P. A. Piroue, and R. L. Sumner. Production of hadrons with large transverse momentum at 200, 300, and 400 GeV. *Phys. Rev. D*, 11:3105, 1975.

[9] K. Dusling, W. Li, and B. Schenke. Novel collective phenomena in high-energy protonproton and protonnucleus collisions. *Int. J. Mod. Phys. E*, 25(01):1630002, 2016.

[10] A. Adare et al. Quadrupole Anisotropy in Dihadron Azimuthal Correlations in Central d+Au Collisions at $\sqrt{s_{NN}}$=200 GeV. *Phys. Rev. Lett.*, 111(21):212301, 2013.

[11] A. Adare et al. Measurement of long-range angular correlation and quadrupole anisotropy of pions and (anti)protons in central d+Au collisions at $\sqrt{s_{NN}}$=200 GeV. *Phys. Rev. Lett.*, 114(19):192301, 2015.

[12] A. Adare et al. Nuclear Modification of $\psi\prime$, χ_c, and J/ψ Production in d+Au Collisions at $\sqrt{s_{NN}}$=200 GeV. *Phys. Rev. Lett.*, 111(20):202301, 2013.

[13] B. B. Abelev et al. Suppression of $\psi(2S)$ production in p-Pb collisions at $\sqrt{s_{NN}}$ = 5. 02 TeV. J. High Energy Phys. **12 (2014)** 073.

[14] E. G. Ferreiro. Excited charmonium suppression in protonnucleus collisions as a consequence of comovers. *Phys. Lett. B*, 749:98, 2015.

[15] X. Du and R. Rapp. In-Medium Charmonium Production in Proton-Nucleus Collisions. J. High Energy Phys. **03 (2019)** 015.

[16] A. Beraudo, A. De Pace, M. Monteno, M. Nardi, and F. Prino. Heavy-flavour production in high-energy d-Au and p-Pb collisions. J. High Energy Phys. **03 (2016)** 123.

[17] U.A. Acharya et al. Measurement of J/ψ at forward and backward rapidity in $p+p$, $p+Al$, $p+Au$, and ^{3}He+Au collisions at $\sqrt{s_{NN}}$ = 200 GeV. 10 2019.

[18] A. Adare et al. Cold Nuclear Matter Effects on J/ψ Yields as a Function of Rapidity and Nuclear Geometry in Deuteron-Gold Collisions at $\sqrt{s_{NN}}$ = 200 GeV. *Phys. Rev. Lett.*, 107:142301, 2011.

[19] A. Adare et al. Transverse-Momentum Dependence of the J/ψ Nuclear Modification in d+Au Collisions at $\sqrt{s_{NN}}$ = 200 GeV. *Phys. Rev. C*, 87(3):034904, 2013.

[20] K. Adcox et al. PHENIX detector overview. *Nucl. Instrum. Methods Phys. Res., Sec. A*, 499:469, 2003.

[21] H. Akikawa et al. PHENIX muon arms. *Nucl. Instrum. Methods Phys. Res., Sec. A*, 499:537, 2003.

[22] S. Adachi et al. Trigger electronics upgrade of PHENIX muon tracker. *Nucl. Instrum. Methods Phys. Res., Sec. A*, 703:114, 2013.

[23] A. Kusina, J.-P. Lansberg, I. Schienbein, and H.-S. Shao. Gluon Shadowing in Heavy-Flavor Production at the LHC. *Phys. Rev. Lett.*, 121(5):052004, 2018.

[24] H.-S. Shao. HELAC-Onia: An automatic matrix element generator for heavy quarkonium physics. *Comput. Phys. Commun.*, 184:2562, 2013.

[25] H.-S. Shao. HELAC-Onia 2.0: an upgraded matrix-element and event generator for heavy quarkonium physics. *Comput. Phys. Commun.*, 198:238, 2016.

[26] J.-P. Lansberg and H.-S. Shao. Towards an automated tool to evaluate the impact of the nuclear modification of the gluon density on quarkonium, D and B meson production in protonnucleus collisions. *Eur. Phys. J. C*, 77(1):1, 2017.

[27] X. Du and R. Rapp. Sequential Regeneration of Charmonia in Heavy-Ion Collisions. *Nucl. Phys. A*, 943:147, 2015.

[28] C. Loizides, J. Nagle, and P. Steinberg. Improved version of the PHOBOS Glauber Monte Carlo. *SoftwareX*, 1-2:13, 2015.

[29] K. J. Eskola, H. Paukkunen, and C. A. Salgado. EPS09: A New Generation of NLO and LO Nuclear Parton Distribution Functions. J. High Energy Phys. **04 (2009)** 065.

[30] Xingbo Zhao and Ralf Rapp. Forward and midrapidity charmonium production at RHIC. *Eur. Phys. J.*,

Journal of Physics: Conference Series **1602** (2020) 012029 doi:10.1088/1742-6596/1602/1/012029

C62:109–117, 2009.
[31] A. Adare et al. Pseudorapidity Dependence of Particle Production and Elliptic Flow in Asymmetric Nuclear Collisions of p+Al, p+Au, d+Au, and ^{3}He+Au at $\sqrt{s_{NN}} = 200$ GeV. *Phys. Rev. Lett.*, 121(22):222301, 2018.

Journal of Physics: Conference Series **1602** (2020) 012030 doi:10.1088/1742-6596/1602/1/012030

PHENIX results on J/ψ polarization in p+p collisions.

Alexandre Lebedev for the PHENIX collaboration
Iowa State University, Ames, IA 50011, USA
E-mail: lebedev@iastate.edu

Abstract. One of the big questions in quarkonia production is how heavy-quark pairs are produced in the initial hard scattering and hadronize into final quarkonia states. This question also affects the estimation of quarkonia breakup cross-sections and color screening in heavy ion collisions. Many existing models describe well general features of quarkonia production like cross-section and transverse momentum distributions. In order to differentiate between various models one has to study quarkonia production in more details. One of such tools is a study of angular distributions of decay leptons, usually called polarization. The PHENIX experiment has measured inclusive J/ψ polar and azimuthal angular decay coefficients in the mid ($|y| < 0.35$) and forward ($1.2 < |y| < 2.2$) rapidity regions in p+p collisions at 200 GeV and 510 GeV. This talk will present the analysis strategy as well as the results in different polarization frames.

1. Introduction

Measurements of quarkonium states in p+p collisions provide a unique tool for testing Quantum Chromodynamics (QCD). The J/ψ is a colorless neutral meson with spin 1, decays with considerable branching ratio into lepton pairs, and is produced predominantly via gluon fusion at Relativistic Heavy Ion Collider (RHIC) energies. Various theoretical models have been developed to describe the general features of J/ψ production such as transverse momentum (p_T) spectra, while consistent predictions of the spin alignment measurements for quarkonia remain a challenge.

The spin alignment of a positively charged lepton from a J/ψ decay, commonly known as "polarization", has been measured at the Tevatron [1], RHIC [2, 3], and the Large Hadron Collider [4, 5, 6]. Measuring spin alignment provides additional tests for the theory and understanding dominant quarkonium production mechanisms in different kinematic regimes. The J/ψ polarization is measured by fitting the angular distribution of a positively charged lepton, shown in Eq. (1), to data and extracting decay angular coefficients.

$$\frac{dN}{d\Omega} \sim 1 + \lambda_\theta \cos^2\theta + \lambda_{\theta\phi} \sin^2\theta \cos 2\phi + \lambda_\phi \sin 2\theta \cos\phi \tag{1}$$

where θ is the polar and ϕ is the azimuthal angles of a positive decay lepton in J/ψ rest frame. The coefficients λ_θ, $\lambda_{\theta\phi}$ and λ_ϕ are determined most commonly in the helicity (HX) frame [9], Collins-Soper (C-S) frame [10] and Gottfried-Jackson (G-J) frame [11] defined in the J/ψ production plane.

Equation (2) shows the two frame-invariant variables defined to characterize the decay angular distribution.

Journal of Physics: Conference Series **1602** (2020) 012030 doi:10.1088/1742-6596/1602/1/012030

$$\tilde{\lambda} = \frac{\lambda_\theta + 3\lambda_\phi}{1 - \lambda_\phi}, \quad F = \frac{1 + \lambda_\theta + 2\lambda_\phi}{3 + \lambda_\theta} \tag{2}$$

Previously a PHENIX J/ψ measurement at $\sqrt{s} = 510$ GeV in p+p at forward rapidity showed largely longitudinal net polarization (negative angular coefficients) [3], while a prior midrapidity PHENIX measurement at $\sqrt{s} = 200$ GeV was consistent with no strong polarization [2]. The present analysis for midrapidity J/ψ production in $\sqrt{s} = 510$ GeV collisions is complementary to both previous measurements.

2. Analysis Procedure

J/ψ's were reconstructed via invariant mass distribution of opposite sign electron pairs. Combinatorial background was subrtacted using same-sign pairs (for high p_T pairs detected in the same PHENIX Central Arm) or mixed events method (for low p_T pairs detected in opposite Arms). Signal and remaining correlated (physics) background were fit by a sum of a Crystal Ball function for signal and an exponential for the background. Number of J/ψ's was then determined by adding up counts in 2.8-3.3 GeV/c^2 mass range (after all background subtraction). An example of such fit is shown in Fig. 1. $\cos\theta - \phi$ phase space was divided into 10 bins in ϕ and 5 bins in $\cos\theta$ and number of J/ψ's were counded for each bin, as is shown in Fig. 2.

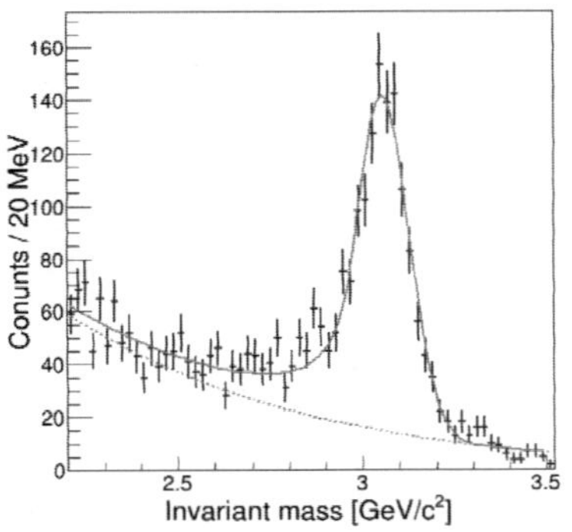

Figure 1. Example of a dielectron invariant mass distribution fit.

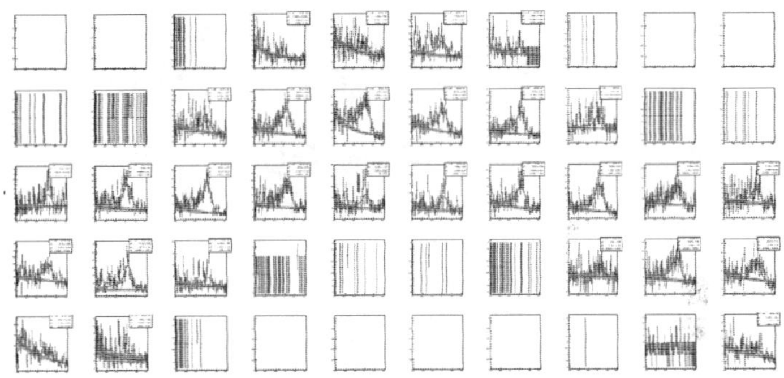

Figure 2. Invariant mass distributions for all $\cos\theta - \phi$ bins in Gottfried-Jackson frame.

The $\cos\theta - \phi$ histogram obtained in this way was then corrected for acceptance times efficiency calculated using full Geant 3 simulation and fit with Eq. 1 An example of such fit is shown in Fig. 3 for Gottfried Jackson frame at mid-rapidity. Color contours show one (red), two (blue) and three (black) sigma uncertainty of the fit. Red point indicates most probable value.

3. Results

The results of J/ψ polarization measurement at mid-rapidity are shown in Fig. 4 for the λ_θ, $\tilde{\lambda}$ and F for the three polarization frames. The results are consistent with no polarization. At low p_T PHENIX acceptance for Collins-Soper frame is very limited, and therefore the measurement was not performed.

The results of J/ψ polarization measurement at forward rapidity are shown in Fig. 5 and Fig. 6 for the λ_θ and $\tilde{\lambda}$ correspondingly. These reasults were published by the PHENIX collaboration

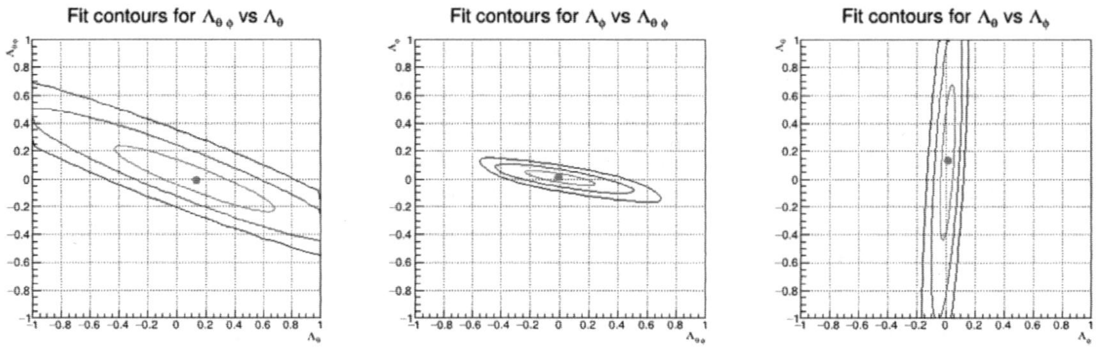

Figure 3. Example of a fit to $\cos\theta - \phi$ distribution.

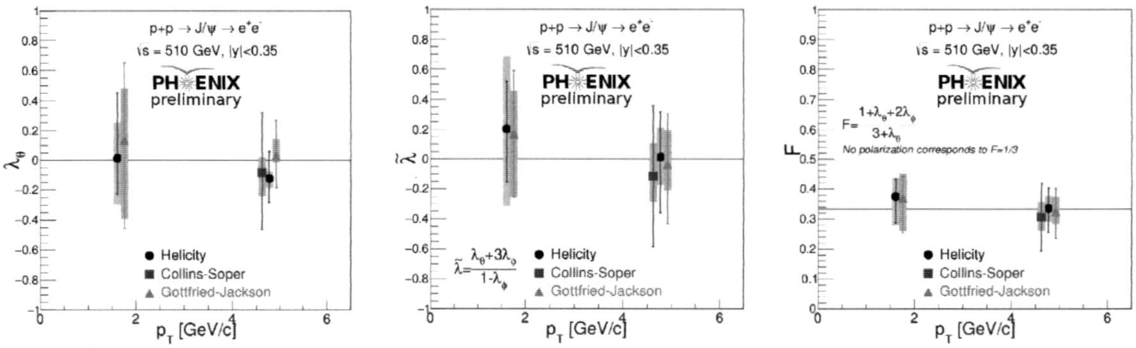

Figure 4. λ_θ (left), $\tilde{\lambda}$ (center) and F (right) as a function of transverse momentum.

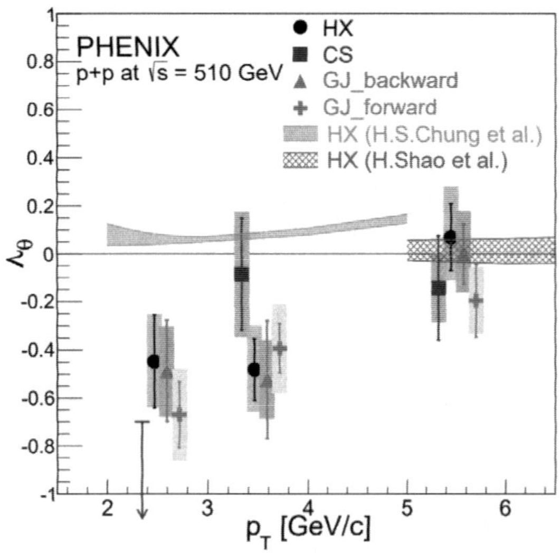

Figure 5. λ_θ as a function of p_T for forward rapidity [3].

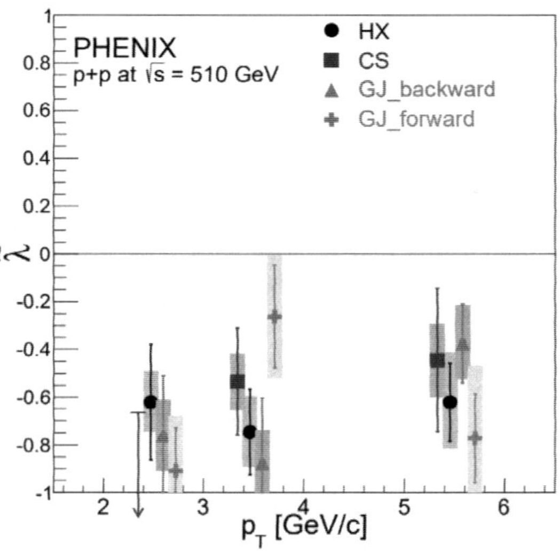

Figure 6. $\tilde{\lambda}$ as a function of p_T for forward rapidity [3].

previously in Ref. [3]. The results indicate negative (longitudinal) polarization, in contrast with two theory predictions by H.S.Chung et al. [12] and H.Shao et al. [13], also shown in Fig. 5.

λ_θ and $\tilde{\lambda}$ dependence on rapidity is shown in Fig. 7 and Fig. 8 correspondingly. The p_T integration range for mid-rapidity measurement is 3-10 GeV/c, while for forward rapidity it is 4-10 GeV/c. λ_θ is consistent with no polarization in all frames, while $\tilde{\lambda}$ seems to indicate a trend to negative (longitudinal) polarization at forward rapidity, although experimental uncertainties are rather large.

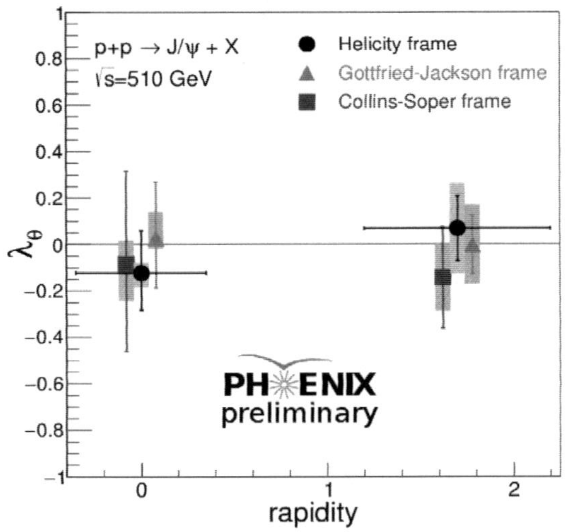

Figure 7. λ_θ as a function of rapidity.

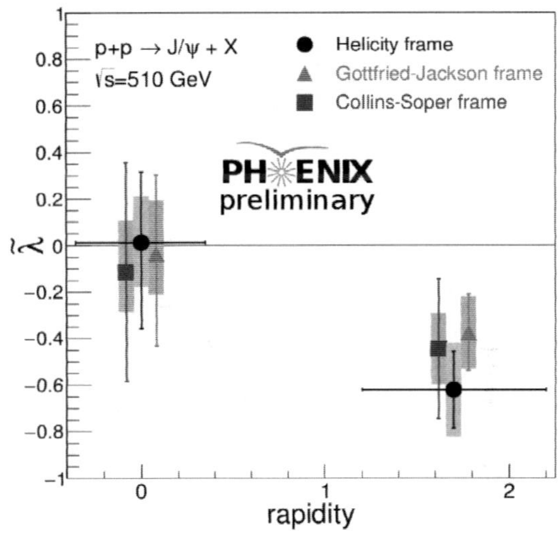

Figure 8. $\tilde{\lambda}$ as a function of rapidity.

Theory comparizons are shown in Fig. 9, Fig. 10, and Fig. 11. Fig. 9 shows λ_θ as a function of $x_T = 2p_T/\sqrt{s}$ in Helicity frame for mid-rapidity at $\sqrt{s} = 200$ GeV and $\sqrt{s} = 510$ GeV. The data are are overlaid with CSM and NRQCD predictions in the Helicity frame from [7]. Both calculations are in desagreement with the data. Fig. 10 shows λ_θ as a function of p_T for Helicity and Collins-Soper frames. The predictions from [7] are also shown for Helicity (red band) and Collins-Soper (blue band) frames. While prediction for Helicity frame disagree with the data, calculations in Collins-Soper frame describe data reasonably well. Fig. 11 shows $\tilde{\lambda}$ as a function of p_T for Heicity frame. alond with teoretical predictions from the same two models. Just like for λ_θ in Fig. 9 theory disagrees with the data.

Fig. 12 shows cross-section times branchinh ratio of J/ψ as a function of transverse momentum. The distribution is fit with a Kaplan function. Fit parameters are $A = 37.6 \pm 2.2$ nb/(GeV/c), $b = 4.33 \pm 0.28$ GeV/c, and $n = 4.61 \pm 0.32$. Green curve shows theory prediction based on full NRQCD at NLO with leading relativistic corrections that include CS and CO states, provided by Butenschön et al. [7] and is in reasonable agreement with the data above $p_T > 2GeV/c$.

It is interesting to note that the shape of the p_T dependence at $\sqrt{s} = 510$ GeV is significantly different from that at $\sqrt{s} = 200$ GeV, measured by the PHENIX experiment previously [2].

The p_T-integrated cross section times branching ratio is shown in Fig. 13 along with the previous PHENIX results at $\sqrt{s} = 200$ GeV [8] and the world results from the LHC and Tevatron. All systematic errors for CDF and ALICE experiments were added in quadrature. A simple logarithmic dependence on the collision energy is seen for J/ψ production at midrapidity, making estimates of J/ψ yield at any $\sqrt{s}$ easy, and inviting the theory community to model the trend. The fit parameters are A = 70.4 nb, and b = 9.27 GeV [1].

Journal of Physics: Conference Series **1602** (2020) 012030 doi:10.1088/1742-6596/1602/1/012030

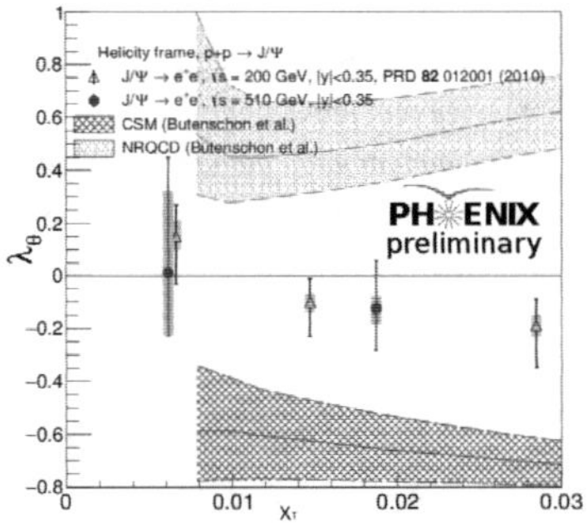

Figure 9. λ_θ as a function of $x_T = 2p_T/\sqrt{s}$.

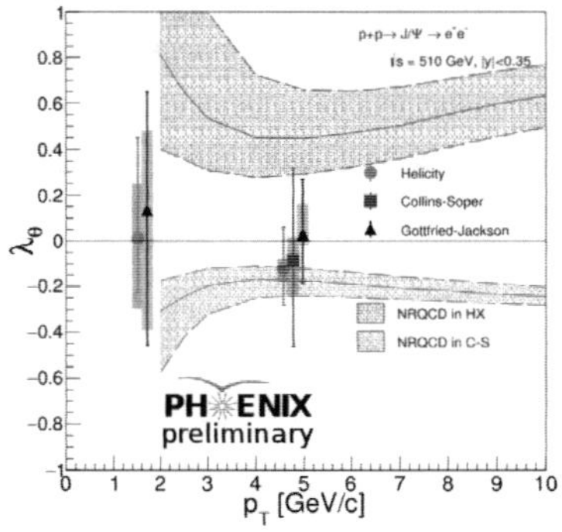

Figure 10. $\tilde{\lambda}$ as a function of p_T.

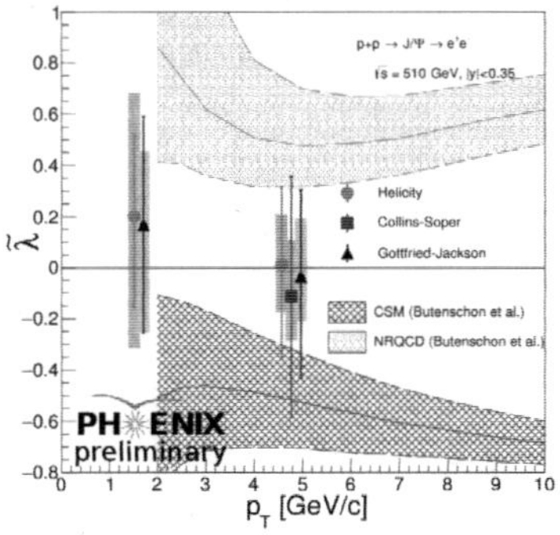

Figure 11. $\tilde{\lambda}$ as a function of p_T.

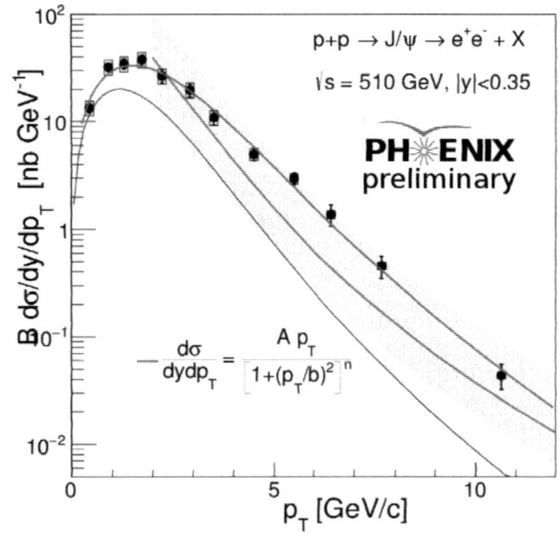

Figure 12. The measured differential J/ψ cross section times branching ratio as a function of p_T. Open rectangles show systematic point-by-point uncertainties. Global normalization uncertainty of 10.1 % is not shown. Green curve shows theory prediction by Butenschön et al. [7]. Light green band indicates theory uncertainty.

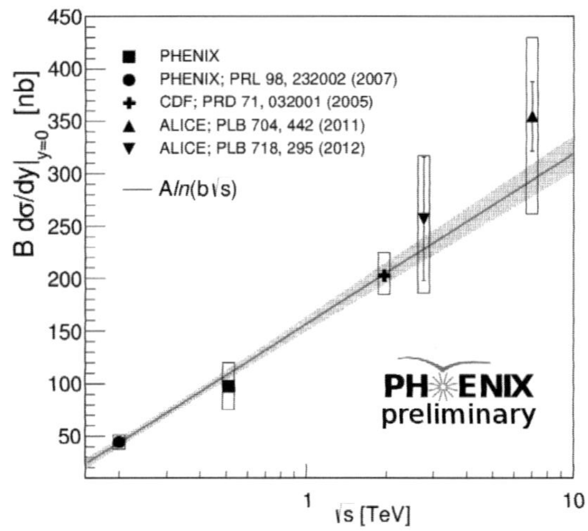

Figure 13. The PHENIX results of p_T-integrated cross section times branching ratio for J/ψ production at midrapidity, shown with the world data. The pink band shows the one-sigma fit uncertainty.

4. Conclusions

The PHENIX experiment has measured J/ψ polarization in p+p collisions at 200 and 510 GeV both at mid- and forward rapidity. The results are consistent with no polarization at mid-rapidity, but there is an indication of negative polarization at forward rapidity with some p_T dependence. Various NRQCD-based predictions can not describe the full set of polarization data, while J/ψ production cross-section as a function of p_T is reasonably well described by NRQCD calculations.

The PHENIX experiment has measured J/ψ p_T distributions and production cross-section in p+p collisions. Cross-section's $\sqrt{s}$ dependence follows a simple logarithmic law.

References

[1] A. Abulencia et al. (CDF Collaboration), Phys. Rev.484Lett.99, 132001 (2007).
[2] A. Adare et al. (PHENIX Collaboration), Phys. Rev. D48682, 012001 (2010).487[13]
[3] A. Adare et al. (PHENIX), Phys. Rev.D95, 092003488(2017), 1612.06807.489[14]
[4] B. Abelev et al. (ALICE Collaboration), Phys. Rev. Lett.490108, 082001 (2012).
[5] S. Chatrchyan et al. (CMS Collaboration), Phys. Lett. B492727, 381 (2013).493[16]
[6] R. Aaij et al. (LHCb Collaboration), Eur. Phys. J. C73,4942631 (2013)
[7] M. Butensch́on and B. A. Kniehl, Phys. Rev. Lett. 108, 172002 (2012).
[8] A. Adare et al. (PHENIX Collaboration), Phys. Rev. Lett. 98, 232002 (2007).
[9] M Jacob and G.C. Wick, Ann. Phys. 7, 404 (1959).
[10] J. C. Collins and D. E. Soper, Phys. Rev. D 16, 2219 (1977).
[11] K. Gottfried and J.D. Jackson, Nuovo Cimento 33, 309 (1964).
[12] H. S. Chung et al., Phys. Rev. D83, 037501 (2011)
[13] H. Shao et al., J. High Energ. Phys. 1505, 103 (2015)

Journal of Physics: Conference Series **1602** (2020) 012031 doi:10.1088/1742-6596/1602/1/012031

Summary of charm results from ALICE

Cristina Terrevoli (for the ALICE Collaboration)
Department of Physics, University of Houston, Texas, USA
E-mail: cristina.terrevoli@cern.ch

Abstract. An overview of the most recent charm measurements from the ALICE Collaboration is presented in this document. The study of heavy-flavour production is important to understand the properties of the hot and dense medium formed in ultra-relativistic heavy-ion collisions. In this document the journey of charm in the medium, from the production to the hadronisation into hadrons, is described. Both open charm and quarkonium measurements are presented, starting from D-meson and J/ψ production cross sections in pp collisions and their comparison with theoretical calculations. Recent measurements of relative abundances of different particle species are discussed, in particular the D_s/D^0 and Λ_c^+/D^0 ratios, that are sensitive to the hadronisation mechanisms and to the strangeness and to the baryon-to-meson enhancement. The energy loss mechanisms and the collective motion in the medium are investigated through the D-meson and J/ψ nuclear modification factor and the elliptic flow measurements. Further investigations on possible medium-like effects observed in high-multiplicity pp and p-Pb collisions are also addressed.

1. Introduction

One of the main goal of the ALICE experiment at the LHC is the characterization of the extremely hot and strongly coupled medium, the quark-gluon plamsa (QGP), formed in high-energy heavy-ion collisions. Heavy quarks, i.e. charm and beauty, are ideal probes of the medium, since, due to their large mass, they are produced in the initial stages of the collisions via hard partonic scattering processes. Therefore they probe the whole evolution of the QGP, loosing energy, and possibly thermalising as a consequence of their interactions via elastic scatterings and gluon radiation in the medium. Precise measurements of the yield modifications of heavy-flavor hadrons in heavy-ion collisions with respect to measurements in proton-proton (pp) collisions, and studies of azimuthal anisotropy for particles with charm quarks, could help to put constraints on the transport coefficients of the QGP, as the heavy-quark diffusion coefficient and its dependence on the temperature (see [1] for a recent review). Heavy quarks are also sensitive to the QGP temperature, that is reflected in the observed suppression of charmonium states, that probe the color screening and regeneration in medium.

Smaller colliding systems such as pp or p-Pb, are not only useful as reference for Pb–Pb measurements, but they are also interesting for the study of Quantum chromodynamics (QCD) and Cold Nuclear Matter (CNM) effects. Furthermore, recent observations reveal the interplay, in particle production, between fragmentation mechanisms in vacuum and quark-coalescence mechanisms in the hot medium [2, 3]. In addition, in high-multiplicity events in small systems, signatures of medium-like effects, that resemble those characteristic to QGP production, have been observed [4, 5]. Heavy-flavor measurements versus event activity play a key role in clarifying the origin of such signatures.

Journal of Physics: Conference Series **1602** (2020) 012031 doi:10.1088/1742-6596/1602/1/012031

2. Heavy flavour hadron production cross sections

The study of the production of hadrons containing heavy quarks in pp collisions is a sensitive test of QCD calculations with the factorisation approach. In this scheme, the p_{T} differential production cross sections of hadrons containing charm quarks are calculated as a convolution of three terms: (i) the parton distribution functions (PDFs) of the incoming protons, (ii) the partonic scattering cross section, calculated as a perturbative series in powers of the strong coupling constant α_{s}, and (iii) the fragmentation function, which parametrises the non-perturbative evolution of a heavy quark into a given species of heavy-flavour hadron [6]. The heavy-flavour p_{T}-differential production cross section is measured by ALICE in different systems and different center-of-mass energies. In Fig. 1 the charmed D^0 meson production cross section measured at central rapidity, on the left, and the preliminary results of the J/ψ production cross section measured at forward rapidity, on the right, are shown, both measured down to $p_{\mathrm{T}} = 0$ in pp collisions at $\sqrt{s} = 5.02$ TeV. They are also compared with models that are based on the factorization approach, with NLO calculations and different schemes [7, 8]. The uncertainties on the theoretical predictions, which are dominated by the choice of the scales of the perturbative calculation are significantly larger than the uncertainties on the measured data points. The precise measurements of the production cross sections down to $p_{\mathrm{T}} = 0$ provide important constraints to pQCD calculations and to low-x gluon PDFs and represent an essential reference for the study of effects induced by cold and hot strongly-interacting matter in the case of proton–nucleus and nucleus–nucleus collisions, respectively.

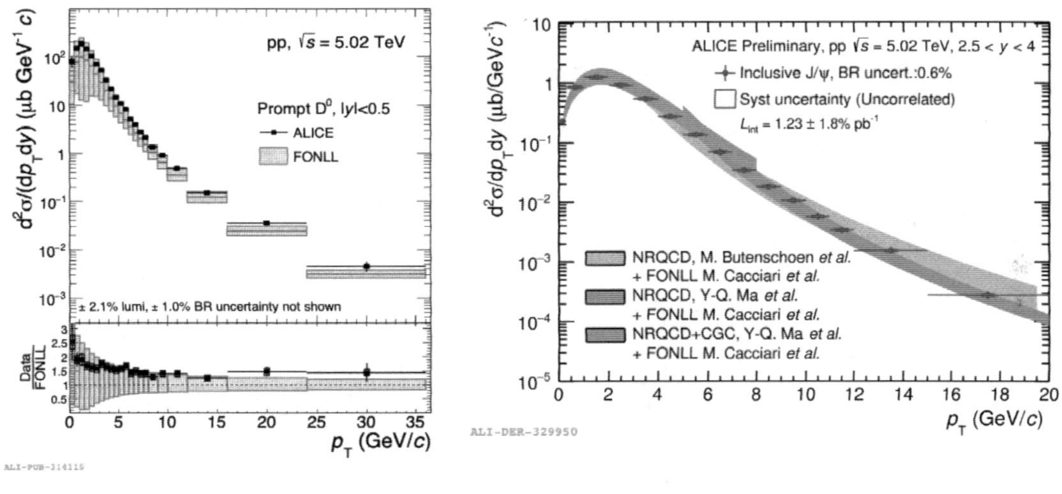

Figure 1. p_T-differential production cross section measured at central rapidity for D^0 (left) and at forward rapidity for J/ψ (right) down to $p_{\mathrm{T}} = 0$, in pp collisions at $\sqrt{s}= 5.02$ TeV. Theoretical calculations based on the QCD calculations describe data within uncertainties [7, 8].

3. Particle specie ratios in pp, p-Pb and Pb–Pb

The relative abundances of open charmed hadron species are sensitive to the fragmentation fractions of the charm quark, and, more in general, test the hadronisation scenario of charm quarks into hadrons, which has been supposed since long to be independent of the collision system and energy.

3.1. D-meson cross-section ratios

In Fig. 2 the ratios of the p_{T}-differential cross sections for prompt $\mathrm{D}^+/\mathrm{D}^0$, on the left, and $\mathrm{D}_s^+/\mathrm{D}^0$, on the right, are reported, respectively, in different collisions systems (pp and Pb–Pb collisions at $\sqrt{s_{\mathrm{NN}}}$=5.02 TeV) and different center-of-mass energies (pp collisions at $\sqrt{s}$=5.02, 7, 13 TeV).

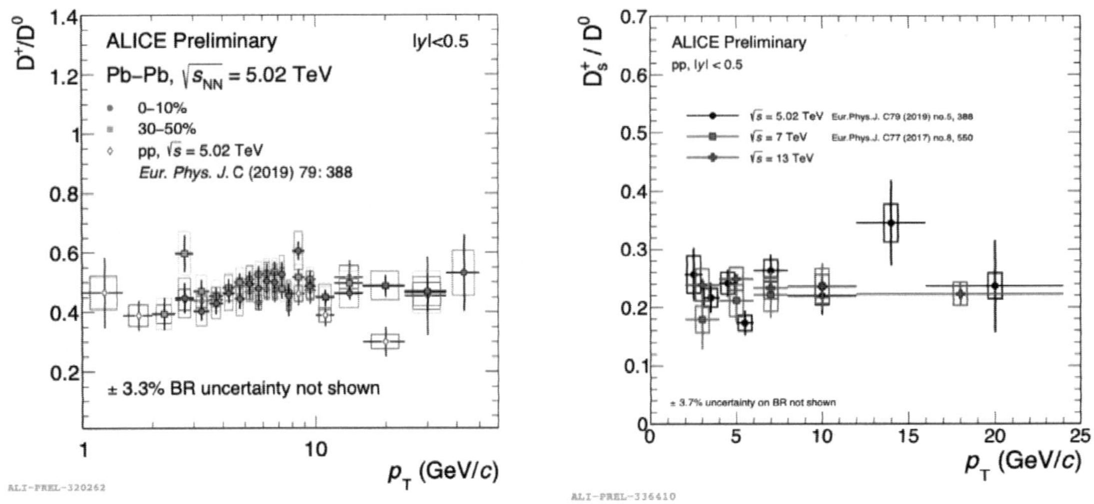

Figure 2. Ratios of $\mathrm{D}^+/\mathrm{D}^0$ production cross sections in different collision systems (pp, Pb–Pb collisions, on the left) and $\mathrm{D}_s^+/\mathrm{D}^0$ in different center-of-mass energy in pp collisions (on the right), as a function of p_{T}.

The measurements do not show a significant p_{T} dependence within the uncertainties. The agreement across different collisions systems suggests no difference between the fragmentation functions of charm quarks to D^0, D^+ in pp and Pb–Pb collisions. No differences are observed in the measurements of D_s^+ mesons in pp collisions at different center-of-mass energies. The ratios of D-meson particle species are also in agreements with measurements done at $\mathrm{e}^+\mathrm{e}^-$ colliders. Within the present uncertainties, these observations suggest that, looking uniquely in the meson sector, the universality of the fragmentation fractions for charm can not be ruled out.

3.2. Charmed baryon-to-meson cross-section ratios in pp collisions

Λ_c production cross section is measured by ALICE in pp, p–Pb and Pb–Pb collisions at central rapidity. In Fig. 3 the Λ_c^+/D^0 ratio is shown as a function of p_{T}, for pp collisions at $\sqrt{s}$= 5.02, 7 TeV, and for p-Pb collisions at $\sqrt{s_{\mathrm{NN}}}$= 5.02 TeV, and a strong p_{T} dependence is evident. In the hadronic collision systems the Λ_c^+/D^0 ratio is larger with respect to measurements done at $\mathrm{e}^+\mathrm{e}^-$ colliders, and it is underestimated by models tuned on $\mathrm{e}^+\mathrm{e}^-$ measurements [9].

Theoretical calculations based on standard PYTHIA tunes with hadronisation processes modelled via simple string fragmentation mechanisms and other MC generators [19, 20] also tuned on $\mathrm{e}^+\mathrm{e}^-$ measurements, fail to describe the data and the p_{T} trend. This could indicate that the fragmentation fractions of charm quarks are non-universal with respect to collision system and energy.

While theoretical calculations based on PYTHIA8.243 with hadronisation processes implemented via color reconnection mechanisms beyond the leading color approximation, and with an enhanced baryon formation [11], and a Statistical Hadronisation Models (SHM) with enhanced set of baryons [10], are closer to the data, and well describe the p_T shape of the measurements.

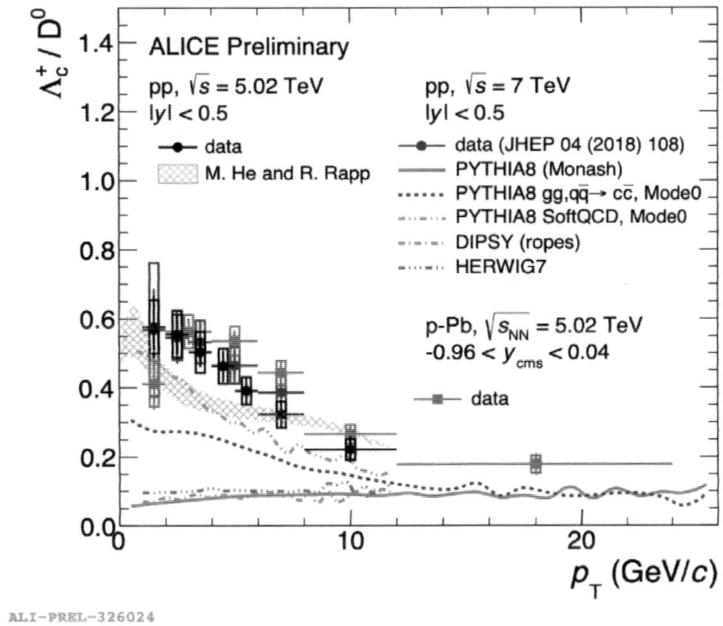

Figure 3. Λ_c^+/D^0 ratio as a function of p_T measured in pp collisions at $\sqrt{s} = 5.02$ and 7 TeV, and in p-Pb collisions at $\sqrt{s_{NN}} = 5.02$ TeV, compared to theoretical models (see text for details).

3.3. Strange to non-strange D-meson and baryon-to-meson ratios in Pb–Pb collisions

The modification of the charm-quark hadronisation mechanisms in the presence of QGP is investigated comparing the ratios between the measured yields of hadronic species in Pb–Pb and pp collisions. A modification is expected for strange D mesons with respect to non-strange D mesons, and baryon over meson states, in particular if the recombination (or coalescence) plays a role in Pb–Pb collisions. In Fig. 4, on the left panel the D_s^+/D^0 ratio is shown in the most central (0–10%) and semi-central (30–50%) collisions in Pb–Pb, in comparison with measurements done in pp collisions. A hint of larger ratio up to $p_T = 8$ GeV/c in Pb–Pb is observed for D_s^+/D^0, and could be explained as different hadronisation processes that occur in Pb–Pb, where a charm quark hadronises via coalescence with a strange quark in the QGP, and the enhancement noticed for D_s^+ is due to the strangeness enhancement in the medium [12].

In Fig. 4, on the right, the baryon-to-meson Λ_c^+/D^0 ratio in Pb–Pb collisions is shown for two centrality classes, and it is compared to the one measured in pp collisions. There is a hint of a higher Λ_c^+/D^0 ratio in Pb–Pb with respect to pp collisions at intermediate p_T for both 0–10% and 30–50% centrality classes, that could be interpreted as an interplay between recombination in medium that favours the formation of baryons, and the radial-flow that pushes the spectra towards harder p_T in Pb–Pb.

In Fig. 5, the double ratio of D_s^+/D^0 in Pb–Pb versus pp collisions, and the Λ_c^+/D^0 ratio are shown in comparisons with models that include fragmentation and coalescence mechanisms in the QGP, and describe the data within uncertainties [14, 16]. For Λ_c^+/D^0, in addition, also a model based on SHM calculations describes the data within uncertainties [15].

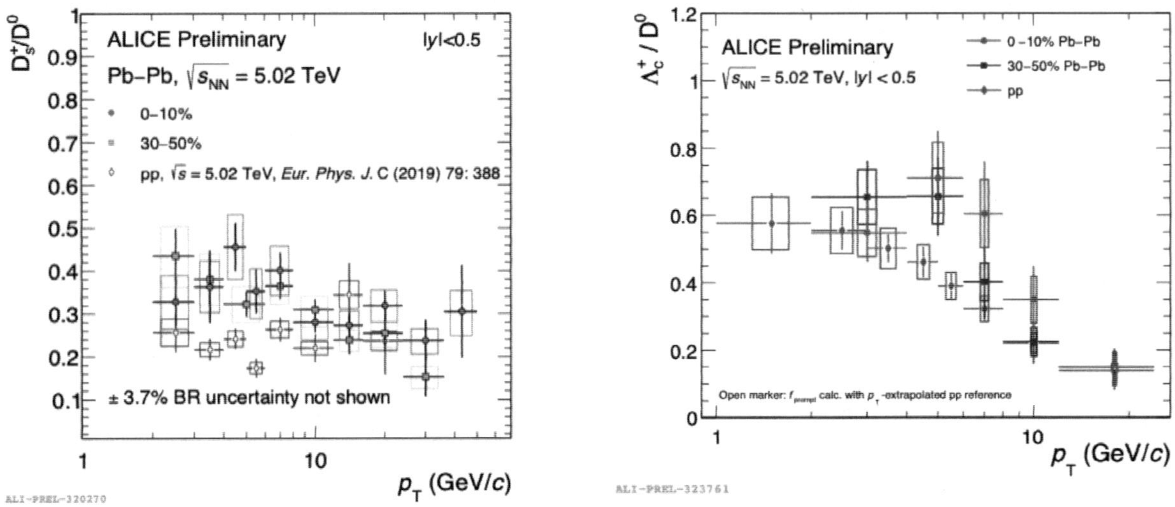

Figure 4. D_s^+/D^0 ratio of prompt yields as function of p_T for pp and Pb–Pb collisions (left). Λ_c^+/D^0 baryon-to-meson ratio in pp and Pb–Pb collisions (right).

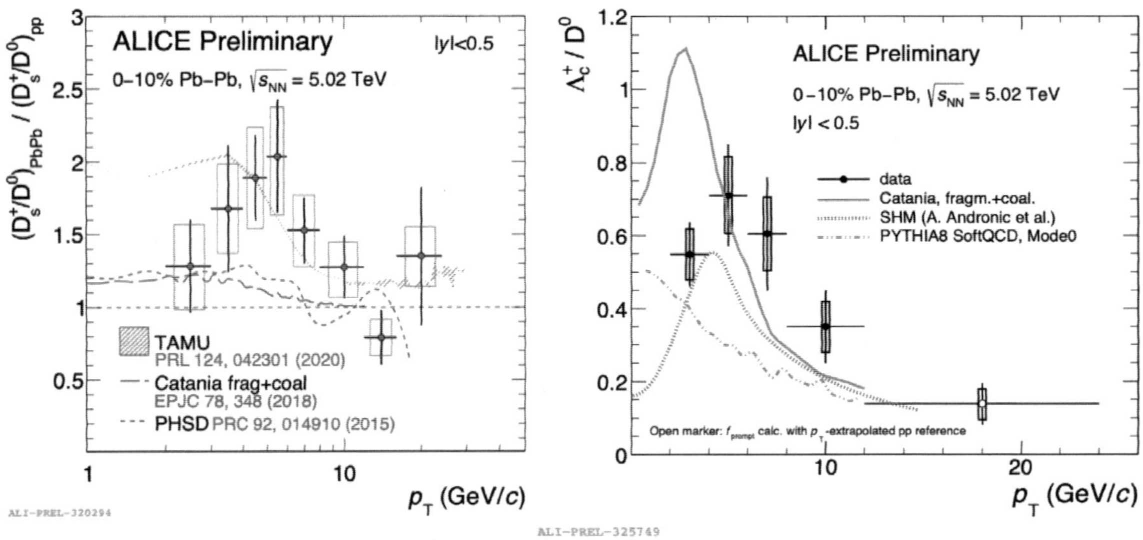

Figure 5. The double ratio D_s^+/D^0 in Pb–Pb and pp collisions in comparison with models that include framgentation and coalescence in Pb–Pb (right) and the baryon-to-meson ratio of Λ_c^+/D^0 compared with models that include fragmentation and coalescence, with statistical hadronisation model and with PYTHIA8 color reconnection tune.

Finally, profiting of the large sample of pp collisions collected at $\sqrt{s}$ = 13 TeV in Run 2, ALICE went a step further in the investigation of the baryon hadronisation mechanisms, with a more differential analysis, measuring the Λ_c^+/D^0 ratio as a function of charged particle multiplicity. The ratios are shown in Fig. 6 as a function of charged particle multiplicity, for different p_T intervals, and across different collision systems, from pp to p–Pb and Pb–Pb. At intermediate p_T, i.e. from $2 < p_T < 8$ GeV/c, focusing on the pp measurements, a hint of increasing ratio from low to high multiplicity ranges is observed, and the higher multiplicity measurements are close to the Pb–Pb measurements. The smooth increasing trend with multiplicity is also observed from pp to Pb–Pb measurements at low $p_{\rm T}$, while the trend is flat at higher $p_{\rm T}$. This could suggest that some peculiar mechanisms, probably in the hadronisation processes, related to the increase of multiplicity is present also in pp collisions, at low-intermediate $p_{\rm T}$, that bring the ratios very close to Pb–Pb measurements. Theoretical calculations based on the color reconnection mechanisms with enhanced processes for baryons formations [11], describe the data in pp versus multiplicity, within the uncertainties.

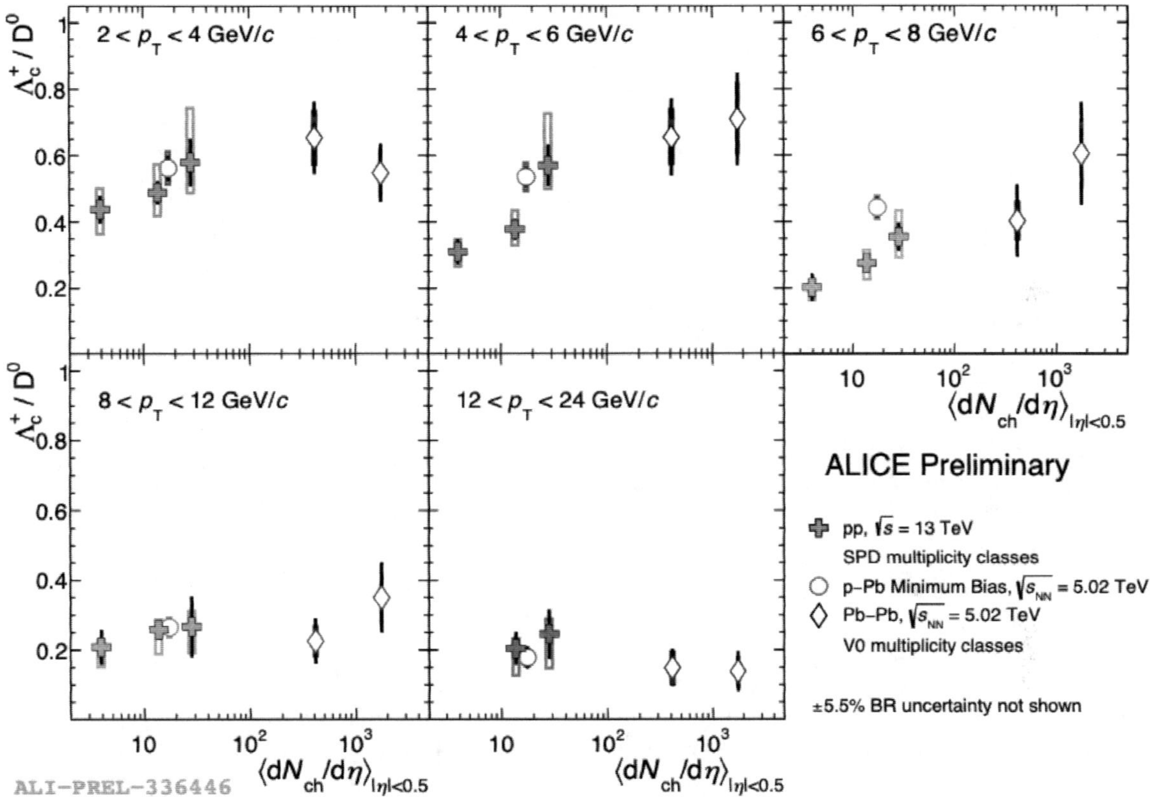

Figure 6. Λ_c^+/D^0 ratio as a function of charged particle multiplicity, from measurements performed in pp, p–Pb, to Pb–Pb, in different $p_{\rm T}$ intervals.

4. Nuclear modification factor in p–Pb and Pb–Pb

The energy loss mechanisms of charm in the QGP are investigated comparing the production of charmed hadron species in Pb–Pb collisions, scaled by the number of binary nucleon-nucleon collisions, with the production in pp collisions, measuring the nuclear modification factor $R_{\rm AA}$. In the left panel of Fig. 7, the D-meson $R_{\rm AA}$ for three different centrality classes is shown. For the first time at LHC, the D-meson spectra is measured down to $p_{\rm T}$=0 in Pb–Pb collisions, in central 0–10% collisions. A large suppression is observed at low-intermediate $p_{\rm T}$, and it is larger for the most central class, than for the semi-peripheral centrality class [22]. The suppression is due to final-state effects, and this is confirmed by the observation of the nuclear modification factor measured in p–Pb collisions, R_{pPb}, that is compatible with unity within uncertainty, as shown in the right panel of Fig. 7. The R_{pPb} is compatible with models that include CNM effects, while models that include the formation of QGP in the p–Pb collisions are disfavoured by the data [21].

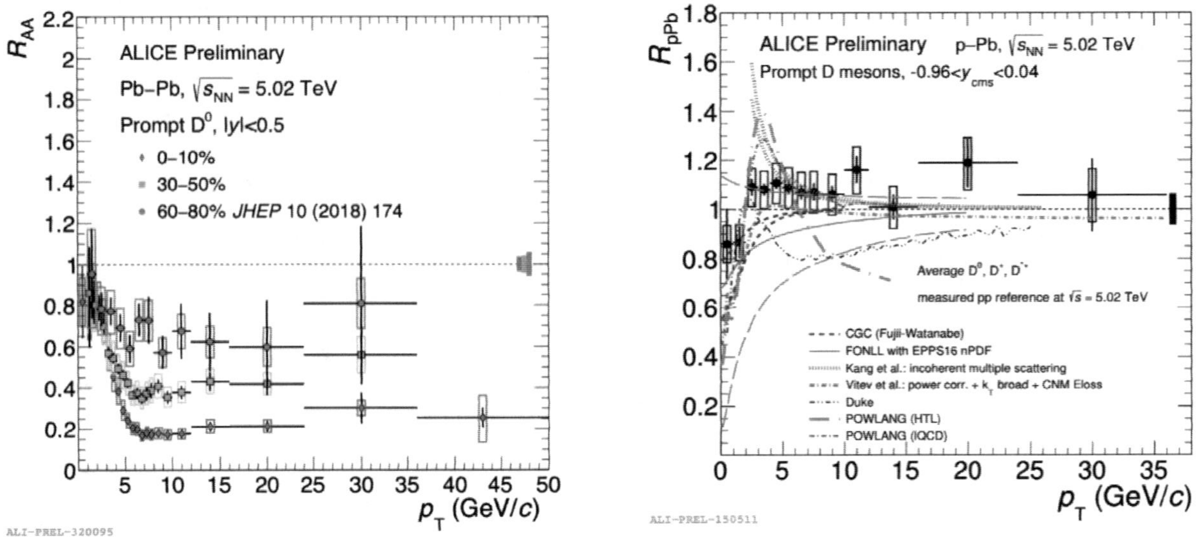

Figure 7. Left panel: D-meson R_{AA} in three different centrality classes in Pb–Pb collisions. Right panel: D-meson R_{pPb} in comparison with models that include CNM effects, and models with QGP [21].

A hierarchy in the nuclear modification factor $R_{\text{AA}}(\text{gluons, u, d}) < \text{R}_{\text{AA}}(\text{c}) < \text{R}_{\text{AA}}(\text{b})$ is expected at low-intermediate p_{T}, that could reflects the one expected for the energy loss of gluons, light and heavy quarks in the medium. The first part of this hierarchy is visible in Fig. 8, where the D-meson R_{AA} is compared with the charged particle R_{AA}. On the left panel: a larger suppression is observed for the light flavour particles at low p_{T}, while a similar suppression between charged particles and D-mesons is measured at high p_{T}. In the same figure, the D_s^+ and Λ_c^+ R_{AA} are also shown, and a hint of smaller suppression is visible, confirming the role of coalescence for strange and baryon states at intermediate p_{T}, that enhances the D_s^+ and Λ_c^+ in Pb–Pb collisions.

In the right panel of the figure 8, the non-prompt D^0 R_{AA}, i.e. the D^0 coming from a beauty meson, is shown in comparison with the prompt D^0 R_{AA}. The observation of a less suppressed non-prompt D^0 yield in the intermediate p_{T} reflects the expected quark mass ordering of the R_{AA}, since a smaller suppression for b than c quarks is predicted.

The J/ψ R_{AA} is presented in Fig. 9, for central and forward rapidity measurements (left panel). A rapidity dependence of the R_{AA} is observed, indicating that the R_{AA} measured at central rapidity is less suppressed in comparison to the one measured at forward rapidity. This observation is consistent with models that include a regeneration mechanisms of the quarkonia states in the medium [23]. In the right panel of the same figure, the J/ψ R_{AA} at central rapidity down to very low p_{T} is compared with a SHM model and a transport model, that agree with data within uncertainties [15, 18].

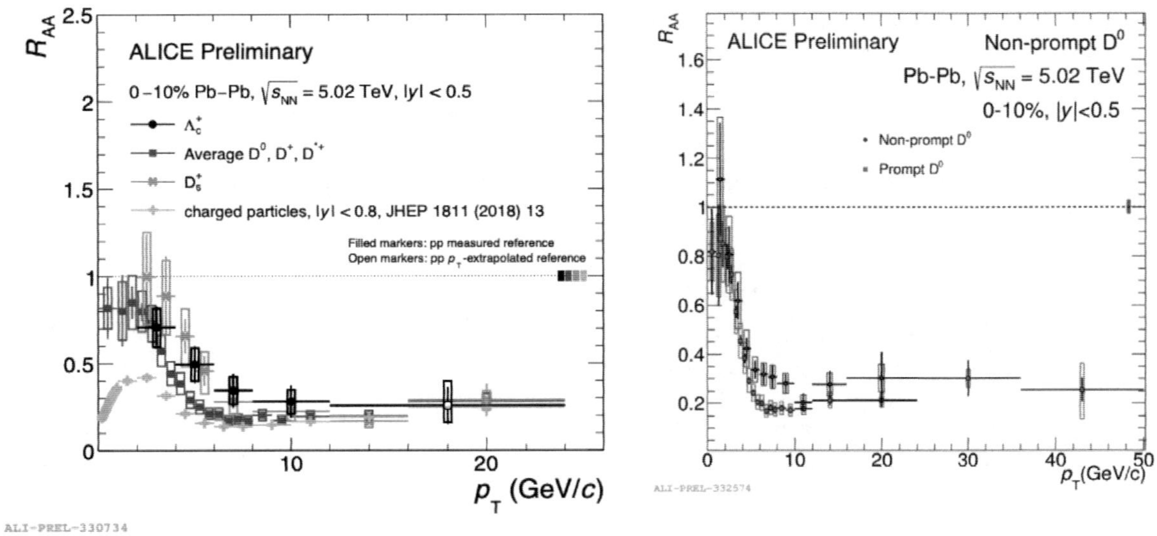

Figure 8. Left panel: Charmed hadron R_{AA} (non-strange D mesons, $\mathrm{D_s^+}$, Λ_c^+) in comparison with charged particle R_{AA} in centrality classes 0-10% in Pb–Pb collisions. Right panel: prompt and non-prompt $\mathrm{D^0}$ R_{AA} in 0-10% centrality class in Pb–Pb collisions.

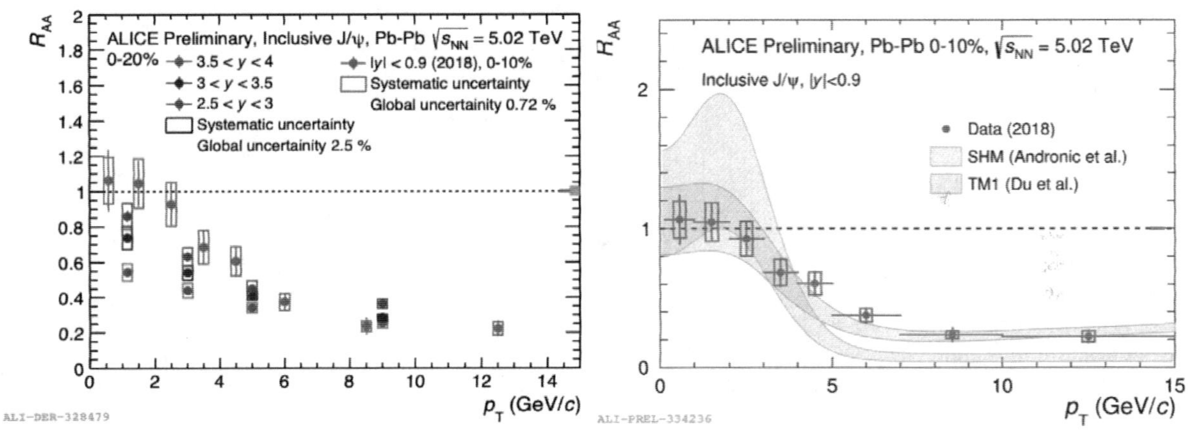

Figure 9. J/ψ R_{AA} in Pb–Pb collisions. Left panel: comparison of J/ψ R_{AA} measured at central and forward rapidity. Right panel: J/ψ R_{AA} at central rapidity, in comparison with models.

5. Nuclear modification factor, elliptic flow and constraint to models

The left panel of Fig. 10 shows the R_{AA} of D-mesons compared to several model calculations with different ingredients and different schemes for the implementation of heavy flavor transport modelling, hadronisation mechanisms, and radiative/collisional processes for energy loss [22]. Models that contain fragmentation and coalescence and both radiative and collisional processes, generally provide better description of the measurements at low and high p_{T}. The v_2 coefficient ("elliptic flow"), that quantifies the azimuthal anisothropy of D-mesons in semi-central Pb–Pb collisions, is shown in the right panel of Fig.10. A positive v_2 is observed for heavy-flavour hadrons, and this suggests that charm quarks largely thermalise in the QGP. To achieve a stronger discriminating power of data over models, a simultaneous description of R_{AA} and v_2 in the whole measured p_{T} range should be achieved. This is still a challenge for the models. With the precision nowadays achieved, the experimental measurements start to provide constraints to the models, for the characterisation of the charm and beauty interaction with the medium. This would be very useful for an improved constraint of the medium transport parameters, such as the heavy-quark diffusion coefficient, one of the main goal of the future prospects with data collected in Run 3 of the LHC [25].

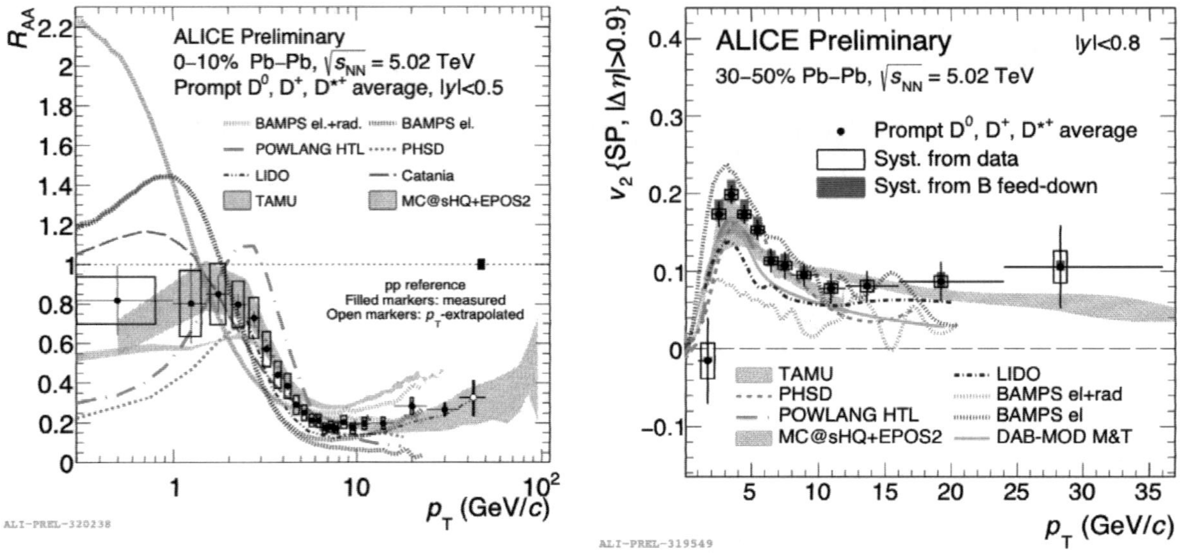

Figure 10. R_{AA} and v_2 in comparison with same set of models [24, ?]

6. Conclusions

An overview of selected recent heavy-flavor results from ALICE in pp, p-Pb and Pb-Pb colliding systems has been presented. Production of charmed mesons in pp collisions are well described by QCD-inspired models within uncertainties. Ratios of different particle species are sensitive to the hadronisation mechanisms in vacuum and in medium. D meson ratios are found to be universal among different collision systems and center of mass energies, while baryon-over-meson ratios are underestimated by theoretical calculations, and different with respect to measurements done at the electron colliders, indicating that baryon-hadronisation mechanisms are not universal, and that they are probably not fully understood. The suppression of charmed non-strange and strange D-mesons, charmed baryons, non-prompt D mesons and charged particles follow the expected hierarchy, suggested by the energy loss expectations of different quarks in the medium. A positive azimuthal anisotropy is observed for charmed mesons, that indicates that charm participates to the collective motion in the medium reaching thermalisation before freeze out. The Run 3 phase of LHC with increased luminosity and detector upgrades will bring the era of precision charmed baryon and beauty measurements and will help to better constraint the properties of the QGP.

References

[1] Prino F and Rapp R 2016 *Journal of Phys. G: Nucl. and Part. Phys.* **43**
[2] Acharya S et al. (ALICE) 2018 *J.High Ener. Phys.* **1810**, 1742019

[3] Acharya S et al. (ALICE) 2019 *Phys. Letters B* **793** 212-223
[4] Abelev B et al. (ALICE) 2013 *Phys. Lett. B* **726** 164–177
[5] Adam J et al. (ALICE) 2017 *Nature Phys* **13**, 535–539
[6] Collins J C, Soper Davison E and Sterman G F 1989 *Adv. Ser. Direct. High Energy Phys.* **5** 1-91
[7] Acharya S et al. (ALICE) 2019 *Eur.Phys.J.C* **79** no.5, 388
[8] Acharya S et al. (ALICE) 2019 *J. High Ener. Phys.* **84**
[9] Acharya S et al. (ALICE) 2018 *J. High Ener. Phys* **04** 108 523
[10] He M and Rapp R, 2019 *Phys. Lett. B* **795** 117
[11] Christiansen R and Skands P Z 2015, *JHEP* **08** 003531
[12] Rapp R et al., 2020 *Phys. Rew. Lett.* **124** 042301
[13] Song T et al., 2015 *Phys. Rev. C* **92** 014910
[14] Plumari S et al. 2018 *Eur. Phys. J. C* **78** no.4, 348
[15] Andronic A. et al., 2019 *Phys. Lett. B* **797** 134836
[16] He M, Fries R J and Rapp R, 2014 *Phys. Lett. B* **735** 445
[17] Acharya S et al. (ALICE) 2020 *Phys. Lett. B* **805** 135434
[18] Du et al. 2011 *Nucl. Phys. A* **859** 114-125
[19] Bahr M et al. 2008 *Eur. Phys. J. C* **58** 639–707
[20] Bierlich C and Christiansen J R, 2015 *Phys. Rev. D* **92** no. 9, 094010
[21] Acharya S et al. (ALICE) 2019 *JHEP* **92**
[22] Acharya S et al. (ALICE) 2018 *JHEP* **1810** 174
[23] Acharya S et al. (ALICE) 2020 *Phys. Lett. B* **805** 135434
[24] Acharya S et al. (ALICE) 2020 *CERN-EP-2020-082*, arXiv:2005.11131
[25] Citron Z, et al, 2019 *CERN-LPCC-2018-07*, arXiv:1812.06772

Two-component source to explain Λ and $\bar{\Lambda}$ global polarization in non-central heavy-ion collisions

A Ayala[1,2], E Cuautle[1], I Domínguez[3], I Maldonado[3], J Salinas[1] and M E Tejeda-Yeomans[4,5]

[1]Instituto de Ciencias Nucleares, Universidad Nacional Autónoma de México, Apartado Postal 70-543, CdMx 04510, Mexico
[2]Centre for Theoretical and Mathematical Physics, and Department of Physics, University of Cape Town, Rondebosch 7700, South Africa
[3]Facultad de Ciencias Físico-Matemáticas, Universidad Autónoma de Sinaloa, Avenida de las Américas y Boulevard Universitarios, Ciudad Universitaria, C.P. 80000, Culiacán, Sinaloa, Mexico
[4]Facultad de Ciencias - CUICBAS, Universidad de Colima, Bernal Díaz del Castillo No. 340, Col. Villas San Sebastián, 28045 Colima, Mexico
[5]Departamento de Física, Universidad de Sonora, Boulevard Luis Encinas J. y Rosales, Colonia Centro, Hermosillo, Sonora 83000, Mexico

E-mail: ivonne.alicia.maldonado@gmail.com

Abstract. The STAR Beam Energy Scan program has found a difference in the global polarization of Λs and $\bar{\Lambda}$s produced in Au + Au collisions. This difference is larger for lower center of mass collision energies. In this work we show that a two-component source, consisting of a high density core and a lower density corona, can describe quantitatively the Λ and $\bar{\Lambda}$ polarizations as a function of collision energy.

1. Introduction

The polarization of hadrons in high energy collisions has been a topic of much interest mainly because this is related to the possibility to understand the role played by spin degrees of freedom during hadronization. In heavy-ion reactions, hadron polarization properties can provide useful information to understand hadron production mechanisms. The polarization of particles along a preferred direction sheds light on how these properties are linked to those of the medium produced in the reaction, which in turn are associated with the initial conditions and the dynamics of the Quark Gluon Plasma (QGP). In these reactions, the possibility to create a vortical fluid can be quantified in terms of the thermal vorticity [1–13].

The Λ polarization was discovered in the mid-70's [14] and it has been studied in different collision systems such as p+p, p+A, A+A, DIS and e^+ + e^- over a wide energy range. In this context, the Λ hyperon is one of the most studied particles given that it is easy to produce and to identify by reconstructing the topology of its decay into two particles of opposite charge. However, the mechanism whereby polarization occurs has not yet been understood, although several models have been developed to try to explain it [15–35]. Furthermore, interest in this topic has been recently renewed since the global polarization can be used to measure the thermal vorticity and perhaps also the electromagnetic field generated within the plasma stage in non-central heavy-ion collisions [36, 37].

Recently the magnitude of the global polarization has been measured by ALICE [38] and the STAR Beam Energy Scan (BES) [39, 40]. The results of the latter have obtained the global polarization of Λs and $\bar{\Lambda}$s as a function of the collision energy. These polarization are at the level of a few percent and decrease as the collision energy increases from $\sqrt{s_{NN}} = 7.7$ to 200 GeV. They also exhibit a clear difference between Λ and $\bar{\Lambda}$ polarizations, the latter being larger for lower collision energies.

In this work we show that the Λ and $\bar{\Lambda}$ polarization in semi-central heavy-ion collisions can be understood when accounting for the abundances of Λs and $\bar{\Lambda}$s coming from a high density core and a less dense corona, using the results of the intrinsic *s*-quark/antiquark polarization as a seed for the $\Lambda/\bar{\Lambda}$ polarization [41].

2. Λ and $\bar{\Lambda}$ from two-component model

In non-central heavy ion collisions, Λs and $\bar{\Lambda}$s are produced in different density regions within the interaction zone [42], as is shown in the Fig. 1. Therefore, their properties can differ depending on whether these particles come from the core, the central region with higher density, or from the corona, the peripheral region with lower density.

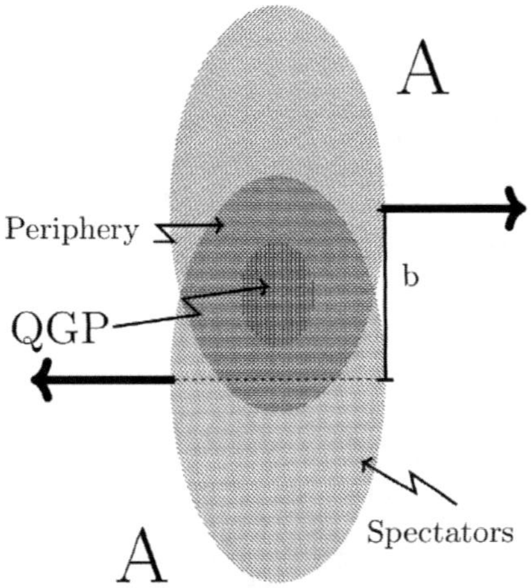

Figure 1. Representation of a non-central heavy-ion collision of a symmetric system with impact parameter b(Taken from Ref. [42]). The Λs and $\bar{\Lambda}$s coming from the core or central region are produced by QGP processes,whereas, those from the periphery, or corona, are produced by n + n kind of reactions.

In this scenario the number of Λs produced in the collision N_Λ can be rewritten in terms of the number of Λs coming from the corona $N_{\Lambda\,\text{QGP}}$ plus the number of Λs originating from the core $N_{\Lambda\,\text{REC}}$, such that

$$N_\Lambda = N_{\Lambda\,\text{QGP}} + N_{\Lambda\,\text{REC}} \tag{1}$$

The subscripts "QGP" and "REC" refer to the kind of processes responsible for the production of these hyperons, that is, QGP and recombination induced processes, respectively. The latter are similar to the polarization producing processes in p+p reactions.

3. Polarization from a two-component model

Considering that the polarization asymmetry of a given baryon species produced in high energy reactions, defined by

$$\mathcal{P}^\Lambda = \frac{N^\uparrow - N^\downarrow}{N^\uparrow + N^\downarrow} \tag{2}$$

namely, the ratio of the difference between the number of baryons with their spin pointing along and opposite to a given direction, to their sum, can be rewritten in terms of the number of Λs or $\bar{\Lambda}$s produced in the different density regions, then the Λ and $\bar{\Lambda}$ polarization is given by

$$\begin{aligned} \mathcal{P}^{\Lambda} &= \frac{(N^{\uparrow}_{\Lambda\,\mathrm{QGP}} + N^{\uparrow}_{\Lambda\,\mathrm{REC}}) - (N^{\downarrow}_{\Lambda\,\mathrm{QGP}} + N^{\downarrow}_{\Lambda\,\mathrm{REC}})}{(N^{\uparrow}_{\Lambda\,\mathrm{QGP}} + N^{\uparrow}_{\Lambda\,\mathrm{REC}}) + (N^{\downarrow}_{\Lambda\,\mathrm{QGP}} + N^{\downarrow}_{\Lambda\,\mathrm{REC}})}, \\ \mathcal{P}^{\overline{\Lambda}} &= \frac{(N^{\uparrow}_{\overline{\Lambda}\,\mathrm{QGP}} + N^{\uparrow}_{\overline{\Lambda}\,\mathrm{REC}}) - (N^{\downarrow}_{\overline{\Lambda}\,\mathrm{QGP}} + N^{\downarrow}_{\overline{\Lambda}\,\mathrm{REC}})}{(N^{\uparrow}_{\overline{\Lambda}\,\mathrm{QGP}} + N^{\uparrow}_{\overline{\Lambda}\,\mathrm{REC}}) + (N^{\downarrow}_{\overline{\Lambda}\,\mathrm{QGP}} + N^{\downarrow}_{\overline{\Lambda}\,\mathrm{REC}})}. \end{aligned} \tag{3}$$

After some algebra, we can express the previous equations as

$$\begin{aligned} \mathcal{P}^{\Lambda} &= \frac{\left(\mathcal{P}^{\Lambda}_{\mathrm{REC}} + \frac{N^{\uparrow}_{\Lambda\,\mathrm{QGP}} - N^{\downarrow}_{\Lambda\,\mathrm{QGP}}}{N_{\Lambda\,\mathrm{REC}}}\right)}{\left(1 + \frac{N_{\Lambda\,\mathrm{QGP}}}{N_{\Lambda\,\mathrm{REC}}}\right)}, \\ \mathcal{P}^{\overline{\Lambda}} &= \frac{\left(\mathcal{P}^{\overline{\Lambda}}_{\mathrm{REC}} + \frac{N^{\uparrow}_{\overline{\Lambda}\,\mathrm{QGP}} - N^{\downarrow}_{\overline{\Lambda}\,\mathrm{QGP}}}{N_{\Lambda\,\mathrm{REC}}}\right)}{\left(1 + \frac{N_{\overline{\Lambda}\,\mathrm{QGP}}}{N_{\overline{\Lambda}\,\mathrm{REC}}}\right)}, \end{aligned} \tag{4}$$

where $\mathcal{P}^{\Lambda}_{\mathrm{REC}}$ and $\mathcal{P}^{\overline{\Lambda}}_{\mathrm{REC}}$ refers to the polarization along the global angular momentum produced in the corona

$$\mathcal{P}^{\Lambda}_{\mathrm{REC}} = \frac{N^{\uparrow}_{\Lambda\,\mathrm{REC}} - N^{\downarrow}_{\Lambda\,\mathrm{REC}}}{N^{\uparrow}_{\Lambda\,\mathrm{REC}} + N^{\downarrow}_{\Lambda\,\mathrm{REC}}}, \tag{5}$$

$$\mathcal{P}^{\overline{\Lambda}}_{\mathrm{REC}} = \frac{N^{\uparrow}_{\overline{\Lambda}\,\mathrm{REC}} - N^{\downarrow}_{\overline{\Lambda}\,\mathrm{REC}}}{N^{\uparrow}_{\overline{\Lambda}\,\mathrm{REC}} + N^{\downarrow}_{\overline{\Lambda}\,\mathrm{REC}}}. \tag{6}$$

We can make some working estimations as follows:

- $\mathcal{P}^{\Lambda}_{\mathrm{REC}} = \mathcal{P}^{\overline{\Lambda}}_{\mathrm{REC}} = 0$ because, although the colliding nucleons in this zone partake of the vortical motion, reactions in cold nuclear matter are less efficient to align the spin in the direction of the angular momentum than in the QGP.
- The intrinsic global Λ and $\bar{\Lambda}$ polarizations z and $\bar{z}$, defined as

$$\begin{aligned} z &= \frac{(N^{\uparrow}_{\Lambda\,\mathrm{QGP}} - N^{\downarrow}_{\Lambda\,\mathrm{QGP}})}{N_{\Lambda\,\mathrm{QGP}}} \\ \bar{z} &= \frac{(N^{\uparrow}_{\overline{\Lambda}\,\mathrm{QGP}} - N^{\downarrow}_{\overline{\Lambda}\,\mathrm{QGP}})}{N_{\overline{\Lambda}\,\mathrm{QGP}}} \simeq \frac{(N^{\uparrow}_{\overline{\Lambda}\,\mathrm{QGP}} - N^{\downarrow}_{\overline{\Lambda}\,\mathrm{QGP}})}{N_{\Lambda\,\mathrm{QGP}}}, \end{aligned} \tag{7}$$

  are non zero and come mainly from the core, since the reactions in this zone are more efficient to align the particle spin to global angular momentum, albeit they are small. Additionally, we assume that $N_{\overline{\Lambda}\,\mathrm{QGP}} \simeq N_{\Lambda\,\mathrm{QGP}}$ in the QGP.
- We expect that the number of Λs produced in the corona is larger than the number of $\bar{\Lambda}$s, since they come from cold nuclear matter collisions and these processes are related to p +

p reactions in which it is more difficult to produce three antiquarks than only one s-quark. Then we can write

$$N_{\overline{\Lambda}\,\mathrm{REC}} \equiv w N_{\Lambda\,\mathrm{REC}} \tag{8}$$

with $w < 1$. Simulations of p + p collisions using the Ultrarelativistic Quantum Molecular Dynamics (UrQMD) generator, show that the value of w is smaller than 0.4 for the energies considered as can be seen in Fig. 2.

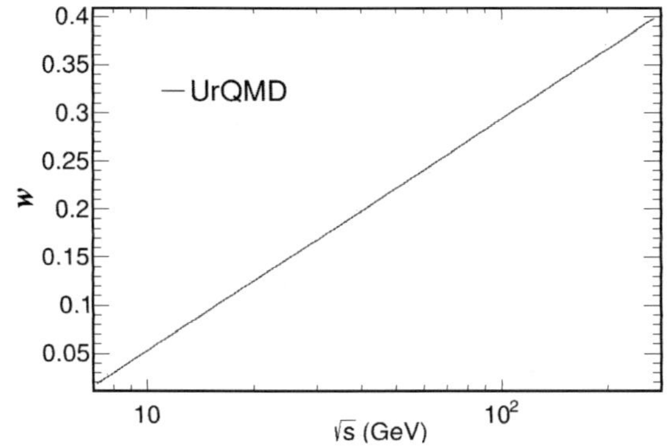

Figure 2. $w = N_{\overline{\Lambda}\,\mathrm{REC}}/N_{\Lambda\,\mathrm{REC}}$ from p + p collisions simulated with UrQMD. w increases with energy. For energy ranges below 200 GeV, the maximum w-value is smaller than 0.4

With the previous assumptions we can rewrite Eqs. 4 as

$$\mathcal{P}^{\Lambda} = \frac{z\frac{N_{\Lambda\,\mathrm{QGP}}}{N_{\Lambda\,\mathrm{REC}}}}{\left(1+\frac{N_{\Lambda\,\mathrm{QGP}}}{N_{\Lambda\,\mathrm{REC}}}\right)}, \tag{9}$$

$$\mathcal{P}^{\overline{\Lambda}} = \frac{\left(\frac{\bar{z}}{w}\right)\frac{N_{\Lambda\,\mathrm{QGP}}}{N_{\Lambda\,\mathrm{REC}}}}{\left(1+\left(\frac{1}{w}\right)\frac{N_{\Lambda\,\mathrm{QGP}}}{N_{\Lambda\,\mathrm{REC}}}\right)} \tag{10}$$

which depend on z and $\bar{z}$ the intrinsic Λ and $\bar{\Lambda}$ polarization, w and the ratio $\frac{N_{\Lambda\,\mathrm{QGP}}}{N_{\Lambda\,\mathrm{REC}}}$. We expect $\bar{z}$ to be smaller than z Nonetheless, the factor $1/w > 1$ amplifies the global $\bar{\Lambda}$ polarization which leads to $\mathcal{P}^{\overline{\Lambda}} > \mathcal{P}^{\Lambda}$. This is shown in Fig. 3 in which we plot $\mathcal{P}^{\overline{\Lambda}}/\mathcal{P}^{\Lambda}$ as a function of w, for different values of the intrinsic polarization and of the ratio of Λ and $\bar{\Lambda}$ polarizations. We can see that in the extreme situation where $\bar{z} = z$ and $N_{\Lambda\,\mathrm{QGP}}/N_{\Lambda\,\mathrm{REC}} = 1$, $\mathcal{P}^{\overline{\Lambda}}/\mathcal{P}^{\Lambda}$ is always larger than 1 for $0 < w < 1$. Considering a more realistic scenario, we take the intrinsic polarization to satisfy $\bar{z} < z$ with $N_{\Lambda\,\mathrm{QGP}}/N_{\Lambda\,\mathrm{REC}}$ smaller than 1 and find that there is still a range of w values for which $\mathcal{P}^{\overline{\Lambda}}/\mathcal{P}^{\Lambda}$ is larger than 1. This region shrinks when $N_{\Lambda\,\mathrm{QGP}}/N_{\Lambda\,\mathrm{REC}} > 1$.

To get the ratio $N_{\Lambda\,\mathrm{QGP}}/N_{\Lambda\,\mathrm{REC}}$ we analyze the production of Λs in the QGP and REC regions. The number of Λs in the QGP region varies as the square of the number of participants $N_{p\,\mathrm{QGP}}$ in the collision at a given impact parameter b

$$\langle s \rangle = N_{\Lambda\,\mathrm{QGP}} = c N_{p\,\mathrm{QGP}}^{2} \tag{11}$$

with $0.001 \leq c \leq 0.005$ [43], for the case when assuming that all the s-quarks hadronize into Λs. Nevertheless, since we know that also other strange hadrons are produced, c becomes smaller.

The number of participants is calculated with

$$N_{p\,\mathrm{QGP}} = \int n_p(\vec{\sigma}, \vec{b})\theta[n_p(\vec{\sigma}, \vec{b}) - n_c] d^2 s \tag{12}$$

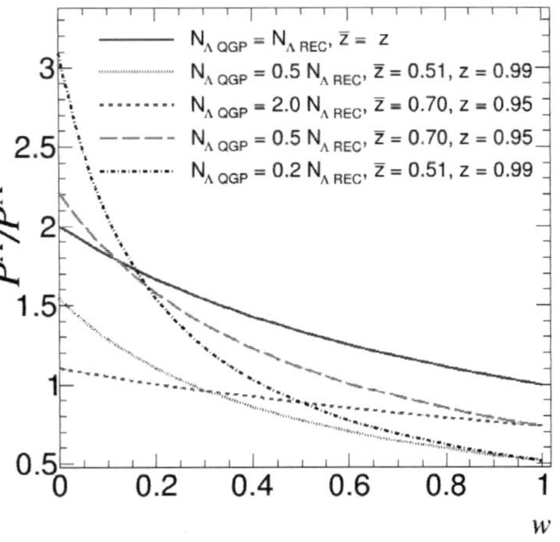

Figure 3. $\mathcal{P}^{\bar{\Lambda}}/\mathcal{P}^{\Lambda}$ as a function of w. We observe that this ratio decreases as w increases. However, there is a range of w-values for which $\mathcal{P}^{\bar{\Lambda}}/\mathcal{P}^{\Lambda} > 1$ whose width depends on the relation between $N_{\Lambda\,\mathrm{QGP}}$, $N_{\Lambda\,\mathrm{REC}}$, z and $\bar{z}$.

in terms of the density of participants per unit transverse area given by

$$n_p(\vec{s},\vec{b}) = T_A(\vec{s})[1 - e^{-\sigma_{NN}T_B(\vec{s}-\vec{b})}] + T_B(\vec{s}-\vec{b})[1 - e^{-\sigma_{NN}T_A(\vec{s})}] \tag{13}$$

with $\vec{b}$ the vector directed along the impact parameter on the nuclei overlap area, σ_{NN} is the collision energy dependent nucleon-nucleon cross-section [44, 45] and $T_A(\vec{s}) = \int_{-\infty}^{\infty} \rho_A(\vec{s},z)dz$ is the thickness function, the nucleon density per unit area in the transverse plane to the collision axis. We take as the nuclear density

$$\rho_A(\vec{r}) = \frac{\rho_0}{1 + e^{\frac{r-R_A}{a}}} \tag{14}$$

a Woods-Saxon profile with a skin depth $a = 0.41$ and a radius for a nucleus with mass number A of $R_A = 1.1A^{1/3}$ fm.

On the other hand, the number of Λs produced in the corona can be written as

$$N_{\Lambda\,\mathrm{REC}} = \sigma_{NN}^{\Lambda} \int d^2s T_B(\vec{b}-\vec{s})T_A(\vec{s})\theta[n_c - n_p(\vec{s},\vec{b})] \tag{15}$$

where the nucleon-nucleon cross-section for Λ production is parameterized using data as $\sigma_{\Lambda}^{NN} = 1.37\ln(\sqrt{s} - 0.94)$. As in the case of the QGP region, the number of Λs depends on the minimal critical density n_c required to produce a QGP. The accepted value is 3.3 fm^{-2} given by the J/Ψ suppression in the threshold model [46]. Figure 4 shows $N_{\Lambda\,\mathrm{REC}}$ and $N_{\Lambda\,\mathrm{QGP}}$ as a function of impact parameter for a collision energy of $\sqrt{s_{NN}} = 7.7$ GeV. Notice that in this case, the ratio $N_{\Lambda\,\mathrm{QGP}}/N_{\Lambda\,\mathrm{REC}}$ becomes smaller than 1 for $b \gtrsim 6$ fm.
We consider that the intrinsic global polarization of Λ and $\bar{\Lambda}$ comes from the corresponding polarization of strange quarks (z) and antiquarks ($\bar{z}$) [47] given by

$$z = 1 - e^{-\frac{t}{\tau}}, \ \bar{z} = 1 - e^{-\frac{t}{\bar{\tau}}} \tag{16}$$

as a function of the formation time (t) of Λ ($\bar{\Lambda}$) and the relaxation time τ ($\bar{\tau}$) required for the alignment between the spin of the quark (antiquark) and the thermal vorticity. We resort to the results of Ref. [48] in which the relaxation time, for the alignment between the spin of quark s or $\bar{s}$ with the thermal vorticity is computed as a function of the collision energy calculated at

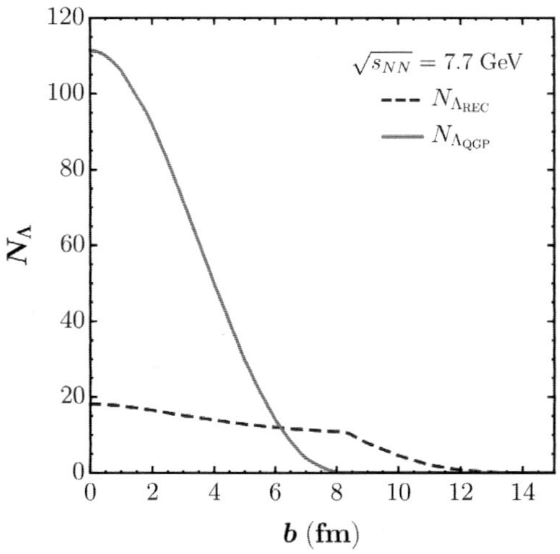

Figure 4. Number of Λs produced in the core and in the corona as a function of impact parameter. The number of Λs produced in the core $N_{\Lambda\,QGP}$ is smaller than the number of Λs produced in the corona for an impact parameter larger than a certain value. In the case of $\sqrt{s_{NN}} = 7.7$ GeV, this occurs for $b \gtrsim 6$ fm.

the freeze-out, when the maximum baryon density is reached, in terms of the temperature T and the baryon chemical potential μ_B as a function of the energy parameterized as

$$\begin{aligned} T(\mu_B) &= 166 - 139\mu_B^2 - 53\mu_B^4 \\ \mu_B &= \frac{1308}{1000 + 0.273\sqrt{s}} \end{aligned} \tag{17}$$

where μ_B and T are expressed in MeV [49].

Figure 5 shows the intrinsic global polarization z for Λ and $\bar{z}$ for $\bar{\Lambda}$ calculated with Eq. (16) for semi-central collisions at an impact parameter $b = 8$ fm and a formation time $t = 5$ fm, as a function of energy. We observe that $\bar{z} < z$ for low collision energies, behavior that is inverted for energy $\sqrt{s_{NN}} > 200$ GeV. For smaller formation times the intrinsic polarization z and $\bar{z}$ are smaller but the trend is similar.

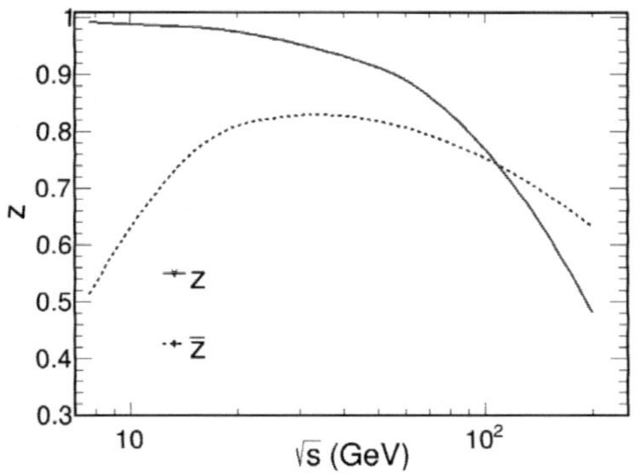

Figure 5. Intrinsic global polarization z for Λ and $\bar{z}$ for $\bar{\Lambda}$ as functions of energy $\sqrt{s_{NN}}$ for semicentral collisions at an impact parameter $b = 8$ fm and $t = 5$ fm. Notice that $\bar{z} < z$ for $\sqrt{s_{NN}} < 200$ GeV, energy for which this behavior is inverted.

Figure 6 shows the global polarization calculated for two different formation times for Λ and $\bar{\Lambda}$ within the QGP, $t = 1, 3$ fm and for $b = 8$ fm. The calculation is compared with data from the STAR-BES. Notice that the data is well described over the entire collision energy range.

4. Summary and Conclusions

The Λ and $\bar{\Lambda}$ global polarization measured by the STAR-BES shows that the polarization of $\bar{\Lambda}$ is larger than Λ at low energies and the difference decreases as the energy increases. In this

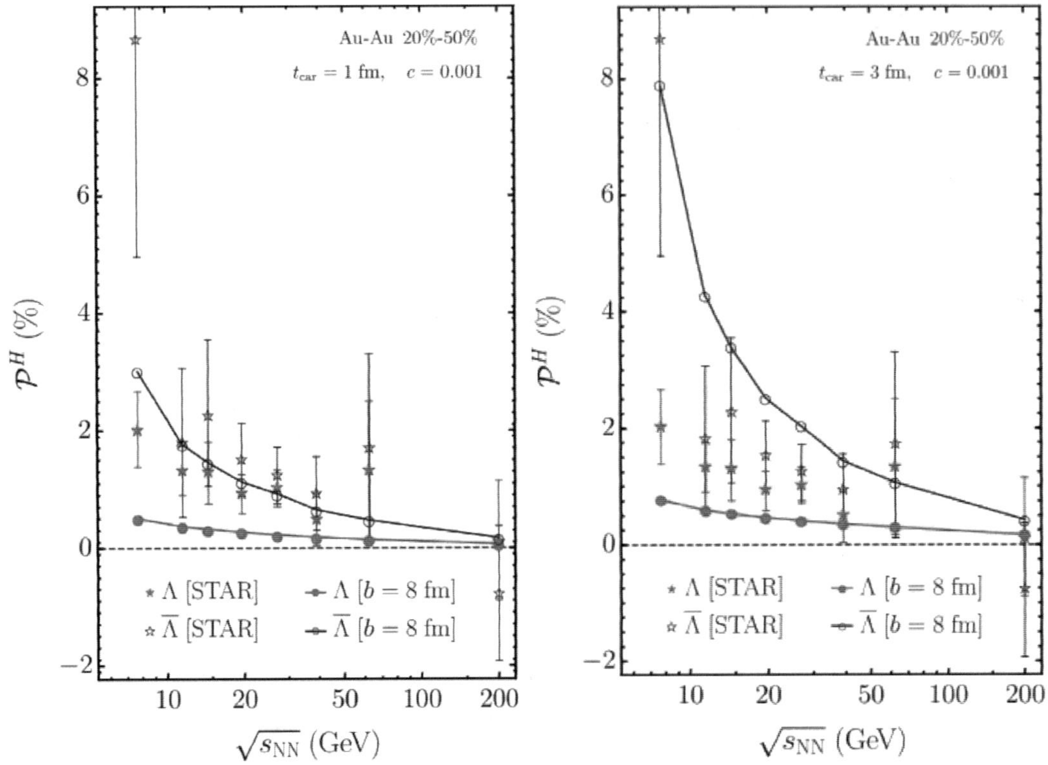

Figure 6. Global Λ and $\bar{\Lambda}$ polarizations compared with data from BES [39] for two different formation times $t = 1$, and 3 fm.

work we have shown that this can be understood in a two-component model where Λs come from a high density core, namely, the region where the QGP is created, and a less dense corona where the production mechanism comes from p+p like processes. We have shown that when the ratio $N_{\bar{\Lambda}\,\mathrm{REC}}/N_{\Lambda\,\mathrm{REC}}$ is smaller than one, an amplification for the global $\bar{\Lambda}$ polarization over that of Λ is obtained, in spite of the intrinsic Λ being larger than the intrinsic $\bar{\Lambda}$ polarization. With reasonable assumptions, this two-component model provides a qualitative and quantitative description of the STAR-BES data. A more detailed analysis has been performed and reported in [41].

Acknowledgments

A.A. thanks F. Wang for helpful comments. Support for this work has been received in part by UNAM-DGAPA-PAPIIT grant number IG100219 and by Consejo Nacional de Ciencia y Tecnología grant numbers A1-S-7655 and A1-S-16215. I.M. acknowledges support from a postdoctoral fellowship granted by Consejo Nacional de Ciencia y Tecnología.

References

[1] Jacob M 1988 *Z. Phys.* C **38**, 273-276 doi:10.1007/BF01574549
[2] Barros C C and Hama Y 2011 *Phys. Lett.* B **699**, 74-77
[3] Ladygin V P, Jerusalimov A P and Ladygina N B 2010 *Phys. Part. Nucl. Lett.* **7**, 349-354
[4] Becattini F, Csernai L and Wang D J 2013 *Phys. Rev.* C **88**, 034905
Erratum: 2016 *Phys. Rev.* C **93**, 069901
[5] Xie Y, Glastad R C and Csernai L P 2015 *Phys. Rev.* C **92**, 064901

[6] Karpenko I and Becattini F 2017 *Eur. Phys. J.* C **77**, 213
[7] Xie Y L, Bleicher M, Stöcker H, Wang D J and Csernai L P 2016 *Phys. Rev.* C **94**, 054907
[8] Jiang Y, Lin Z W and Liao J 2017 *Phys. Rev.* C **94**, 044910
Erratum: 2017 *Phys. Rev.* C **95**, 049904
[9] Shi S, Li K, Liao J 2019 *Phys. Lett.* B **788**, 409-413
[10] Li H, Pang L G, Wang Q and Xia X L 2017 *Phys. Rev.* C **96**, 054908
[11] Karpenko I and Becattini F 2017 *Nucl. Phys.* A **967**, 764-767
[12] Xia X L, Li H, Tang Z B and Wang Q 2018 *Phys. Rev.* C **98**, 024905
[13] Suvarieva D, Gudima K and Zinchenko A 2018 *Phys. Part. Nucl. Lett.* **15**, 182-188
[14] Bunce G *et al.* 1976 *Phys. Rev. Lett.* **36** 1113
[15] Groom D E and Marshall J H 1967 *Phys. Rev.* **159**, 1213-1219
[16] Yen W L, Ammann A c, Carmony D D, Eisner R L, Garfinkel A F, Gutay L J, Lakshmi R V, Miller D H and Tautfest G W 1969 *Phys. Rev. Lett.* **22**, 963–966
[17] Andersson B, Gustafson G, and Ingelman G 1979 *Phys. Lett.* B **85**, 417–420
[18] DeGrand T A and Miettinen H I 1981 *Phys. Rev.* D **23**, 1227 *ibid* 1981 **24**, 2419
Fujita T and Matsuyama T 1988 *ibid* **38**, 401
DeGrand T A 1988 *ibid* **38**, 403
[19] Panagiotou A D 1990 *Int. J. Mod. Phys.* A **5**, 1197
[20] Soffer J and Tornqvist N A 1992 *Phys. Rev. Lett.* **68**, 907-910
[21] Gustafson G and Hakkinen J 1993*Phys. Lett.* B **303**, 350-354
[22] Ellis J R, Kharzeev D and Kotzinian A 1996 *Z. Phys.* C **69**, 467-474
[23] Ellis J R, Kotzinian A, Naumov D and Sapozhnikov M 2007 *Eur. Phys. J.* C **52**, 283-294
[24] Jaffe R L 1996 *Phys. Rev.* D **54**, R6581-R6585
[25] Kotzinian A, Bravar A, and von Harrach D 1998 *Eur. Phys. J.* C **2**, 329-337
[26] de Florian D, Stratmann M and Vogelsang W 1998 *Phys. Rev.* D **57**, 5811–5824
[27] de Florian D, Stratmann M and Vogelsang W 1998 *Phys. Rev. Lett.* **81**, 530–533
[28] Boros C and Liang Z T 1998 *Phys. Rev.* D **57**, 4491-4494
[29] Anselmino M, Boglione M and Murgia F 2000 *Phys. Lett.* B **481**, 253-262
[30] Anselmino M, Boer D, D'Alesio U and Murgia F 2001 *Phys. Rev.* D **63**, 054029
[31] Ma B Q, Schmidt I, Soffer J and Yang J J 2000 *Eur. Phys. J.* C **16**, 657-664
[32] Alikhanov I and Grebenyuk O 2008 *Eur. Phys. J.* C **54**, 123-127
[33] Yang Y, Lu Z and Schmidt I 2017 *Phys. Rev.* D **96**, 034010
[34] Sun Y and Ko C M 2017 *Phys. Rev.* C **96**, 024906
[35] Han Z Z and Xu J 2018 *Phys. Lett.* B **786**, 255-259
[36] Becattini F, Piccinini F and Rizzo J 2008 *Phys. Rev.* C **77**, 024906
[37] Becattini F, Karpenko I, Lisa M, Upsal I and Voloshin S 2017 *Phys. Rev.* C **95**, 054902
[38] Acharya S *et al.* (ALICE Collaboration) 2020 *Phys. Rev.* C **101**, no.4, 044611
[39] Adamczyk L *et al.* (STAR Collaboration) 2017 *Nature* **548**, 62-65
[40] Adam J *et al.* (STAR Collaboration) 2018 *Phys. Rev.* C **98**, 014910
[41] Ayala A *et. al.*, 2020 "Core meets corona: a two-component source to explain Lambda and anti-Lambda global polarization in semi-central heavy-ion collisions," *arXiv:*hep-ph/2003.13757
[42] Ayala A, Cuautle E, Herrera G and Montaño L 2002 *Phys. Rev.* C **65**, 024902
[43] Letessier J, Rafelski J and Tounsi A 1996 *Phys. Lett.* B **389**, 586
[44] Nakamura K *et al.* (Particle Data Group) 2010 *J. Phys.* G **37**, 075021
[45] Abelev B *et al.* (ALICE Collaboration) 2013 *Phys. Rev.* C **88**, 044909
[46] Blaizot J P and Ollitrault J Y 1996 Phys. Rev. Lett. **77**, 1703-1706
[47] Ayala A, Quark spin – thermal vorticity alignment and the Λ, $\overline{\Lambda}$ polarization in heavy-ion collisions, 36th Winter Workshop on Nuclear Dynamics, Puerto Vallarta, Mexico, March 2020.
https://indico.cern.ch/event/841247/contributions/3740432/attachments/1998415/3334657/Ayala.pdf
[48] Ayala A, de la Cruz D, Hernández L A and Salinas J 2020 *arXiv:*hep-ph/2003.06545
[49] Randrup J and Cleymans J 2006 *Phys. Rev.* C **74**, 047901

AUTHOR INDEX

Institute of Physics
Dirac House, Temple Back
Bristol BS1 6BE UK

ISSN: 1742-6588
ISBN 978-1-7138-5749-5

9 781713 857495

CURRAN ASSOCIATES INC.
proceedings.com